# 40 YEARS OF BEREZINSKII–KOSTERLITZ–THOULESS THEORY

# 40 YEARS OF BEREZINSKII–KOSTERLITZ–THOULESS THEORY

Editor

## Jorge V José
Indiana University, USA

NEW JERSEY • LONDON • SINGAPORE • BEIJING • SHANGHAI • HONG KONG • TAIPEI • CHENNAI

*Published by*

World Scientific Publishing Co. Pte. Ltd.
5 Toh Tuck Link, Singapore 596224
*USA office:* 27 Warren Street, Suite 401-402, Hackensack, NJ 07601
*UK office:* 57 Shelton Street, Covent Garden, London WC2H 9HE

**British Library Cataloguing-in-Publication Data**
A catalogue record for this book is available from the British Library.

**40 YEARS OF BEREZINSKII–KOSTERLITZ–THOULESS THEORY**

Copyright © 2013 by World Scientific Publishing Co. Pte. Ltd.

*All rights reserved. This book, or parts thereof, may not be reproduced in any form or by any means, electronic or mechanical, including photocopying, recording or any information storage and retrieval system now known or to be invented, without written permission from the Publisher.*

For photocopying of material in this volume, please pay a copying fee through the Copyright Clearance Center, Inc., 222 Rosewood Drive, Danvers, MA 01923, USA. In this case permission to photocopy is not required from the publisher.

ISBN 978-981-4417-62-4
ISBN 978-981-4417-63-1 (pbk)

Printed by FuIsland Offset Printing (S) Pte Ltd Singapore

# Contents

Introduction and Overview
    J. V. José     vii

1. Early Work on Defect Driven Phase Transitions
    J. M. Kosterlitz and D. J. Thouless     1

2. Duality, Gauge Symmetries, Renormalization Groups and the BKT Transition
    J. V. José     69

3. Berezinskii–Kosterlitz–Thouless Transition through the Eyes of Duality
    G. Ortiz, E. Cobanera and Z. Nussinov     93

4. The Berezinskii–Kosterlitz–Thouless Transition in Superconductors
    A. M. Goldman     135

5. Berezinskii–Kosterlitz–Thouless Transition within the Sine-Gordon Approach: The Role of the Vortex-Core Energy
    L. Benfatto, C. Castellani and T. Giamarchi     161

6. The Two-Dimensional Fully Frustrated XY Model
    S. Teitel     201

7. Charges and Vortices in Josephson Junction Arrays
    R. Fazio and G. Schön     237

8. Superinsulator–Superconductor Duality in Two Dimensions and Berezinskii–Kosterlitz–Thouless Transition
    V. M. Vinokur and T. I. Baturina     255

9. BKT Physics with Two-Dimensional Atomic Gases
    Z. Hadzibabic and J. Dalibard     297

10. Vortex Physics in the Quantum Hall Bilayer
    *H. A. Fertig and G. Murthy*    325

Index    349

# Introduction

More than 40 years ago, Vladimir Berezinskii[1] (1971) and the team of Michael Kosterlitz and David Thouless[2] (1973) (BKT) published the results of their independent investigations into the puzzling question of whether or not long-range order exists in two-dimensional systems with continuous symmetry. This work would have a seminal impact upon condensed matter physics, and other areas in physics, an impact we explore in the present volume. Heuristic arguments and rigorous mathematical theorems had long led to the general belief that there could not be a stable thermodynamic ordered phase at low temperatures. However, experimental and theoretical evidence had indicated that there was something else going on with this problem. BKT introduced the idea of a *topological phase transition*, where pairs of bound vortex excitations unbind at a critical temperature "$T_{\text{BKT}}$". The nature and characteristics of the BKT transition are different in several respects from the more common second order phase transitions. While there is no long-range order with a finite order parameter for all temperatures, as shown by the earlier theorems, we now know that there is, however, a continuous line of critical points below $T_{\text{BKT}}$. At low temperatures, the two-point correlation function decays algebraically, with a temperature dependent exponent up to $T_{\text{BKT}}$. Above $T_{\text{BKT}}$, the correlation function decays exponentially in distance but has a correlation length that diverges exponentially with temperature, rather than as a power law. The seminal ideas of BKT have had an impact in many areas of theoretical and experimental physics, some examples of which are presented in this volume. Given that there are thousands of papers where BKT work has been applied, in this 40th Anniversary volume, we provide only a small sampling of the ways in which BKT ideas have been used. It is interesting to note that the number of citations to the KT and BKT work is comparable to the number of citations to the seminal 1958 paper by Phil Anderson on localization theory.[3]

Soon after I started my PhD thesis in 1974, under the advice of Leo Kadanoff, I attended my first summer school conference on *The Helium Liquids* held at St. Andrews University in Scotland. In that meeting Ann Eggington presented a review of the experimental and theoretical situation about whether or not long-range order exists in restricted dimensionalities, in particular, in Helium films.[4] In spite of the theorem proofs, valid in the thermodynamic limit, there was puzzling experimental evidence suggesting the existence of superfluidity in thin $^4$He films. She mentioned the KT work at the end of her review as one possible resolution to this controversy. It was my first introduction to the theories of BKT, as at that time, my thesis work was unrelated to that problem. I started a postdoctoral position with Leo after I finished my PhD thesis in 1976. He was interested in trying to understand this conundrum since some respected investigators had pointed out difficulties with the low order approximations used in BKT theory, in particular, the complete decoupling of spin-waves from vortex excitations.

Although the BKT theory was very physically appealing, alternative theories to the BKT theory were offered during that time. For example, Johannes Zittartz did a careful perturbative self-consistent analysis of the 2D-XY or planar model Hamiltonian, believed to describe superfluid Helium (1976).[5] He did not find evidence or justification for the discontinuous jump in the two-point correlation function exponent, one of the landmark predictions of BKT theory. He obtained evidence for two different low temperature phase transitions and clearly stated that he believed that KT theory was incorrect. Alan Luther and Doug Scalapino also reanalyzed the classical Planar model through a mapping to a one-dimensional spin-1/2 fermion gas (1977).[6] That type of mapping was very popular at the time since it could lead to exact solutions to higher dimensional classical models. This paper generally did not agree with the BKT results either. They found a low temperature power law decay, with a non-universal correlation function exponent of $1/\sqrt{8}$, instead of the universal $1/4$ found by KT at criticality.

In the first chapter of the current volume, Kosterlitz and Thouless offer some historical perspective on the development of and responses to their theory. (Unfortunately, Berezinskii is no longer with us to present his perspective). KT offer a detailed analysis of how they came up with their topological phase transition idea. It is always interesting to trace back what actually went through people's minds and what influenced their thoughts and what trajectory led them to their conclusions. This is often not mentioned in a formal publication. In their essay here, KT also review how things have evolved since they published their paper. They make some corrections and improvements to their original ideas and analysis: for example, their initial conclusion that their theory would not apply to charged superconductors.

They also mention that five years after the publication of their first paper, it had only had three citations. The situation has changed dramatically after 1977.

The mid-1970s were a very exciting period for physicists since there were a number of important developments that took place at that time. Some of those developments contributed to providing a more definitive basis for the correctness of the BKT scenario.

Within the area of high energy physics, Gerard't Hooft had just proven that the Yang–Mills, Weinberg–Salam U(1) × SU(2) non-Abelian gauge theory, was indeed renormalizable.[7] The theory led to the formulation of the Standard Model, based on the Higgs mechanism. Simultaneously, Kenneth Wilson developed the Renormalization Group theory (RG),[8] based in part on Kadanoff's block spin transformations.[9] This theory was providing quantitative answers to long-standing questions in critical phenomena. This was also the time when Wilson introduced the lattice Gauge theory formulation of U(1) × SU(2) to handle the strong coupling limit of the theory.[10,11] Soon after, Alexander Migdal developed an approximate renormalization-group approach to study Abelian and non-Abelian lattice gauge theories.[12] A number of these developments, together with an extension of the duality transformations concept to Abelian gauge theories, have led to a fresh look and reanalysis of the two-dimensional XY model. The results from using a new approach have led to strong support for the validity of the BKT theory and an alternative derivation of many of their previous results. In Chapter 2, I discuss the impact of these developments upon the analysis of 2-XY model that we engaged in during my postdoctoral fellowship with Leo Kadanoff. Our analysis resulted in a 1977 paper written with Leo, Scott Kirkpatrick and David Nelson (JKKN).[13]

In Chapter 3, Gerardo Ortiz, Emilio Cobanera and Zohar Nussinov review recent work and controversies on duality transformations applied to the 2-D XY and p-clock models. The p-clock model was introduced and analyzed in the JKKN paper. There are still many open questions about the critical structure of this family of models, in particular, the different critical structures for the p-clock model when $p$ is bigger or smaller than 5. In their analysis, they use a new Transfer Matrix formulation of these models, together with an analysis of their corresponding topological excitations.

One of the initial conclusions from the KT paper was that their topological phase transition would not occur in superconductors. This was based on the assumption that the vortex–vortex interactions would decay with a $1/r$ power-law. In Chapter 4, Allen Goldman points out that, in fact, the vortex–vortex interaction decays logarithmically in a two-dimensional superconducting film, or in a planar Josephson junction array (JJA), up to what turned out to be large screening lengths. The logarithmic nature of the vortex–vortex interactions was, however, first derived by Judea Pearl in his 1964 PhD thesis.[14] Once it was recognized that BKT theory could apply to superconductors, a series of extensions of the theory were formulated. Goldman provides a review of applications of BKT to superconducting films and JJA. In Chapter 5, Clara Benfatto, Claudio Castellani and Thierry Giamarchi (BCG) present a critical review of the theoretical and experimental applications of BKT theory to superconductors to date. They point out that BKT theory was developed having the 2D-XY model in mind, which may not always work perfectly in superconductors. BCG use the Sine–Gordon representation in their analysis, pointing out that one of the hallmarks of KT predictions, i.e. the jump in the superfluid density at criticality,[15] has just recently been measured experimentally.

In contrast to neutral superfluids, superconductors and JJA respond in a very special way to the presence of external magnetic fields. In JJA the ratio of the magnetic flux, $\Phi$, penetrating a basic unit cell divided by the quantum of flux $\Phi_0$, known as the frustration parameter, "$f = \frac{\Phi}{\Phi_0}$", can lead to very interesting and complex types of phase transitions: $f$ can take, in principle, any rational to irrational real number values. In Chapter 6, Steve Teitel reviews the present theoretical and experimental situation of the "fully frustrated", $f = 1/2$, 2-D-XY model. He introduced this model, together with Ciriyam Jayaprakash, in 1983.[16] It has led to a number of interesting theoretical and experimental physics results, some of which have been tested in JJAs. He reviews the present situation for the fully frustrated $f = 1/2$ case, which even after many years of study has not yet been fully understood. Teitel also reviews briefly the work pertaining to other "$f$" values, including rational and irrational numbers, for which there is also a very large literature.

Josephson Junction Arrays (JJA) are also the focus of Rosaria Fazio and Gerd Schön's contribution to this volume in Chapter 7. JJA became a very good laboratory to study BKT transitions and its extensions under different types of conditions. Each JJ junction can be characterized by a quantum parameter, $\alpha$, given by the ratio of the Josephson capacitive charging energy, $E_C$, divided by the Josephson energy, $E_J$. Depending on the value of $\alpha$, we can be in a regime dominated either by thermal or quantum fluctuations. Fazio and Schön review the case when $\alpha$ can be large and quantum fluctuations can become dominant. This occurs when the JJ sizes are in the submicron regime. In this case, not only the thermally excited vortices are important but also capacitively induced charges as well. They discuss another type of duality transformation between the classical and the quantum fluctuations regimes.

Along the same lines, in Chapter 8, Valerii Vinokur and Tatyana Baturina also discuss the competition between classical and quantum fluctuations in a 2-D superconducting system. They consider the interplay of classical and quantum fluctuations via the existence of a superinsulator–superconductor duality relationship having a charge BKT transition dually related to a vortex induced BKT transition. They make a series of proposals for having practical device applications using this type of duality transformation.

A recent application of BKT ideas is in ultra-cold, quasi-two-dimensional, trapped atomic gases. In Chapter 9, Zoran Hadzibabic and Jean Dalibard provide a critical review of the recent remarkable results on the competition between the BKT type transitions against Bose–Einstein condensation. This is a relatively recent application of the BKT scenario and a number of issues need to be resolved, both experimentally and theoretically, in order to have a more comprehensive understanding of the BKT type transition present in these systems.

Finally in Chapter 10, Herb Fertig and Ganpathy Murthy look at a further extension of the BKT ideas in a quantum Hall bilayer with filling factor $\nu = 1$, which maps into a 2-D superfluid system, having charged vortices they call *merons* and in the presence of disorder. There is significant interest in this system both from the theoretical and experimental points of view, and a number of questions remain about the zero and very low temperature phases which are dominated by quantum fluctuations and disorder.

As the various chapters within this book make clear, the applications, developments and extensions of the BKT seminal ideas continue unabated some thirty-five years after the ideas gained broad currency. The chapters here cover only a small portion of the areas in which BKT theory has had and is having an impact. Most of the chapters are reviews in nature. We hope that each one of them provides a view of the present state of affairs in the areas they cover. We cannot yet discern where BKT may take us in the decades to come, but are confident that the full impact of these ideas has yet to be described. One area which is not covered in detail here but which is currently very active relates to the applications of BKT theory to the problem of confinement in high energy physics, in particular, within the context of lattice gauge theory. A recent series of reviews to applications to the quark confinement problem have been discussed in a recent volume.[17]

*Jorge V. José*

Department of Physics and Department of Integrative and Cellular Physiology, Indiana University, IN 47405, USA

## References

1. V. L. Berezinskii, Destruction of long range order in one-dimensional and two-dimensional systems with a continuous symmetry group. II. Quantum systems, *Zh. Eksp. Teor. Fiz.* **61**, 1144 (1971); *Sov. Phys. JETP* **34**, 610 (1972).
2. J. M. Kosterlitz and D. J. Thouless, Ordering, metastability and phase transitions in two-dimensional systems, *J. Phys. C* **6**, 1181 (1973).
3. P. W. Anderson, Absence of diffusion in certain random lattices, *Phys. Rev.* **109**, 1492 (1958).
4. A. Eggington, Superfluidity in restricted geometries, in *The Helium Liquids, Proc. Fifteen Scottish University Summer School in Physics*, 1974. eds. J. G. M. Armitage and I. E. Farquhar (Academic Press, 1975), pp. 175–209.
5. J. Zittartz, Phase transition in the two-dimensional classical XY-model, *J. Physik B* **23**, 63 (1976).
6. A. Luther and D. J. Scalapino, Critical properties of the two-dimensional planar model, *Phys. Rev. B* **16**, 1153 (1977).
7. G. 't Hooft, Renormalization of massless Yang–Mills fields, *Nucl. Phys. B* **33**, 173 (1971).
8. K. G. Wilson, Renormalization group and critical phenomena. Renormalization group and the Kadanoff scaling picture, *Phys. Rev. B* **4**, 3174, 3184 (1971).
9. L. P. Kadanoff, Scaling laws for Ising models near Tc, *Physics* **2**, 263 (1966).
10. K. G. Wilson, Quarks and strings on a lattice, *Phys. Rev. D* **10**, 2445 (1974).
11. K. G. Wilson, Confinement of quarks, *Phys. Rev. D* **10**, 2445–2459 (1974).
12. A. A. Migdal, Recursion equations in gauge field theories, *Z. Exper. Teoret. Fiz.* **69**, 810 (1975); *Sov. Phys. JETP* **42**, 743 (1975).
13. J. V. José, L. P. Kadanoff, S. Kirkpatrick and D. R. Nelson, (JKKN) Renormalization, vortices, and symmetry-breaking perturbations in the two-dimensional planar model, *Phys. Rev. B* **16**, 1217 (1977).
14. J. Pearl, Appl. Current distributions in superconducting films carrying quantized fluxoids, *Phys. Lett.* **5**, 65 (1964).
15. D. R. Nelson and J. M. Kosterlitz, Universal jump in the superfluid density of two-dimensional superfluids, *Phys. Rev. Lett.* **39**, 1201 (1977).
16. S. Teitel and C. Jayaprakash, Josephson junction arrays in transverse magnetic fields, *Phys. Rev. Lett.* **51**, 1999 (1983).
17. J. Greensite, An introduction to the confinement problem, *Lecture Notes in Physics* **821** (2011).

# Chapter 1

# Early Work on Defect Driven Phase Transitions

J. Michael Kosterlitz

*Department of Physics, Brown University, Providence, RI 02912, USA*

David J. Thouless

*Department of Physics, Box 351560, University of Washington, Seattle, WA 98195, USA*

> This article summarizes the early history of the theory of phase transitions driven by topological defects, such as vortices in superfluid helium films or dislocations and disclinations in two-dimensional solids. We start with a review of our two earliest papers, pointing out their errors and omissions as well as their insights. We then describe the work, partly done by Kosterlitz but mostly done by other people, which corrected these oversights, and applied these ideas to experimental systems, and to numerical and experimental simulations.

## 1.1. Introduction

The idea that certain phase transitions might be driven by the occurrence of an equilibrium concentration of defects such as dislocations in solids or quantized vortices in superfluids is almost as old as the recognition that such defects have an important role in the mechanical or electrical properties of such materials. Indeed, in Onsager's incredibly brief initial note on the existence of quantized vortices in superfluid helium, the possibility of a vortex-driven transition from the superfluid phase to the normal fluid phase is mentioned,[1] and this possibility was later presented by Feynman[2] in a much more accessible form. There are also discussions of the idea that the spontaneous formation of an equilibrium density of dislocations in solids could lead to the transition from a rigid solid to a liquid with viscous flow. Such models involve handling the statistical mechanics of line defects in a three-dimensional medium, and such problems are notoriously difficult to handle.

## 1.2. One-Dimensional Ising Model

The first exposure of one of us (DJT) to the ideas that were to prove important in the theory of defect driven phase transitions happened as a result of a visit to Bell Labs. There I was told by Philip Anderson of the work with his student Gideon Yuval and his colleague Don Hamann on the solution of the Kondo problem (for a magnetic impurity in a low temperature metal), by means of a transformation of the problem to a one-dimensional Ising model with a long-range interaction between the spins falling off with distance like $1/r^2$.[3-6] They recognized that such an inverse square law for the interactions had a slower fall-off with distance than those known to give no equilibrium magnetization,[7] and faster than those known to have magnetization at sufficiently low temperatures.[8] Anderson asked me if I knew anything about this intermediate case, so I thought about it, and, on the train back to Stony Brook, realized that the argument on the last page of the 1958 translation of Landau and Lifshitz, *Statistical Physics*,[9] could be adapted to give a plausible answer to this problem, so I tried it out on some of the statistical mechanics faculty and visitors at SUNY Stony Brook, such as Bob Griffiths and Barry McCoy.

The basic idea of the Landau and Lifshitz argument is to consider the statistical mechanics of the domain walls between blocks of spin up and spin down. If $a$ is the spacing between spins, then $a$ is also the length occupied by each domain wall. The entropy is the same as the entropy in a one-component, one-dimensional hard "sphere" gas, so that, in a system of length $L$ with $p$ domain walls, the entropy is

$$S \approx k_B p \ln(L/pa), \tag{1.1}$$

and so it is logarithmically divergent in the limit of large $L$, fixed $p \neq 0$. In the case of a finite range interaction the energy per domain wall is finite in the limit of large $L$, so that at any nonzero temperature the entropy dominates, there is a nonzero concentration of domain walls, and zero equilibrium magnetization, in this limit.

In the particular case of a $Ja^2/r^2$ interaction between spins, the energy of a single domain wall at $x$ can be approximated as

$$E = 2J \int_{-L/2}^{x-a/2} \int_{x+a/2}^{L/2} \frac{1}{(x_2-x_1)^2} dx_2 dx_1 = 2J \ln \frac{(L+a)^2/4 - x^2}{La} \approx 2J \ln \frac{L}{4a}, \tag{1.2}$$

for $x$ in the interior, well away from the boundaries at $x = \pm L/2$. At low temperatures the free energy $F = E - TS$ is dominated for large $L$ by the energy, so there are no free dislocations in equilibrium, but for $T > J/k_B$ the entropy dominates, free domain walls are thermodynamically stable, and there is no net magnetization.

The argument that was published in 1969[10] had a few more refinements. In particular, although isolated domain walls are forbidden at low temperatures in the thermodynamic limit, pairs of domain walls, which provide a nonzero magnetization density, are not forbidden. A pair of domain walls a distance $na$ apart contains $n$ spins in the reverse direction, and has energy

$$E_n \approx 4Ja^2 \ln n. \tag{1.3}$$

A concentration $c_n$ of such pairs of domain walls reduces the magnetization $M$ by

$$-\delta M = \sum_n 2nc_n, \tag{1.4}$$

and this reduces the magnetization from its zero-temperature value of unity as the temperature is increased. This has the effect of reducing the coefficient of the $\ln L$ term in Eq. (1.2) for the domain wall energy by a factor approximately equal to $\delta M(T)$, and so the inequality that determines the possibility of isolated domain walls, or of the dissociation of domain wall pairs into free domain walls, becomes

$$k_B T \geq JM. \tag{1.5}$$

I was startled by the conclusion that the magnetization would decrease monotonically as the temperature was raised, but that, rather than decreasing continuously to zero, must jump to zero when the inequality (1.5) is satisfied. This is not a first order transition, since the energy is a continuous function of $T$.

This result was confirmed in more careful theoretical work by Dyson,[11] by Fröhlich and Spencer[12] and by Aizenman et al.[13] Initially, apart from its association with the scaling theory of the Kondo effect,[6] it did not seem much more than a curiosity, but its wider significance was soon appreciated.

## 1.3. Vortex Driven Transitions in Superfluid Films

In 1971 two things came together for us. Thouless decided to give a graduate course on superfluidity and superconductivity, and, in explaining that the energy of a quantized vortex in a thin film of superfluid depends logarithmically on the area of the film, realized that the same entropy–energy balance that occurs in the $1/r^2$ Ising model should also occur in the natural vortex model for a neutral superfluid film. At about the same time, Kosterlitz left Torino and arrived in Birmingham in his second post-doctoral position; he began looking for an area of physics outside elementary particle theory to try out. He was enticed into condensed matter physics by our colleague Eric Canel, and the two of us started talking about vortex-driven transitions in superfluid films. This was the start of our collaboration to explore the possibility of a new sort of phase transition in two-dimensional systems.

The flow velocity in a superfluid condensate is $\hbar/M_A$ times the gradient of the phase of the condensate wave function, where $M_A$ is the atomic mass (or pair mass in the case of fermion superfluidity), and so the flow velocity is always the gradient of a potential function. The appearance of uniform rotation in superfluid helium was a paradox resolved by Onsager's suggestion that it was the effect of an approximately uniform distribution of quantized vortex lines in the rotating liquid, with the microscopic velocity being a sum of terms $\hbar \nabla \phi_i / M_A$, where $\phi_i$ is the azimuthal angle referred to the core of a particular vortex $i$. This general picture was confirmed by the experimental work of Hall and Vinen in 1956.[14]

In three dimensions the energy and entropy of interacting vortex lines are complicated to calculate, and the needed approximations are unconvincing. In a two-dimensional system, such as a thin film of superfluid helium, the vortices are quantized, with circulation an integer multiple of $h/M_A$.[1,2] The only degrees of freedom of a quantized vortex are the two coordinates of the position of the vortex core, which, in a classical theory of vortex dynamics, such as the one formulated by Kirchhoff,[15] are conjugate dynamical variables, and which should therefore be governed by an uncertainty relation in a quantized theory. This uncertainty relation quantizes the area $A = \pi R^2$ into cells of area $\pi a_1^2 = M_A/\rho_s$, and this gives the entropy per vortex as

$$S_v \approx k_B \ln(L^2/a_1^2). \tag{1.6}$$

The vortices can have positive or negative circulation $\pm h/M_A$. The energy of an isolated vortex near the center of a disk of superfluid of radius $R$ can be written as

$$E_V \approx 2\pi \int_{a_0}^{R} \rho_s \left(\frac{\hbar}{M_A r}\right)^2 r\, dr = 2\pi \rho_s (\hbar/M_A)^2 \ln(R/a_0). \tag{1.7}$$

Here $a_0$ is the vortex core radius, the healing distance over which the superfluid density is suppressed by the high speed near the core, and $\rho_s$ is the superfluid density per unit area, which is *defined* and measured in terms of the energy density for a given superfluid flow velocity. If the isolated vortex is not close to the center of the system, but is close to one of the boundaries, the energy is reduced by the backflow produced by the boundary, just as the energy of a free electrostatic charge is reduced by the charge density induced in a neighboring conducting surface, so that at distance $l$ from the boundary it becomes

$$E_V(l) \approx 2\pi \int_{a_0}^{l} \rho_s \left(\frac{\hbar}{M_A r}\right)^2 r dr = 2\pi \rho_s \left(\frac{\hbar}{M_A}\right)^2 \ln(l/a_0). \quad (1.8)$$

There is obviously a close analogy between Eqs. (1.1) and (1.2) for the one-dimensional Ising model with $1/r^2$ interactions, and Eqs. (1.6)–(1.8) for vortices in a superfluid thin film. In analogy with Eq. (1.3) for the Ising model, the energy of a pair of oppositely rotating vortices with circulation $\pm \hbar/M_A$ separated by a distance $d$ is given as

$$E_{vp}(d) = 2\pi \rho_s \left(\frac{\hbar}{M_A}\right)^2 \ln \frac{d}{a_0}. \quad (1.9)$$

Comparison of Eqs. (1.6) and (1.7) shows that the condition for stability of the superfluid against the appearance of isolated vortices is

$$k_B T \leq \pi \rho_s (\hbar/M_A)^2. \quad (1.10)$$

These arguments were given in our 1972 and 1973 papers.[16,17] However we got an incorrect factor of $M_A/M^*$, where $M^*$ is an effective mass, in our derivation. A proper derivation was given by Nelson and Kosterlitz.[18] The corrected result is important, because the parameter-free relation between the critical temperature and superfluid density can be, and was, experimentally verified. A straightforward explanation for this robust relation between the maximum temperature for stable superfluidity comes from the fact that the coefficient of the logarithm in the expression (1.7) depends on the average flow induced on a circuit at large distance from the vortex core by the $2\pi$ twist of the phase angle, and the superfluid density is also defined by the energy induced by a twist of the phase angle imposed over a large area. Therefore, even if nonuniformity of the substrate causes nonuniformity of the superfluid film, it is the *same* average that comes into the expression for superfluid density and for vortex energy.

For a short while we could congratulate ourselves on discovering completely new physics. It did not last long, because, on a trip to Paris, Thouless met Paul Martin, who was spending a sabbatical there, and he said that he had heard the story before, from some Russian visitor who knew of Berezinskii's work. This was too early for us to have the translations available, we had little knowledge of Russian between us, and we cannot remember why we did not get a colleague to translate the papers for us. We cited two papers by Berezinskii;[19,20] the first of these is irrelevant, but the second anticipated our published work by a year. For reasons we do not know, and about which it is inappropriate for us to speculate, our work had more impact than Berezinskii's, and was much more often quoted. There seem to have been nearly 2200 papers published since 1972 that mention our work in the title or abstract, only 3 of which were published in the first five years. Of these articles, 1700 are classed as purely theoretical and 500 as experimental, according to Inspec.

We have our own stories of missed opportunities. One of us (DJT) went down to the University of Sussex to give a seminar to Douglas Brewer's group there. I knew that he and his collaborators had done good work on superfluid films, and hoped that he would follow up our result that the superfluid density should have a discontinuity at the superfluid–normal critical temperature. Douglas was politely interested, and told me that Isadore Rudnick's group at UCLA had done some experiments on superfluid films, and had found such a discontinuity,[21,22] but that a careful reanalysis of their data had made this unexpected discontinuity disappear. I failed to look up the UCLA paper, and Douglas did not follow up on our ideas. There are two morals in this story. One is that theorists should read the relevant experimental literature carefully, and the other is that experimentalists should not stop checking their results just because they are consistent with existing theory.

## 1.4. Other Systems with Defect-Mediated Transitions

At that time the main subject of interest for theorists in statistical mechanics was the "universal" nature of critical behavior, pioneered by Cyril Domb, Michael Fisher,[23] Leo Kadanoff and many others, which led to the renormalization procedure to determine critical exponents, described in the papers of Kenneth Wilson[24] and Wilson and Fisher.[25] The argument was that, as the critical point is approached in a transition such as the liquid–vapor transition, length scales get larger and larger, and the details of the mechanism become irrelevant. Close to a critical point only broad features, such as the dimensionality of the order parameter, are important. It was therefore natural to extend the arguments to other two-dimensional systems which have similar defects with a similar long-range interaction between them. In some cases we did this successfully, in other cases we got important details wrong, and many other cases we did not think about.

In the mid 1930s Peierls had produced general arguments[26,27] that showed that in two-dimensional systems with an order parameter with more than one degree of freedom, such as a superfluid, or a magnet with magnetism isotropic in two- or three-dimensional space, long-wavelength fluctuations of the azimuthal angle of the magnetization would result in the average magnetization tending to a zero limit as the system got large. Twenty years later this result was derived more carefully by Mermin and Wagner[28] and by Hohenberg.[29] Although, later authors had proved that the average magnetization must be zero, a number of people pointed out that it was not proved whether higher order correlations of the spins could have a nonzero limit at low temperatures, but a vanishing limit at higher temperatures. This would still allow for a critical point at the temperature where this transition occurred.

In this section we describe the theory roughly as we saw it in the early 1970s, and point out what we missed and where we went astray. In Sec. 1.6 we describe the essential modifications of the work described in this section that were made in the late 1970s.

### 1.4.1. *Two-dimensional magnetic systems*

Magnetic systems are usually regarded by theorists as the standard for classifying the universality classes of other systems of transitions with critical points or lines. They are not always readily available experimentally, since most real magnetic systems are solids, and the lower symmetry of a solid as compared with an isotropic liquid, either because of the intrinsic symmetry of an ideal crystal, or because of the local effects of impurities, may complicate the analysis of results. Furthermore, ferromagnetic systems are more complicated than antiferromagnetic systems, since the long-range interaction between the aligned magnetic dipoles gives rise to the formation of domain walls. The basic example of the Ising model universality class has a nearest neighbor interaction between the $z$ components of the spins on a regular lattice, in two, or three dimensions, but a weaker interaction between the other two components of spin does not change the behavior near the critical temperature and close to zero magnetic field. There is good evidence that the order–disorder transition in a binary alloy and the gas–liquid transition in a one-component fluid are also in this universality class.[30] In the classical Heisenberg model the spin can point in any direction in space, and the interaction energy of any neighboring pair of spins is proportional to the scalar product of the two neighboring spins. In the planar spin model there is an interaction between neighboring spins which is stronger for the $x, y$ components than it is for the $z$ components, and is unchanged by rotations of the spins about the $z$ axis. It is this planar spin model in a two-dimensional system which is thought to be in the same universality class as the superfluid film; the real and imaginary parts of the local value of the condensate wave function correspond to the $x$ and $y$ components of the spin. The spin system is two-dimensional if it is thin enough that spin directions on opposite faces of the film are strongly correlated. The analog of a vortex is a point in the two-dimensional space around which the projection of the spin direction in the plane rotates by $\pm 2\pi$, and we often use the term 'vortex' to describe such a structure in a spin system.

At large distances from the vortex core the energy per nearest neighbor bond is proportional to

$$|J_x|(1 - \cos \delta\phi) \approx \frac{|J_x|}{2}(\delta\phi)^2, \qquad (1.11)$$

where $J_x = J_y$ is the strength of the in-plane coupling, and $\delta\phi$ is the difference between the azimuthal angles of the two spins. Summation of this over the bonds surrounding a vortex will give close approximations to the integrals shown in Eqs. (1.7) and (1.8). In Sec. 1.2 the result of such a modification in the domain wall energy led to an increase in the magnetic susceptibility with rising temperature, while in Sec. 1.3 the decrease in the vortex energy leads to an decrease in the superfluid density as the temperature rises, as is discussed in more detail in Sec. 1.5. For the planar magnetic system the quantity that corresponds to the magnetization in the one-dimensional inverse square law magnet and to the superfluid density in the two-dimensional superfluid is the *spin-wave stiffness*, which is a measure of the energy associated with a progressive in-plane twisting of the magnetization direction. This is analogous to the superfluid density, because the superfluid density measures the extra energy density associated with a steady supercurrent, and the superfluid velocity, by definition, is given by $\hbar/m_B$ times the gradient of the phase of the condensate, where $m_B$ is the boson mass.

### 1.4.2. *Isotropic Heisenberg model*

Although analysis of the high temperature series by Stanley and Kaplan[31] seemed to show a critical temperature both for the planar spin model and for the Heisenberg model in two dimensions, slightly later work by Stanley[32] and by Moore[33] showed that the indication of a phase transition was much clearer for planar spins than for the isotropic Heisenberg model. This is consistent with the argument that we gave, that, for the Heisenberg model, there should only be a finite energy barrier separating topologically different states. A more careful argument of this sort was given by Belyavin and Polyakov.[34]

The argument is that for the planar spin model and its analogs there is a topological invariant characterizing a vortex-like configuration, which is the winding number of the polar angle $\phi_s$ of the planar spin direction,

$$n_w = \frac{1}{2\pi} \oint d\phi_s . \qquad (1.12)$$

Since $\phi_s$ is single valued modulo $2\pi$, $n_w$ must be integer. Furthermore, for many of the systems we discuss, at low temperatures there is an energy barrier between configurations with different winding numbers whose magnitude diverges logarithmically with the linear dimensions of the system, so that spontaneous fluctuations of the winding numbers are suppressed. The simplest way of adding a free 'vortex' is to add it close to the boundary of the two-dimensional space in which the system resides, in which case, as Eq. (1.8) shows, the energy is small. When this new vortex is pulled away from the surface, its energy increases logarithmically with the distance from the surface. Alternatively, two vortices of opposite sign can be created close to one another, and then pulled away from one another, so the energy increases logarithmically as their separation increases.

For the isotropic Heisenberg model there is no loop integral defining a topological invariant, since any loop on the 2-sphere that defines the directions of the spins, characterized by the polar angles of the spin directions, can be shrunk continuously to a point. If the spin direction on the distant boundary of the two-dimensional space tends to $\theta = 0$ then there is an invariant integral over all space

$$n_S = \left(\frac{1}{2\pi}\right) \int dx \int dy \left[\frac{\partial \cos\theta}{\partial x}\frac{\partial \phi}{\partial y} - \frac{\partial \cos\theta}{\partial y}\frac{\partial \phi}{\partial x}\right] = \frac{\Delta\phi}{2\pi}, \qquad (1.13)$$

where $\Delta\phi/2\pi$ is the winding number of $\phi$ at the distant boundary at which $\theta = \pi/2$. This measures the number of times the mapping of the $xy$ space onto the sphere defined by the polar angles covers the $\theta\phi$ sphere. This topological invariant was known to Gauss. The excitation away from the $\theta = 0$ fully aligned ground state, to a state in which $n_S = 1$, is known as a *skyrmion*, in honor of our former colleague and boss, who introduced this as a model for an elementary particle.[35] For an interaction that can be approximated as proportional to $|\nabla S|^2$ this excitation has a finite, scale-independent, energy, so that skyrmions can be excited with nonzero density at arbitrarily low temperatures, and there is no phase transition to a magnetized state at a nonzero temperature.

Theoretical arguments were advanced for the importance of skyrmions, rather than spin waves, as excitations giving the response of certain quantum Hall materials to magnetic fields.[36,37] Experimental evidence in support of this has accumulated,[38] and there is a recent review by Ezawa and Tsitsishvili.[39]

### 1.4.3. *Two-dimensional Coulomb plasma*

The discussion of the interaction of vortices and of the importance of the logarithmic dependence of the energy of a vortex pair on the distance between the two, which was discussed in Sec. 1.3, suggested that a useful model for the statistical mechanics of these vortices would be a two-dimensional plasma of positive and negative charges. The interaction energy between the charges $q_1, q_2$ has to be proportional to $q_1 q_2 \ln |\mathbf{r}_1 - \mathbf{r}_2|$, and this is not what happens for real electric charges confined to a thin layer. We therefore do not know of a real system of charges that behave like this, but we can use some of the insights gained from work on three-dimensional systems, such as ionic solutions.[40]

For a dilute plasma, such as occurs in the upper atmosphere, the ions may be bound together in pairs as neutral diatomic molecules, whose diameter depends on the sizes of the two ions, or the ions may be separated from their opposite partners. In response to an electric field, either external, or from one of the free ions, the surrounding plasma is polarized, and this has the effect of screening out the long-range effect of the field due to an ion. In analogy with the dynamics of vortices, for which the $x$ and $y$ coordinates are conjugate variables (a result which Horace Lamb[15] quotes from a textbook on Mechanics by Kirchhoff), the kinetic energy is constant, and quantized in units of two-dimensional area. In the case of a logarithmic dependence of the energy on separation, this screening is incomplete, and we had to adapt renormalization group methods developed by Anderson, Yuval and Hamann[3,4] and by Wilson and Fisher[24,25] to solve this problem. Our initial efforts to make these methods fit our problem were inadequate,[17,41] and later developments are discussed in more detail in Sec. 1.5.

### 1.4.4. *Two-dimensional crystals*

In our early papers[16,17] we discussed the analogy between the vortex-driven superfluid to normal transition in helium films and an analogous melting transition between solid and liquid phases in thin films. We got part of this right, but missed some essential features. These missing features were pointed out and corrected by Halperin and Nelson[42] and by Peter Young.[43]

There are two important differences between liquids and solids. The most obvious one is that an ideal solid is rigid and does not yield to a small stress, but recovers its original shape when the small stress is removed, while a liquid will flow without any memory of its original shape. The second difference is that a liquid is isotropic, with no preferred directions in space, while most solids are crystalline, even if their crystal structure is not immediately obvious to the eye, and a single crystal of a solid has different mechanical, electrical and optical properties in different directions. This crystal structure was first studied in detail by X-ray crystallography, which was developed a hundred years ago by Laue and his collaborators and by the Braggs. In fact, the possible periodic arrays that could be formed by identical atoms were identified, and classified by the use of group theory, and by the examination of the symmetries of real crystals, during the last half of the nineteenth century.

Despite these sharp differences between typical solids and typical liquids, there are actually various intermediate examples known. For example, ordinary silica-based glass has no orientational order, but glasses have existed for hundreds or even thousands of years without showing any signs of losing their shape because of flow induced by gravitational forces. In contrast there are materials, now known as liquid crystals, which flow freely like liquids, but which have local orientational order. Since the 1930s it has been known that it is the presence of *dislocations* in imperfect metals that allows them to flow, or *creep*, under a shearing stress, although this is not like the flow of a liquid, whose rate is proportional to the applied stress, but the rate of creep is very slow under a low stress. This led to the idea that the melting transition in a solid might be due to the spontaneous appearance of thermodynamically stable dislocations at the melting temperature.[44]

In a regular triangular lattice, even if it is disordered by thermally excited sound waves, each atom has six immediate neighbors. Even when thermal excitation of vacancies and interstitials is taken into account, the shape remains stable against stress,[45] and the anisotropy revealed by the X-ray diffraction pattern remains.[46] In 1934 papers appeared more or less simultaneously, written by E. Orowan,[47] M. Polanyi,[48] and G. I. Taylor,[49] which described how what are now known as *dislocations* can greatly reduce the energy barriers to the flow of a solid in response to stress. In the ideal two-dimensional case, shearing stress can only be relieved by a whole line of atoms slipping over an adjacent line, and this involves an energy barrier proportional to the length of the line. If there are dislocations in the solid, which involve partial extra (or missing) lines of atoms that terminate at the

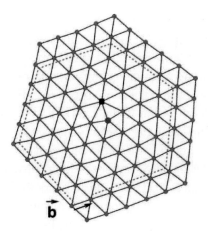

Fig. 1.1. A dislocation in a triangular lattice with an extra line of particles inserted. The dashed line is a Burgers circuit which would close in a perfect lattice but fails to close by a Burger's vector **b**. Note that the dislocation core has a five- and a seven-fold coordinated particles one lattice spacing apart and the Burger's vector, **b**, is perpendicular to the bond joining these. These particles can be regarded as two disclinations separated by one lattice spacing. Reprinted from Fig. 27 of Ref. 50, with permission from Elsevier.

dislocation core inside the solid, as shown in Fig. 1.1, the dislocation can respond to shear stress, with the help of thermally activated vacancy sites and interstitials, by changing its track one or two atoms at a time, so the energy barrier is much lower. The presence of such dislocations in most solids, particularly in metals, explains why plastic flow does actually occur in them, particularly at high temperatures; certain metals, such as copper, lead and gold are quite soft when their purity is high, because their dislocations are free to move.

For each dislocation in a plane lattice there is an extra row of atoms which starts at the site of the dislocation and terminates at the boundary of the crystal. As a result, a path from atom to atom which would be a polygon around the site of the dislocation fails to meet by an amount known as the Burgers vector of the dislocation, as is shown in Fig. 1.1. This vector is perpendicular to the extra row of atoms. There can be bound pairs of dislocations with opposite Burgers vectors, which correspond to extra (or missing) finite rows of atoms between the positions of the two opposite dislocations.

We argued that, since the dislocation in two dimensions is a point defect, like the vortex in a superfluid, and the energy of an isolated dislocation depends logarithmically on the size of the system, there should also be a defect driven melting transition in a two-dimensional solid at the temperature

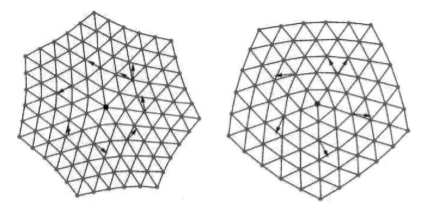

Fig. 1.2. $\pm\pi/3$ disclinations in an underlying triangular lattice. On a circuit enclosing the disclination cores, there is an orientational mismatch by $\pm\pi/3$: $\oint_C d\theta = 2\pi \pm \pi/3 = 2\pi \pm \pi/3$. Reprinted from Fig. 28 of Ref. 50, with permission from Elsevier.

where the energy and entropy terms balance one another.[16,17] There were two significant errors in this argument, which Halperin and Nelson,[42] and Young[43] pointed out. The first error is that the coefficient of the interaction energy of a dislocation pair is not isotropic, like the interaction between vortices, but depends on the angle between the displacement of one dislocation from the other and the direction of the Burgers vector (the normal to the line of extra atoms). The more important correction is that the presence of free dislocations is not enough to destroy the orientational order of the solid, but a further transition must occur.

The point defects that destroy orientational order are known as *disclinations*,[44] which in a two-dimensional triangular lattice are groups of nearest neighbors that form a pentagon or heptagon rather than the hexagon of the typical group of neighbors. In Fig. 1.1 one can see that the core of the dislocation is made up of one pentagon and one adjacent heptagon, which can be regarded as a pair of oppositely charged disclinations in nearest neighbor positions.[44] Isolated disclinations are shown in Fig. 1.2. Along a loop that winds once round such a disclination, parallel transport of a coordinate frame oriented with the local crystal axes would rotate it by an angle $-2\pi/5$ or $+2\pi/7$, and $\pm\pi/3$ relative to an undistorted, triangular reference lattice, according to the sign of the disclination. A random set of positive and negative dislocations does not destroy orientational order in a solid, and so the transition from ordered, rigid solid to disordered fluid can take place through two distinct defect-driven phase transitions.[42,43] At the first transition, the dislocation pairs that form at low temperatures become dissociated to form an oriented fluid known as a *hexatic liquid crystal*. Disclinations play no role in the melting of an elastic solid because there is true long range orientational order[56] and the energy of an isolated disclination is proportional to the area of the elastic system.[45] At the higher temperature transition, the disclination pairs that form a free dislocation dissociate to create an unoriented fluid of the usual type. This is discussed in more detail in Secs. 1.6.3 and 1.7.2.

Shortly after we had written this up and distributed preprints we received a preprint from Feynman which contained the same argument, but with a different expression for the dependence of the energy of a free dislocation on the elastic moduli. Feynman's disagreement with us was worrying, but we found to our relief that we had got it right. Feynman did not actually publish his preprint, but an account of it was included in a paper by Elgin and Goodstein.[51]

### 1.4.5. *Thin film superconductors*

A sufficiently thin film of superconductor can be regarded as two-dimensional and can be described by Ginsburg-Landau free energy[52] for a charged condensate of Cooper pairs,

$$F[\Psi] = \int d^2 r dz \left\{ \frac{1}{2m^*} \left| \left( -i\hbar \nabla - \frac{e^*}{c} \mathbf{A} \right) \Psi \right|^2 \right.$$

$$\left. + \frac{1}{2} r(T) |\Psi|^2 + \frac{1}{4} u |\Psi|^4 + \frac{B^2}{8\pi} - \frac{\mathbf{H} \cdot \mathbf{B}}{4\pi} \right\}. \quad (1.14)$$

The field $\Psi(\mathbf{r})$ is a coarse grained Cooper pair condensate wave function which is finite in the plane of the film, $\mathbf{A}(\mathbf{r}, z)$ is the magnetic vector potential, $\mathbf{B}(\mathbf{r}, z) = \nabla \times \mathbf{A}(\mathbf{r}, z)$. The parameter $r(T)$ is negative below the critical temperature of the film in the absence of a magnetic field, and $u$ is positive and assumed to be temperature independent. Here, $\mathbf{r} = (x, y)$ is a point in the film which lies in the plane $z = 0$, the effective mass $m^* = 2m_e$, and $e^* = 2e$. The important difference between the uncharged superfluid of $^4$He and a charged superfluid of a superconductor is that a vortex in a superconductor has a circulating electric current producing an associated magnetic field, $\mathbf{B}(\mathbf{r}, z)$, which exists in the space outside the film at $(\mathbf{r}, 0)$. The behavior of a superconducting film was first discussed by Pearl[53] who found that the circulating current of a vortex at $\mathbf{r} = 0$ falls off as $1/r^2$, instead of exponentially as in a bulk superconductor. Also, the screening length in a 2D superconducting film is $\lambda_{2d} = \lambda_0^2/d$, where the London penetration depth $\lambda_0$ is of order 1 $\mu$m, and $d$ is the film thickness. To ensure that a real film can be approximated by a 2D film, one must have $d \ll \lambda_0$ so that there is no variation over the film thickness.

On scales larger than the superconducting coherence length, $\xi_0 = \sqrt{\hbar^2/(m^*|r(T)|)}$, which is the length over which the order parameter grows from zero at a vortex core to its mean value, $|\Psi_0| = \sqrt{-r(T)/u}$, well away from the core, the phase only approximation $\Psi(\mathbf{r}, z) = \Psi_0 e^{i\theta(\mathbf{r})}$ is valid, and the free energy of Eq. (1.14) becomes

$$F = \int d^2 r\, dz \left\{ \frac{\rho_s^0}{2}\left(\frac{\hbar}{m^*}\right)^2 \left(\nabla\theta - \frac{e^*}{\hbar c}\mathbf{A}\right)^2 + \frac{1}{8\pi}(\nabla\times\mathbf{A})^2 \right\}$$
$$+ \text{constant}, \qquad (1.15)$$

where $\rho_s^0 = m^*|\Psi_0|^2$ is the mass density of the Cooper pairs. From the variational equation $\delta F/\delta \mathbf{A} = 0$, one obtains

$$\nabla\times(\nabla\times\mathbf{A}) = \nabla\times\mathbf{B} = \frac{4\pi \mathbf{J}}{c} = 4\pi|\Psi_0|^2 \frac{e^*\hbar}{m^* c}\left(\nabla\theta - \frac{e^*}{\hbar c}\mathbf{A}\right),$$

$$\Rightarrow \mathbf{J}(\mathbf{r}, z) = e^*|\Psi_0|^2 \frac{\hbar}{m^*}\left(\nabla\theta - \frac{e^*}{\hbar c}\mathbf{A}\right). \qquad (1.16)$$

To see the effect of the magnetic field $\mathbf{B}$ due to a set of vortices in the superconducting plane at $z = 0$, we follow the analysis of Nelson's book[54] by assuming that the superconducting current is in the plane $z = 0$, so that

$$\mathbf{J}(\mathbf{r}, z) = \frac{c\phi_0}{8\pi^2\lambda_{\text{eff}}}\left(\nabla\theta(\mathbf{r}) - \frac{2\pi}{\phi_0}\mathbf{A}_{2d}(\mathbf{r})\right)\delta(z), \qquad (1.17)$$

where $\lambda_{\text{eff}} = \lambda_0^2/d$ is the effective penetration depth in the superconducting film,[53] and $\phi_0 = 2\pi\hbar c/e^*$ is the flux quantum. Now we can use the standard equations of magnetostatics to solve for the current due to a vortex of strength $m$ at $\mathbf{r} = 0$, so that

$$-\nabla^2 \mathbf{A}(\mathbf{r}, z) + \frac{1}{\lambda_{\text{eff}}}\mathbf{A}_{2d}(\mathbf{r})\delta(z) = \frac{m\phi_0}{2\pi\lambda_{\text{eff}}}\frac{\hat{\mathbf{z}}\times\mathbf{r}}{r^2}\delta(z), \qquad (1.18)$$

since $\theta(\mathbf{r})$ due to the vortex at the origin $\mathbf{r} = 0$ is $\nabla\theta(\mathbf{r}) = 2\pi m(\hat{\mathbf{z}}\times\mathbf{r})/r^2$, and $\mathbf{A}_{2d}(\mathbf{r}) = \mathbf{A}(\mathbf{r}, z=0)$. This can be solved by taking Fourier transforms, with the result that the current in the superconducting plane behaves like a neutral superfluid for $r \ll \lambda_{\text{eff}}$ with

$$\mathbf{J}_{2d}(\mathbf{r}) = \frac{mc\phi_0}{8\pi^2 d\lambda_{\text{eff}} r}\hat{\theta}, \qquad (1.19)$$

and, for $r \gg \lambda_{\text{eff}}$,

$$\mathbf{J}_{2d}(\mathbf{r}) = \frac{mc\phi_0}{4\pi^2 d r^2}\hat{\theta}. \qquad (1.20)$$

This translates into a $\ln r$ vortex–vortex interaction as in a neutral superfluid for $r < \lambda_{\text{eff}}$, and a $1/r$ interaction for $r > \lambda_{\text{eff}}$.

Since the interaction between vortices in a superconducting film falls off faster than $\ln r$, this implies that there is no transition to a true superconducting state in the thermodynamic limit.[17] However, since the effective penetration depth $\lambda_{\text{eff}}$ can be as large as $\mathcal{O}(10^{-2})$m, which is larger than the typical film, the behavior of the system at low temperatures will be indistinguishable from that of a true phase transition rounded by finite size effects. The most common measurement on a superconducting film is the *IV* characteristics at various temperatures $T$. The theory discussed here is not adequate to predict the *IV* characteristics because the voltage $V$ produced by the applied current $I$ is due to the unbinding of vortices, whose subsequent motion produces the measured voltage $V$.[55]

## 1.5. Scaling Theory

Kosterlitz and Thouless did have one advantage over other more knowledgeable workers who were investigating these low dimensional systems in which fluctuations dominate the behavior. Thouless had always been a bit of a maverick who loved unusual physics problems and Kosterlitz was too ignorant of statistical mechanics to realize that the problem of phase transitions in $2D$ magnets, superfluids and crystals was either absurd or too difficult to attempt. It might be absurd because of some well known rigorous theorems about the absence of long range order at any finite temperature in such systems.[28,29,56] Almost all physicists took these theorems to mean that a phase transition could not exist in these systems, implying that the problem was absurd. However, careful reading of Mermin's papers revealed that he did not exclude the possibility of a transition between different states at a finite temperature, but he did prove the absence of true long range order at any $T > 0$. The conventional wisdom at that time was based on Lev Landau's mean field theory,[57] in which a transition between low and high $T$ states involves the destruction of long range order, which was assumed to characterize the low $T$ state, and to be absent in the high $T$ state. Hence, the absence of long range order for all finite temperatures meant that the idea of a phase transition was unlikely, although Imry and Gunther had pointed out that the formation of a two-dimensionally ordered phase would lead to a characteristic signal in the X-ray scattering. Even if a transition did exist, it was clear that large, strongly interacting fluctuations were involved. There was a major dearth of techniques to handle such strongly interacting degrees of freedom, with the exception of perturbation expansions which were technically difficult and whose meaning was obscure, so the chances of success were vanishingly small.

Undeterred by the general view that the problem was either absurd or impossible, Kosterlitz, out of ignorance, and Thouless, out of curiosity, went ahead and essentially solved the problem. In their first paper on superfluids and solids in two dimensions,[16] the concept of topological order was introduced for systems which have no conventional long range order at any finite temperature. There it was argued that the important quantity which distinguishes an ordered phase from a disordered one is the rigidity of the system, and this is controlled by the presence or absence of free dislocations or vortices. In the ordered phase, these topological excitations are bound in pairs, but, when they unbind and become free to move, the rigidity vanishes and the two-dimensional solid responds to a shear stress like a fluid. Exactly analogous arguments can be constructed in terms of the unbinding of vortex pairs into free vortices to discuss the destruction of superfluidity in a two-dimensional film of superfluid $^4$He. The argument is as follows: the energy of an isolated vortex in a $2D$ $XY$ magnet of size $L$ with nearest neighbor coupling constant $J$ is $2\pi J \ln(L/a)$ and the associated entropy is $k_B \ln(L^2/a^2)$ so that the free energy is just $\Delta F = (\Delta E - T\Delta S) = 2(\pi J - k_B T)\ln(L/a)$. For a very large system $L \to \infty$, $\Delta F \to +\infty$ for $k_B T < \pi J$, which corresponds to no free vortices and finite rigidity, while, for $k_B T > \pi J$, $\Delta F \to -\infty$, the probability of thermally activated vortices approaches unity. This means that the generalized rigidity vanishes or that the system is disordered.

Once the mechanism and the relevant excitations had been identified, Kosterlitz and Thouless decided to attempt to make their ideas more quantitative. This took them into completely unexplored regions of statistical mechanics. The first essential step was to write the Hamiltonians for the superfluid film and the $2D$ solid in terms of the relevant excitations. We knew that the lowest energy excitations, spin waves in the magnet and phonons in the solid, although they destroyed the conventional long range order in two-dimensional $XY$ magnets, superfluids and solids,[28,29,56] these excitations did not induce true disorder with the characteristic exponential decay of correlation functions in a high temperature phase, nor did they reduce the rigidity of the systems. Define the phase angle $\theta(\mathbf{r})$ of the order parameter $\psi(\mathbf{r})$ by

$$\psi(\mathbf{r}) = |\psi|e^{i\theta(\mathbf{r})}, \quad \theta(\mathbf{r}) = \phi(\mathbf{r}) + \sum_{\mathbf{R}} n(\mathbf{R})\Theta(\mathbf{r},\mathbf{R}), \quad (1.21)$$

where $\phi(\mathbf{r})$ corresponds to the smooth variations of the phase angle $\theta(\mathbf{r})$, and $\Theta(\mathbf{r},\mathbf{R}) = \tan^{-1}[(y-Y)/(x-X)]$ is the contribution to $\theta(\mathbf{r})$ of a vortex unit circulation at the point $\mathbf{R}$ on the sites of the dual lattice. The source field

$n(\mathbf{R})$ is $\sum_i n_i \delta(\mathbf{R} - \mathbf{R}_i)$ where $n_i$ is the quantized circulation of the vortex at $\mathbf{R}_i$. After some algebra, we obtained an expression for the energy of a configuration of phase angles in the presence of an arbitrary set of vortices $n(\mathbf{R})$ in Eq. (50) of Ref. 17,

$$\frac{H}{k_B T} = \frac{H_{sw}}{k_B T} + \frac{H_v}{k_B T},$$

$$\frac{H_{sw}}{k_B T} \approx \frac{1}{2} K_0(T) \int d^2\mathbf{r} (\nabla \phi(\mathbf{r}))^2, \qquad (1.22)$$

$$\frac{H_v}{k_B T} \approx -\pi K_0(T) \int_{|\mathbf{R}-\mathbf{R}'|>a} d^2\mathbf{R} d^2\mathbf{R}' n(\mathbf{R}) n(\mathbf{R}') \ln \frac{|\mathbf{R}-\mathbf{R}'|}{a}$$

$$- \ln y_0 \int d^2\mathbf{R} n^2(\mathbf{R}).$$

We have eliminated the last term in the expression we gave in Eq. (50) of Ref. 17, which was proportional to the net charge and to $\ln(L/a)$, because the correct expression for the energy of a nonzero net charge is actually sensitive to boundary conditions. For the case of a Coulomb gas enclosed by a conducting boundary, circular or polygonal, each charge is accompanied by image charges, which attract it to the boundary, so the $\ln(L/a)$ should be replaced by a quantity of order unity. The same happens for the vortex gas in a superfluid. We therefore have a good reason to ignore this term.

This expression describes the classical $XY$ magnet with the parameter $K_0(T)$ equal to $J/k_B T$, were $J$ is the nearest neighbor interaction energy, and $K_0(T)$ is $(\hbar/m)^2(\rho_s^0/k_B T)$ for a superfluid film; $\rho_s^0$ is the bare superfluid mass per unit area. The parameter $y_0 = \exp(-E_c/k_B T)$ is the vortex fugacity whose exact value turns out to be unimportant. The expression for $H_{sw}$ ignores both the nonlinear interactions of the spin waves with one another, and their scattering by the vortices. The vortex core energy, $E_c$, can be regarded as the contribution of the region close to the vortex core, $r \leq a$, where the magnitude $|\psi(\mathbf{r})|$ is varying rapidly and $\mathbf{v}_s = \nabla \theta$ is very large, so that the quadratic form of the action is no longer a good description. The short distance cutoff, $a$, can be regarded as the size of a vortex core and $L \gg a$ is the linear size of the system.

Note that, in the planar magnet, the exchange constant $J$ determines both $K_0(T)$ and the fugacity $y_0$ so that the two parameters are actually not independent, although they are treated as independent parameters. This is convenient as the calculations involve a perturbation expansion in $y_0(T)$ which violates the relation between the fugacity and $K_0(T)$. This does not break the $O(2)$ symmetry of the original problem and turns out not to affect the quantities of interest, as treating $y_0$ as an independent variable is just altering an irrelevant variable which is known not to affect physical quantities. In 1972, Kosterlitz and Thouless did not understand Wilson's seminal theory[24,30] of critical phenomena and had no idea of the renormalization group classification of relevant, irrelevant and marginal operators, but this approach just felt right. Also, we were very influenced by the similarity to the $1D$ Ising ferromagnet with $1/r^2$ interactions, solved by Anderson, Yuval and Hamann.[6] Both problems contain point defects with logarithmic interactions between them, and the treatment by Anderson, Yuval and Hamann also involved a fugacity and an interaction strength which scale differently, but whose initial values are related in an irrelevant fashion.

Smooth variations of the phase angle are controlled by $H_{sw}$ in Eq. (1.22) and lead to algebraic decay of the two-point correlation function,[20]

$$\langle e^{i(\phi(\mathbf{r})-\phi(0))}\rangle \sim r^{-1/2\pi K_0(T)} . \tag{1.23}$$

From this, we see that fluctuations destroy long range order in $2D$, since the magnetization $m = \langle e^{i\phi(\mathbf{r})}\rangle \sim L^{-1/4\pi K_0(T)}$, where $L$ is the linear size of the system. This approximation is insufficient to induce a transition to a true disordered phase in which one expects this correlation function to decay exponentially. To proceed further, we were faced with the daunting task of considering the problem with the Hamiltonian of Eq. (1.22) where $H_v$ is the Hamiltonian for a set of point vortices $\{n(\mathbf{R})\}$ interacting with a $\ln r$ Coulomb potential. We assume that the system has no net charge in the absence of an explicit external electrostatic potential.

We considered[17] the problem of a neutral set of charges $n(\mathbf{R}) = \pm 1$ since such charges have the largest fugacity and are the most probable, and ignored larger charges because these are less probable. We considered a pair of opposite charges, a distance $r$ apart, and concluded that they would be screened by the polarization of smaller pairs which were screened by yet smaller pairs. The main result of this paper is that there is a scale dependent dielectric constant $\epsilon(r) = K_0/K(r)$ where the force between a pair of opposite unit charges separated by $r$ is $2\pi K(r)/r$. The potential energy $U_{\text{eff}}(r)$ of this pair is

$$U_{\text{eff}}(r) = 2\pi \int_a^r dr' \frac{K(r')}{r'} \equiv 2\pi f(r) \ln\left(\frac{r}{a}\right), \tag{1.24}$$

and the results of Ref. 17 are equivalent to

$$K^{-1}(r) = K_0^{-1} + 4\pi^3 y_0^2 \int_a^r \frac{dr'}{a} \left(\frac{r'}{a}\right)^{3-2\pi f(r')} . \tag{1.25}$$

We were unable to solve this and made an unfortunate and unnecessary approximation by replacing $f(r')$ by $K(r)$ in Eq. (1.25) and solving self-consistently for $K(r)$. This replacement of $f(r')$ by $K(r)$ was justified by the smallness of their difference, but it did lead to some strange and incorrect predictions.[17]

In an important paper[59] the approximation was shown to be unnecessary and a correct procedure given. Define a scale dependent fugacity by

$$y(r) = y_0 \left(\frac{r}{a}\right)^{2-\pi f(r)}$$
$$= y_0 \exp\left(2\ln(r/a) - \pi \int_a^r dr' \frac{K(r')}{r'}\right), \quad (1.26)$$
$$K^{-1}(r) = K_0^{-1} + 4\pi^3 \int_a^r \frac{dr'}{r'} y^2(r'),$$

so that, by differentiating these with respect to $\ln r$,

$$\frac{dy(r)}{d\ln r} = (2 - \pi K(r))y(r),$$
$$\frac{dK^{-1}(r)}{d\ln r} = 4\pi^3 y^2(r), \quad (1.27)$$

which are exactly the recursion relations derived earlier by one of us.[41] However, even if we had not made our unnecessary approximation and derived these recursion relations, it is not clear that we would have known what to do with them as renormalization group equations for scale dependent coupling constants were unknown to the majority of physicists in 1972.

Kosterlitz was not happy with the results of our self-consistent calculations, and had been given a preprint of the paper by Anderson, Yuval and Hamann,[6] which opened his eyes to strange ideas like scale dependent parameters, scaling and the renormalization group derived by performing a partial trace in a partition function for the 1D Ising model with $1/r^2$ interactions between spins. There seemed to be a definite connection between our 2D Coulomb gas problem of Eq. (1.22) and the domain wall representation of the 1D Ising model with $1/r^2$ interactions.[5,6] This connection motivated him to spend some months reading and rereading the papers until he understood exactly what the recursion relations meant and how they were derived. This seemed like a worthwhile exercise because the method reduced the impossible task of computing the exact partition function for the $1/r^2$ Ising model to the more feasible task of deriving and solving a few differential equations, which gave most of the interesting information contained in the inaccessible partition function. If one could reduce the vortex representation of the 2D

$XY$ system of Eq. (1.22) to a few analogous differential equations, the few months of effort would be well spent. Finally, Kosterlitz was able to derive the recursion relations of Anderson et al.[6] and, more importantly, understand the physical meaning of the mathematical manipulations involved. He was then able to apply these ideas to the problem of interest, which is the partition function of the 2D Coulomb gas,

$$Z(K_0, y_0) = Z_0(K_0) \sum_{n=0}^{\infty} \frac{y^{2n}}{n!^2} \int \prod_{i=1}^{2n} d^2\mathbf{R}_i \exp\left[+\pi K_0 \sum_{i\neq j}^{2n} n_i n_j \ln\left(\frac{|\mathbf{R}_i - \mathbf{R}_j|}{a}\right)\right],$$

$$Z_0(K_0) = \int \mathcal{D}\phi \, \exp\left(-\frac{K_0}{2} \int d^2\mathbf{r}(\nabla\phi(\mathbf{r}))^2\right).$$

(1.28)

Here, $Z_0(K_0)$ is the trivial Gaussian spin wave partition function, and, in the Coulomb gas part of the partition function, the integrations over the vortex positions $\mathbf{R}_i$ are restricted by the hard core condition $|\mathbf{R}_i - \mathbf{R}_j| > a$, where the short distance cut-off $a$ can be taken as a lattice spacing or the diameter of a vortex core. It turns out that this cut-off length $a$ is, to a large degree, arbitrary, but it cannot be set to zero before the end of the calculations. Similarly, the infrared cut-off, $L$ the size of the system, must be kept finite until the end of the calculations to avoid unphysical divergences. We note that both infrared and ultra-violet cut-offs have clear physical interpretations in this case, and there is no temptation to take the limits $a \to 0$ or $L \to \infty$ unless it is obvious that such limits make physical sense. In fact, these cut-offs appear only in the combination $L/a$ and the appropriate limit is $L/a \to \infty$ which has an obvious physical meaning.

The original derivation of the renormalization group equations[41]

$$\frac{dK^{-1}}{dl} = 4\pi^3 y^2 + \mathcal{O}(y^4),$$

$$\frac{dy}{dl} = (2 - \pi K)y + \mathcal{O}(y^3), \quad (1.29)$$

$$\frac{df}{dl} = 2f + 2\pi y^2 + \mathcal{O}(y^4),$$

was carried out in the most complicated way possible by integrating out the short distance degrees of freedom in the partition function of Eq. (1.28) where the lattice spacing has been rescaled by $a \to ae^{\delta l} = a(1+\delta l)$. At length scale $a(l) = ae^l$ the effective interaction of a fugacity and free energy density are $K = K(l)$, $y = y(l)$, and $f = f(l)$ respectively. Later work derived these recursion relations by technically easier but more sophisticated methods.[18,60]

The most straightforward method[18] starts with the superfluid momentum density correlation function,

$$C_{\alpha\beta}(\vec{q}, K_0, y_0) = \langle g^s_\alpha(\vec{q}) g^s_\beta(-\vec{q})\rangle = A(q)\frac{q_\alpha q_\beta}{q^2} + B(q)\left(\delta_{\alpha,\beta} - \frac{q_\alpha q_\beta}{q^2}\right). \tag{1.30}$$

Hohenberg and Martin[61] showed that the renormalized or measured superfluid density $\rho_s^R(T)$ is given by

$$K_R(T) = \frac{\hbar^2 \rho_s^R(T)}{M_A^2 k_B T} = \lim_{q\to 0}(A(q) - B(q)) = K_0 - 4\pi^2 K_0^2 \lim_{q\to 0}\frac{\langle n(\vec{q})n(-\vec{q})\rangle}{q^2}$$

$$= K_0 + \pi^2 K_0^2 \int d^2\mathbf{r}\, r^2 \langle n(\mathbf{r})n(0)\rangle. \tag{1.31}$$

The vorticity–vorticity correlation function, $\langle n(\mathbf{r})n(0)\rangle$, is calculated as a power series in the fugacity $y_0$ and, to lowest order, one easily obtains

$$K_R^{-1} = K_0^{-1} + 4\pi^3 y_0^2 \int_a^\infty \frac{dr}{a}\left(\frac{r}{a}\right)^{3-2\pi K_0} + \mathcal{O}(y_0^4), \tag{1.32}$$

which is essentially identical to Eq. (1.25) obtained from the self-consistent theory.[59] The perturbative treatment makes explicit that this is correct to lowest order in the vortex fugacity $y_0$. The RG equations are obtained from Eq. (1.32) by writing

$$\int_a^\infty = \int_a^{a(1+\delta l)} + \int_{a(1+\delta l)}^\infty, \tag{1.33}$$

and defining a rescaled fugacity $y(l+\delta l)$ and interaction $K(l+\delta l)$ by

$$y(l+\delta l) = (2 - \pi K(l))y(l)\delta l,$$
$$K^{-1}(l+\delta l) = K^{-1}(l) + 4\pi^3 y^2(l)\delta l, \tag{1.34}$$

where $K(l)$ and $y(l)$ are the coupling constant and fugacity at scale $l$, when the cut-off $a$ has been rescaled to $ae^l$. At this scale, the effective Hamiltonian of the system $\mathcal{H}(K(l), y(l))$ of Eq. (1.22) describes fluctuations on length scales larger than $ae^l$ while the original physical Hamiltonian $\mathcal{H}(K_0, y_0)$ describes fluctuations on all scales larger than $a$. This is precisely Wilson's renormalization group idea[24,30] as applied to the two-dimensional Coulomb gas. Writing Eq. (1.34) as a pair of differential equations, we obtain the recursion relations

$$\frac{dK^{-1}(l)}{dl} = 4\pi^3 y^2(l) + \mathcal{O}(y^4),$$
$$\frac{dy(l)}{dl} = (2 - \pi K(l))y(l) + \mathcal{O}(y^3), \tag{1.35}$$

which are identical to those of Eq. (1.27) except that this treatment explicitly displays that the recursion relations are a perturbation expansion in the fugacity $y(l)$. This derivation, using the renormalized stiffness constant $K_R$, has the big advantage that the RG equations are derived using the perturbation expansion of $K_R(K, y)$, so that one can immediately write the exact relation,

$$K_R(K_0, y_0) = K_R(K(l), y(l)). \tag{1.36}$$

This is a disguised version of the Josephson scaling relation[62] which, in $d$ dimensions, reads

$$K_R(K_0, y_0) = e^{(2-d)l} K_R(K(l), y(l)). \tag{1.37}$$

All the physics of the transition is contained in Eqs. (1.35) and (1.36) and we now have to solve these simple differential equations. It is clear that transition occurs at the fixed point $y(l) = 0$ and $\pi K(l) = 2$, so we write $\pi K(l) = 2 + x(l)$ so that, to lowest order in $x(l)$ and $y(l)$, Eq. (1.35) becomes

$$\begin{aligned} \frac{dx}{dl} &= -16\pi^2 y^2, \\ \frac{dy}{dl} &= -xy. \end{aligned} \tag{1.38}$$

The solutions $x(l)$ and $y(l)$ are related by

$$x^2(l) - 16\pi^2 y^2(l) = C(t), \tag{1.39}$$

where $C(t)$ is a constant of integration, which depends on $t = (T - T_c)/T_c$. As the scale $l$ increases, $x(l)$ and $y(l)$ sweep out a set of hyperbolae in the $y > 0$ half plane. Physically, since $y$ is the vortex fugacity, the density of vortices is proportional to $y$ and $y < 0$ is unphysical. We identify the line $y = 0$ as the spin wave phase since there are no vortices or dislocations in the system if $y = 0$. This phase has finite rigidity and is identified as the low $T$ ordered phase. On the other hand, a finite fugacity $y$ implies a finite density of free defects in the system, which means that the rigidity vanishes at long length scales. We identify this situation as a high temperature disordered phase.

Having identified the possible behaviors of the system by physical arguments, it remains to demonstrate that the renormalization group flows of Eq. (1.35) and the approximate form of Eq. (1.38) yield this expected behavior. The integration constant $C(t)$ can be taken to be zero at $T = T_c$ and it is straightforward to convince oneself that the $x, y$ plane splits into three different regions: (I) $C(t) \geq 0$, $x(l) \geq 4\pi y(l) \geq 0$, (II) $C(t) < 0$,

$|x(l)| < 4\pi y(l)$, and (III) $C(t) > 0$, $x(l) < 0$, $|x(l)| \geq 4\pi y(l)$. Regions (I)–(III) require separate solutions of Eq. (1.38). We choose $C(t) = -b^2 t$ where $t = (T - T_c)/T_c$ is the fractional deviation of $T$ from $T_c$. This linear approximation for the constant $C$ is adequate for $|t| \ll 1$ as there is every reason to expect that $C(t)$ is an analytic function of $t$ and that, to lowest order in $t$, it is linear. We identify $C(t) = 0$ as corresponding to $T = T_c$ because the trajectories $x(l; C)$ and $y(l; C)$ behave differently in the three regions.

In region I, $x(l) > 0$ and $C(t) > 0$, and we have

$$\frac{dx}{dl} = -16\pi^2 y^2,$$

$$\frac{dy}{dl} = -xy,$$

$$x^2 - 16\pi^2 y^2 = C > 0 \Rightarrow \int_{x_0}^{x(l)} dx \left( \frac{1}{x - \sqrt{C}} - \frac{1}{x + \sqrt{C}} \right) \quad (1.40)$$
$$= -2l\sqrt{C},$$

$$x(l) = \sqrt{C} \, \frac{x_0 + \sqrt{C} + (x_0 - \sqrt{C})e^{-2l\sqrt{C}}}{x_0 + \sqrt{C} - (x_0 - \sqrt{C})e^{-2l\sqrt{C}}}.$$

To examine the behavior of $x(l)$ in various limits, we can take the deviation from $T_c$ extremely small so that $x_0 \gg \sqrt{C}$ and $2l\sqrt{C} > 1$ or $2l\sqrt{C} < 1$ in different regimes which allows us to define a correlation length $\xi_-(t)$ separating the regimes. We have

$$\xi_-(t) = e^{\frac{1}{2\sqrt{C(t)}}} = e^{1/2b\sqrt{|t|}},$$
$$x(l) = \frac{x_0}{1 + x_0 l} \quad \text{for } e^l \ll \xi_-(t). \quad (1.41)$$

For $e^l > \xi_-(t)$, we immediately see that

$$x(l) = \sqrt{C}\left(1 + 2e^{-2l\sqrt{C}}\right) \quad \text{and} \quad y(l) = \frac{\sqrt{C}}{2\pi} e^{-l\sqrt{C}}. \quad (1.42)$$

This result gives the behavior of the renormalized stiffness constant $K_R(T)$ when $T$ is slightly below $T_c$. From Eq. (1.36), we have

$$K_R(T) = K_R(K(l), y(l)) = K_R(K(\infty), 0)$$
$$= K(l = \infty) = \frac{2}{\pi} + b|t|^\nu, \quad (1.43)$$

with $\nu = 1/2$. Here, we have used $y(\infty) = 0$ when the effective Hamiltonian is a Gaussian with coupling constant $\pi K(\infty) = 2 + x(\infty) = 2 + \sqrt{C}$, from

Eq. (1.42). Thus, we can deduce the behavior of the measurable stiffness constant $K_R(T)$ in the whole region, $T \leq T_c$. For $T \ll T_c$, we expect that $K_R(T)$ will decrease linearly from its $T = 0$ value as $T$ increases until a few percent below $T_c$. At this temperature, we expect that vortex fluctuations will have some effect and, in the final approach to $T_c^-$, the stiffness $K_R(T)$ plunges to its universal value of $2/\pi$ as $\sqrt{|t|}$.[18]

In region II, $x(l)$ can have either sign and $C(t) = -b^2 t < 0$, so that
$$16\pi^2 y^2 = (x^2 + |C|),$$
$$-l = \int_{x_0}^{x(l)} \frac{dx}{x^2 + |C|} = |C|^{-1/2} \left( \tan^{-1}\left(\frac{x(l)}{\sqrt{|C|}}\right) - \tan^{-1}\left(\frac{x_0}{\sqrt{|C|}}\right) \right). \quad (1.44)$$

Here, one chooses the upper limit of integration to be at $l$ so that $x(l) = -x_0$, and we note that we can choose the deviation from $T_c$ such that $\sqrt{|C(t)|} = b\sqrt{t} \ll x_0$, so that Eq. (1.44) reads $l = \pi |C(t)|^{-1/2}$, from which we can define a correlation length
$$\xi_+(t) = e^{\pi/b\sqrt{t}}, \quad \xi_-(t) = e^{1/2b\sqrt{t}}. \quad (1.45)$$

This theory of the $2D$ planar rotor model and of a superfluid film has two length scales $\xi_-(t)$ and $\xi_+(t)$, both diverging exponentially at the critical point. The length scales $\xi_-$ and $\xi_+$ can be interpreted as controlling the distribution of the defects or vortices in the system. At low $T$, neutral pairs of equal but opposite vorticity are bound together as neutral pairs and the length scale $\xi_-(t)$ is interpreted as the maximum size of a vortex pair. From Eq. (1.42), the density of vortex pairs separated by more than $\xi_-(t)$ is essentially zero since the fugacity $y(l)$ is essentially zero for $e^l > \xi_-(t)$ and is identically zero when $l \to \infty$. We can interpret the absence of unbound vortices as the low $T$ phase where the phase of the complex order parameter has a finite stiffness to twists. This is analogous to a finite spring constant. Diverging correlation lengths at continuous transitions were already well known and accepted, but this exponential divergence was not known to most physicists in the early 1970's, although, for the exactly soluble F model limit of the eight vertex model,[63,64] the correlation length displays a similar exponential divergence.

Finally, in region III, where $C > 0$ and $x(l) = -|x(l)| < 0$, we can integrate the recursion relations exactly as for region I, except we obtain
$$|x(l)|^2 - 16\pi^2 y^2 = C,$$
$$\frac{d|x|}{dl} = |x|^2 - C. \quad (1.46)$$

Since $|x(l)| \geq \sqrt{C}$, this means that $|x(l)|$ increases with $l$ and, when $|x(l)| \gg \sqrt{C}$,

$$|x(l)| = \frac{|x(l^*)|}{1 - |x(l^*)|(l - l^*)} \qquad (1.47)$$

where $l^*$ is the value of $l$ where the trajectory becomes indistinguishable from that with $C = 0$. Note that the physical interpretation of a system in region III is somewhat obscure and this region is unphysical[60] as it is not accessible for the 2D planar rotor model. The RG flows for initial Hamiltonians in this region are attracted to the line $x(l) + 4\pi y(l) = 0$ and flow to large $y(l)$, just as trajectories in region II. We assume that all initial Hamiltonians in regions II and III flow to the same disordered fixed point and that all the Hamiltonians correspond to the same high $T$ disordered phase.

It appears that the system behaves differently if the initial Hamiltonian defined by parameters $K_0(T)$ and $y_0(T)$ lies in region I, II or III. It is clear that region I can be interpreted as a low temperature ordered phase since the vortex fugacity $y(l)$ flows to zero as the length scale, parametrized by $l$, increases. This means that the vortex fugacity $y(l)$ vanishes as $l \to \infty$ which is interpreted to mean that the system contains no free, unbound vortices at length scales larger than $\xi_-(T)$ which, in turn, means that the measured stiffness constant can be computed in region I by exploiting Eq. (1.36) where we have used Eq. (1.39) to obtain $x(\infty) = \sqrt{C(t)} = b\sqrt{|t|}$. In region II, it is natural to assume that $K_R(t > 0) = 0$ which can be derived by treating the unbound vortices on scales $e^l > \xi_+(t)$ by Debye–Hückel theory.[40] Note that Eq. (1.36) predicts that the ratio $\rho_s^R(T)/T$ has a finite and universal value at $T_c^-$[18]

$$\frac{\rho_s^R(T_c^-)}{T_c} = \frac{2M_A^2 k_B}{\pi \hbar^2} \qquad (1.48)$$

which has the approximate value $3.491 \times 10^{-9}\,\mathrm{g\,cm^{-2} K^{-1}}$ for liquid helium.

This result has been verified experimentally[66] to about 10% accuracy, by measuring the variation with temperature of the moment of inertia of a cylindrical roll of Mylar film coated with a film of liquid helium. At high temperatures the normal liquid rotates as the Mylar film rotates, but at low temperatures the superfluid component of the liquid helium ceases to move with the substrate. This method of measuring superfluid density was developed in the middle of the last century by Andronikashvili. The universal jump in superfluid density of Eq. (1.48) is an inescapable prediction of the theory. If experiment had produced a different number, then the whole structure of the vortex theory would be proved incorrect and theorists would have been back to the drawing board. As a historical note, Bishop and Reppy performed the measurements of the superfluid density of their $^4$He films without knowledge of the theoretical prediction, and, without knowledge of the theoretical value, they measured the universal slope $\rho_s^R(T_c^-)/T_c$.[66] Of course, measuring the superfluid density is not as straightforward as it appears. An important feature of this, or almost any other conceivable experiment, is that to get enough sensitivity to distinguish the moment of inertia of the helium film from that of the Mylar substrate it is necessary to use a significantly high frequency for the measurement, which was 2500 Hz in this experiment, and so the static theory described in this section is inadequate. A nonzero frequency theory was developed by Ambegaokar, Halperin, Nelson and Siggia (AHNS).[67] This and related experiments are discussed in more detail in Sec. 1.7.1.

To show that $\rho_s^R = 0$ for $T > T_c$, we follow the treatment of Nelson[65] by defining the generalized stiffness constant at finite $q$ as

$$K_R(q, K, y) = K_0 - 4\pi^2 K_0^2 \frac{\langle n(q)n(-q) \rangle}{q^2}. \tag{1.49}$$

We can use Eq. (1.37) to write

$$K_R(q, K, y) = K_R(qe^l, K(l), y(l)) \tag{1.50}$$

up to the scale $e^{l^*} = \xi_+$ to obtain

$$K_R(q, K, y) = K(l^*) - \frac{4\pi^2 K^2(l^*)}{(qe^{l^*})^2} \langle n(qe^{l^*})n(-qe^{l^*}) \rangle \tag{1.51}$$

where the average must be calculated with the Coulomb gas ensemble with couplings $K(l^*)$ and $y(l^*)$. The vortex Hamiltonian is given by

$$\frac{H_V(l^*)}{k_B T} = \frac{1}{2} \int d^2k \left( \frac{4\pi^2 K(l^*)}{k^2} + B(l^*) \right) n(k)n(-k). \tag{1.52}$$

Now, we can choose $e^{l^*} = \xi_+(T)$ at which scale the system has a dense set of vortices with all integer values of $n(\mathbf{r})$. In such a case, it is a reasonable approximation to integrate rather than sum over $n(q)$ when evaluating the expectation value

$$\langle n(q\xi_+)n(-q\xi_+) \rangle_{l^*} = \frac{1}{4\pi^2 K(l^*)/(q\xi_+)^2 + B(l^*)}. \tag{1.53}$$

Note, $B(l^*) \approx -\ln y(l^*) = \mathcal{O}(1)$ but the exact value is not important. Using this in Eq. (1.49), we obtain

$$K_R(q, T) = \frac{B(l^*)K(l^*)}{B(l^*) + 4\pi^2 K(l^*)/(q\xi_+)^2} \sim (q\xi_+)^2, \tag{1.54}$$

as $q\xi_+(T) \to 0$. This shows that $K_R(T) = 0$ when $T > T_c$, as expected.

Other thermodynamic quantities can be calculated from the renormalization group equations, especially when $T \leq T_c$ because the system is mapped into a simple Gaussian model when the vortex fugacity $y(l) \to 0$. For example, the two-point correlation function $G(r, K, y)$ was computed at $T_c$ by Kosterlitz[41] who found that,

$$G(r, K, y) = \langle e^{i(\theta(\mathbf{r}) - \theta(0))} \rangle \sim r^{-1/4} (\ln r)^{1/8}, \tag{1.55}$$

when $T = T_c$. An improved derivation is sketched below from the renormalization group equations. First, we note the identity, for $d = 2$,

$$\frac{\partial G(l)}{\partial l} = \eta(l) G(l) \quad \text{where} \quad \eta(l) = \frac{1}{2\pi K(l)} = \frac{1}{4} - \frac{x(l)}{8} + \mathcal{O}(x^2),$$

$$G(0) = G(l) \exp\left(-\int_0^l \eta(l') dl'\right) \tag{1.56}$$

where

$$G(l) = G(re^{-l}, K(l), y(l)).$$

We want the correlation function $G(r, K, y)$ in terms of the physical coupling constants. To use the above equation, we need to know the correlation function $G(l)$ which is calculated from the scaled parameters $K(l)$ and $y(l)$. Since our recursion relations are valid as long as the vortex fugacity $y(l)$ remains small, we can choose $l$ in Eq. (1.56) at our convenience. A sensible choice is $re^l = 1$, so we need $G(re^l = 1, K(l), y(l)) = \mathcal{O}(1)$ and the singular part of the correlation function is just

$$G(r, K, y) = \exp\left(-\int_0^{\ln r} \eta(l) dl\right). \tag{1.57}$$

To evaluate this, we use $\eta(l) = 1/(2\pi K(l)) = 1/4 - x(l)/8 + \mathcal{O}(x^2(l))$ where $x(l)$ is given by Eq. (1.42) and we obtain

$$\int_0^l dl' x(l') = \sqrt{C} l + \ln\left[\frac{1 - ae^{-2\sqrt{C}l}}{1 - a}\right] \quad \text{where} \quad a = \frac{x_0 - \sqrt{C}}{x_0 + \sqrt{C}} = 1 - \frac{2\sqrt{C}}{x_0},$$

$$\approx \sqrt{C} l + \ln(1 + x_0 l) \quad \text{when} \quad l \ll \frac{1}{2\sqrt{C}} = \ln \xi_-(T),$$

$$\approx \sqrt{C} l + \ln \frac{x_0}{2\sqrt{C}} \quad \text{when} \quad l \gg \frac{1}{2\sqrt{C}} = \ln \xi_-(T). \tag{1.58}$$

Thus, for $T \leq T_c$, we obtain

$$G(r, t) \sim r^{-\eta(T)} (\ln r)^{1/8} \quad \text{when} \quad 1 \ll r \ll \xi_-(t),$$

$$G(r, t) \sim r^{-\eta(T)} \quad \quad \quad \quad \text{when} \quad 1 \ll \xi_-(t) \ll r, \tag{1.59}$$

where $\eta(T) = 1/(2\pi K_R(T))$. Note that, when $r \gg \xi_-(t)$, a pure power law decay with a temperature dependent exponent, $\eta(T) = 1/(2\pi K_R(T))$, is obtained, corresponding to an effective Gaussian model with a renormalized coupling constant $K_R(T) = K(l = \infty)$. For $r \ll \xi_-(t)$, we obtain the same power law decay with a universal logarithmic correction of $(\ln r)^{1/8}$.

The forms of $G(r,t)$ of Eq. (1.59) are true for $t \le 0$ for all values of $r$, and also for $t > 0$ when $r < \xi_+(t)$. However, above $T_c$, we expect that the correlation function $G(r,t) \sim \exp(r/\xi_+)$ when $r \gg \xi_+(t)$. To show this, we have to be more careful in our use of the RG equation, Eq. (1.56). In this situation, we must choose $l^* = \ln \xi_+$ because this is the limit of validity of the recursion relations, and we obtain

$$G(r, K, y) = G(r/\xi_+, K(l^*), y(l^*)) \exp\left(-\int_0^{l^*} dl\, \eta(l)\right). \quad (1.60)$$

In contrast to our treatment for $r < \xi_+$, when we could choose $e^l = r$, we are left with the problem of calculating the effective correlation function $G(r/\xi_+)$, which controls the $r$ dependence of the correlation function. Equation (1.60) summarizes the content of our RG equations, relating the correlation function of the original system to the correlation function, $G(r/\xi_+)$, of an effective system where the minimum length scale is $\xi_+$, but at a very high temperature. An explicit calculation was done in Ref. 60, using a duality transformation, to obtain the expected result for $r \gg \xi_+$,

$$G(r,t) \sim \exp\left(-\frac{r}{\xi_+(t)}\right). \quad (1.61)$$

## 1.6. Scaling Theory in Analogous Systems

### 1.6.1. *Duality and the roughening of crystal facets*

Interestingly, and it surprised us, the 2D planar rotor model is intimately related to models for the roughening of equilibrium crystal facets as the temperature $T$ is varied. This analogy was pointed out by Chui and Weeks.[74,75] The lattice points of an ideal facet are represented as a two-dimensional lattice, and smooth deviations away from the ideal lattice can be represented as a function of the position on the ideal facet.

If the equilibrium state of a given crystal facet is a smooth plane, that facet may appear in a crystal grown at this temperature, but it is also possible that a stable facet may not occur because it grows too slowly in comparison with competing facets. In particular, it has been known for a long time that a crystal face with screw dislocations in it may grow much faster than an ideal crystal face. There has been a hope that quantum crystals, such as solid $^4$He surrounded by the superfluid liquid, may equilibrate fast enough for the equilibrium facets to appear.

It is not difficult to see that the partition function of the $2D$ Coulomb gas representation of the $2D$ planar rotor model of Eq. (1.22) can be written as[60]

$$Z(K,y) = \sum_{n(\mathbf{r})} \exp\left(-2\pi^2 K \sum_{\mathbf{r},\mathbf{r}'} G(\mathbf{r},\mathbf{r}')n(\mathbf{r})n(\mathbf{r}') - \ln y \sum_{\mathbf{r}} n^2(\mathbf{r})\right)$$

$$= \sum_{n(\mathbf{r})} \left(\prod_{\mathbf{r}} \int_{-\infty}^{+\infty} \frac{d\phi(\mathbf{r})}{2\pi}\right) \exp\left(-\frac{1}{2K}\sum_{\mathbf{r},\mathbf{r}'}(\phi(\mathbf{r})\right.$$

$$\left. - \phi(\mathbf{r}'))^2 + 2\pi i \sum_{\mathbf{r}} n(\mathbf{r})\phi(\mathbf{r}) - \ln y \sum_{\mathbf{r}} n^2(\mathbf{r})\right),$$

where $\quad G(\mathbf{r},\mathbf{r}') = \int \frac{d^2q}{(2\pi)^2} \frac{e^{i\mathbf{q}\cdot\mathbf{r}}}{q^2} = \frac{1}{2\pi}\ln\frac{|\mathbf{r}-\mathbf{r}'|}{L}.$ (1.62)

At this point, we can make the approximation for the vortex fugacity of $y=1$ and $\sum_{-\infty}^{+\infty} e^{2\pi i n\phi} = \sum_{-\infty}^{+\infty} \delta(\phi-h)$ when we obtain the discrete Gaussian model,[60] or we can take $y \ll 1$ so that the partition function becomes

$$Z(K,y) = \int \prod_{\mathbf{r}} \frac{d\phi(\mathbf{r})}{2\pi} \exp\left[-\frac{1}{2K}\sum_{\langle \mathbf{r},\mathbf{r}'\rangle}(\phi(\mathbf{r})-\phi(\mathbf{r}'))^2\right][1+2y\cos(2\pi\phi(\mathbf{r}))]$$

$$= \int \prod_{\mathbf{r}} \frac{d\phi(\mathbf{r})}{2\pi} \exp\left[-\frac{1}{2K}\sum_{\langle \mathbf{r},\mathbf{r}'\rangle}(\phi(\mathbf{r})-\phi(\mathbf{r}'))^2 + 2y\sum_{\mathbf{r}}\cos(2\pi\phi(\mathbf{r}))\right].$$

(1.63)

The last step to convert this into the sine-Gordon roughening model[74,75] is to define the height variable by $\phi(\mathbf{r}) = h(\mathbf{r})/b$ where $h(\mathbf{r})$ is the local facet height above a reference plane in steps of the interplane distance $b$. We finally obtain the sine-Gordon representation of the roughening problem as

$$H(\gamma,u) = H_0 + H_u = \frac{\gamma}{2}\int d^2\mathbf{r}(\nabla h(\mathbf{r}))^2 - u\int \frac{d^2\mathbf{r}}{a^2}\cos\left[\frac{2\pi h(\mathbf{r})}{b}\right], \quad (1.64)$$

where $u = 2y$ when $y \ll 1$ and $a$ is the short distance cut off.

One can derive renormalization flow equations for the parameters $\gamma(l)$ and $u(l)$[74] in a similar way to $K(l)$ and $y(l)$ in the vortex system by defining a renormalized surface stiffness $\gamma_R(T)$ as,

$$F(v) - F(0) = \frac{1}{2}\Omega\gamma_R(T)v^2, \quad (1.65)$$

$$\mathbf{v} = \frac{1}{\Omega}\int d^2\mathbf{r}\langle\nabla h(\mathbf{r})\rangle,$$

where $F(v)$ is the free energy of a surface of area $\Omega$, and $\mathbf{v}$ is the average gradient of $h(\mathbf{r})$. Writing $h(\mathbf{r}) = \mathbf{v}\cdot\mathbf{r} + h'(\mathbf{r})$, the sine-Gordon Hamiltonian is

$$H(\mathbf{v}) = \frac{1}{2}\Omega v^2 + H_0[h'(\mathbf{r})] + H_u[h'(\mathbf{r}) + \mathbf{v}\cdot\mathbf{r}]. \quad (1.66)$$

Expanding $F(\mathbf{v}) = -k_BT\ln\text{Tr}\exp(-H(\mathbf{v})/k_BT)$ to $\mathcal{O}(v^2)$, we obtain

$$\gamma_R(T) = \gamma + \frac{1}{2\Omega}\sum_{\alpha\beta}\frac{\partial^2}{\partial v_\alpha \partial v_\beta}\left(\langle H_u\rangle_0 - \frac{1}{2k_BT}[\langle H_u^2\rangle_0 - \langle H_u\rangle_0^2]\right) + \cdots$$

$$= \gamma + \frac{k_BT}{8}\left(\frac{2\pi}{b}\right)u^2\int\frac{d^2\mathbf{r}}{a^2}\left(\frac{r}{a}\right)^2\langle\cos\left[\frac{2\pi}{b}(h'(\mathbf{r}) - h'(0)\right]\rangle_0, \quad (1.67)$$

$$K_R^{-1} = K^{-1} + \pi^3 u^2 \int_a^\infty \frac{dr}{a}\left(\frac{r}{a}\right)^{3-2\pi K},$$

where $K = k_BT/(\gamma b^2)$ and we have used $\langle\cos[(2\pi/b)(h(\mathbf{r}) - h(0))]\rangle_0 = (r/a)^{-2\pi K}$.

Equations (1.67) and (1.32) have identical forms, and flow equations for the parameters $K(l)$ and $u(l)$ can be written down immediately,

$$\frac{dK^{-1}}{dl} = \pi^3 u^2 + \mathcal{O}(u^4),$$
$$\frac{du}{dl} = (2 - \pi K)u + \mathcal{O}(u^3). \quad (1.68)$$

The only difference is in the temperature, $T$, dependence of the parameter $K(T)$. For the roughening model, $K(T) = k_BT/(\gamma b^2)$, while for the $^4$He film, $K(T) = (\hbar/m)^2\rho_s^0/(k_BT)$. This inversion of the temperature is to be expected because the roughening model is dual to the planar rotator model, and the low $T$ phase of one becomes the high $T$ phase of the other.[60] Thus, in the high $T$ phase of the roughening model, the parameter $u(l) \to 0$, and the Hamiltonian becomes $H/(k_BT) = (K_R(T)/2)\int d^2\mathbf{r}(\nabla h)^2$. The integer spacing between the crystal planes becomes irrelevant at large scales and the interface becomes rough. For $T < T_R$, where $T_R$ is the roughening temperature, the parameter $u(l)$, which is analogous to the fugacity $y(l)$, is relevant, leading to a flat or faceted interface.

Another result from the planar rotor model, which can be directly taken over to the roughening case, is the universal relation,

$$K_R(T_R^+) = \frac{k_B T_R}{\gamma_R(T_R^+) b^2} = \frac{2}{\pi}. \tag{1.69}$$

This result, which follows directly from our results for the planar rotor model and duality, implies that facets with the largest lattice spacing $b$ will have the highest roughening temperature $T_R$, assuming that the surface stiffness $\gamma$ does not depend strongly on $b$. To convert these theoretical predictions into terms of measurable quantities, we must remember that these facets are faces of a finite 3D crystal, so the equilibrium theory of a crystal of fixed volume is needed. The basic idea is that the crystal shape is determined by the condition that the total surface free energy be minimized at fixed total volume.[76] A flat facet with coordinate $-x_0 \leq x \leq +x_0$, in terms of reduced coordinates $\tilde{x} = x/L$, $\tilde{y} = y/L$ and $\tilde{h} = h/L$, obeys

$$\begin{aligned}\tilde{h}(\tilde{x},0) &= \tilde{h}_0 - A(|\tilde{x}| - \tilde{x}_0)^{3/2} \quad \text{for } |\tilde{x}| > \tilde{x}_0, \\ \tilde{h}(\tilde{x},0) &= \tilde{h}_0 \quad \text{for } |\tilde{x}| < \tilde{x}_0,\end{aligned} \tag{1.70}$$

where the reduced facet size, $\tilde{x}_0 \sim b/\xi_+(t) \sim \exp(-B/\sqrt{|T-T_R|})$, and $\tilde{h}_0 = f_0/\gamma$, with $f_0$ the free energy per unit area of the facet.

For $T > T_R$, the facet has disappeared and the reduced interface height is,

$$\tilde{h}(\tilde{x},0) = \tilde{h}_0 - \frac{\gamma}{2\gamma_R}\tilde{x}^2, \tag{1.71}$$

so that the curved rough surface has a reduced curvature, $K_R(T) = (L/R)(k_B T/\gamma b^2) = 2/\pi$, at $T = T_R^+$, which jumps discontinuously to zero at $T = T_R^-$. Here, $L$ is the system size and $R$ is the radius of curvature.[76,77] This is the analogue of the universal jump in the superfluid density $\rho_s^R(T_c^-)/T_c$ in a 2D superfluid film. Of course, measuring this predicted jump in the reduced curvature of a disappearing facet is not easy and, to the best of our knowledge, the universal predictions have not yet been experimentally verified, particularly that the facet size vanishes as $\xi_+^{-1}(T)$,[73,78] but our speculation is that the discrepancy is due to lack of equilibration of the crystal.

### 1.6.2. Substrate effects

The development of the theory was motivated mainly by some very early experiments on thin films of $^4$He adsorbed on a substrate. This system is directly related to the 2D XY model and the predictions for the latter are directly applicable to $^4$He films, provided one can argue that the substrate has no effect or is irrelevant in the RG sense. Substrates are essential for any two-dimensional system as it cannot exist without being supported by a substrate. To our knowledge, the only systems which do not need to be supported everywhere are thin films of liquid crystals[79,80] and a monolayer of graphene,[81] since these are sufficiently strong to survive being suspended from two edges. An isolated thin film of $^4$He or of a magnet cannot exist in the absence of a substrate. In view of this, it is vital to assess the effect of possible perturbations due to a substrate on the ideal two-dimensional system of Sec. 1.5. Before discussing the important effects of the substrate on 2D XY systems, we can immediately see why experiments on superfluidity in $^4$He films can be understood in terms of the idealized theory of Sec. 1.5. This ignores the far from perfect Mylar substrate which can be modeled by a random potential coupling to the local film density $|\psi(\mathbf{r})|^2$. Since the order parameter field is the quantum mechanical condensate wave function, $\psi(\mathbf{r}) = |\psi(\mathbf{r})|e^{i\theta(\mathbf{r})}$, the substrate potential cannot couple to the phase of the condensate wave function but only to its amplitude and is, thus, irrelevant. This argument justifies the use of the ideal theoretical model for superfluid $^4$He films, and partly explains why theoretical predictions agree so well with experiments.

Many other systems are influenced by the substrate, an example being a magnetic system which is heavily influenced by crystal fields with the symmetry of the substrate crystal structure. Consider a 2D XY ferromagnet on a periodic lattice described by a Hamiltonian,

$$\beta H = \frac{1}{2} K(T) \int d^2\mathbf{r}(\nabla\theta)^2 - h_p \int d^2\mathbf{r} \cos(p\theta), \quad (1.72)$$

where $h_p$ is the strength of the $p$-fold symmetry breaking field. This sort of symmetry breaking term represents a crystal field in an XY magnet, and its significance is immediately seen when $|h_p| \gg 1$ in Eq. (1.72). For $h_p > 0$, the free energy is minimized when $\cos p\theta(\mathbf{r}) = +1$ or $\theta(\mathbf{r}) = 2n\pi/p$ and, when $h_p < 0$, $\theta(\mathbf{r}) = (2n+1)\pi/p$, where $n = 0, 1, \ldots, p-1$. These models are known as $p$-state clock models and correspond to the very well known nearest neighbor Ising ferromagnet when $p = 2$ and to the 3-state Potts model for $p = 3$. For larger values, $p = 4, 5, \ldots$, there are many variants depending on the strengths and signs of the higher harmonics of the anisotropy, $\cos[np\theta(\mathbf{r})]$, and of the interaction on a lattice,

$$\beta H = \sum_{\langle \mathbf{r},\mathbf{r}'\rangle} \sum_{n=1}^{[p/2]} K_n[1-\cos(n(\theta(\mathbf{r})-\theta(\mathbf{r}')))] - \sum_{\mathbf{r}} \sum_{n=1}^{[p/2]} h_{pn} \cos(np\theta(\mathbf{r})). \quad (1.73)$$

Remarkably, one can show that this Hamiltonian contains many models in two dimensions which have been of great interest in statistical mechanics, such as the Ising model, various clock models, the $p$-state Potts models, etc. For example, the 4-state Potts model is obtained when $h_4 \to \infty$ which restricts $\theta(\mathbf{r}) = 0, \pi/2, \pi, 3\pi/2$ and $K_1 = 2K_2$. There exists an extensive literature on the connections between these models and the Coulomb gas representations of Eq. (1.73).[82–85]

In the simplest case with only the $n = 1$ terms in Eq. (1.73), we can replace the term in the partition function,[86]

$$e^{h_p \cos p\theta} \to \sum_{m=-\infty}^{+\infty} e^{ipm\theta} e^{(\ln y_p) m^2}, \qquad (1.74)$$

where $y_p \approx h_p/2$, so that the Coulomb gas representation of the partition function for an $XY$ model in a $p$-fold symmetry breaking field becomes[60]

$$Z \propto Z_c(K, y, y_p) = \sum_{n(\mathbf{r})} \sum_{m(\mathbf{r})} e^{-\mathcal{H}_c},$$

$$-\mathcal{H}_c = \pi K_0(T) \sum_{\mathbf{R},\mathbf{R}'} n(\mathbf{R}) n(\mathbf{R}') \ln\left(\frac{|\mathbf{R} - \mathbf{R}'|}{a}\right) + \ln y \sum_{\mathbf{R}} n^2(\mathbf{R})$$

$$+ \frac{p^2}{4\pi K} \sum_{\mathbf{r},\mathbf{r}'} m(\mathbf{r}) m(\mathbf{r}') \ln\left(\frac{|\mathbf{r} - \mathbf{r}'|}{a}\right) + \ln y_p \sum_{\mathbf{r}} m^2(\mathbf{r})$$

$$+ ip \sum_{\mathbf{R},\mathbf{r}} n(\mathbf{R}) m(\mathbf{r}) \tan^{-1}\left(\frac{Y - y}{X - x}\right). \qquad (1.75)$$

Here, the integer vortices $n(\mathbf{R})$ lie on the dual lattice with sites at $\mathbf{R} = (X, Y)$, and the symmetry breaking charges $m(\mathbf{r})$ are on the original lattice with sites at $\mathbf{r} = (x, y)$. Both $n(\mathbf{R})$ and $m(\mathbf{r})$ obey neutrality constraints $\sum_{\mathbf{R}} n(\mathbf{R}) = 0 = \sum_{\mathbf{r}} n(\mathbf{r})$ and the two Coulomb gases are coupled by the angular factor $ip\Theta(\mathbf{R} - \mathbf{r})$. Interestingly, Eq. (1.75) shows a remarkable duality relation[60] under the relabeling of the summation variables $n(\mathbf{R}) \leftrightarrow m(\mathbf{r})$, the replacements $y \leftrightarrow y_p$, and $K \leftrightarrow p^2/(4\pi^2 K)$ the partition function is unchanged, so that

$$Z(K, y, y_p) = Z\left(\frac{p^2}{4\pi^2 K}, y_p, y\right). \qquad (1.76)$$

It is straightforward to write down the recursion relations as expansions in the fugacities $y$ and $y_p$ which read, to lowest order,

$$\frac{dK^{-1}}{dl} = 4\pi^3 y^2 - \pi p^2 K^{-2} y_p^2,$$

$$\frac{dy}{dl} = (2 - \pi K)y, \quad (1.77)$$

$$\frac{dy_p}{dl} = \left(2 - \frac{p^2}{4\pi K}\right) y_p.$$

These are consistent with the duality relation of Eq. (1.76) and were first derived from this.[60]

From these RG equations, we see that the charge fugacity $y_p$ is relevant at temperatures for which $K^{-1} < 8\pi/p^2$ while the vortex fugacity $y$ is relevant for $K^{-1} > \pi/2$. Thus, when $p > 4$, there is a band of temperatures,

$$\frac{2}{\pi} \leq K_R(T) \leq \frac{p^2}{8\pi}, \quad (1.78)$$

for which both fugacities, $y$ and $y_p$, are irrelevant when the system reduces to a Gaussian model, and is the same as the pure $XY$ model at low temperature. At lower temperatures, $K_R^{-1}(T) < 8\pi/p^2$, the anisotropy term $h_p \cos p\theta$ becomes relevant, the $XY$ spins are restricted to one of $p$ directions, and the system becomes a discrete $Z_p$ clock model. Note that, for $p = 2$ and $p = 3$, there is no intermediate region where both fugacities are irrelevant, so that one expects that the model has a transition at $T_c$ in the 2D Ising class when $p = 2$, and the 3-state Potts class when $p = 3$. For the special case, $p = 4$, the multicritical point, $K = 2/\pi$ and $y = y_4 = 0$, is the meeting point of three critical lines separating a high $T$ disordered phase and two low $T$ ordered phases with two different spin orientations of $\theta(\mathbf{r}) = n\pi/2$ or $\theta(\mathbf{r}) = (2n+1)\pi/4$, with $n = 0, 1, 2, 3$. This point is identical to the critical point of the $F$ model,[63] and the three critical lines of continuously varying exponents meeting there, are in the universality class of the Baxter model.[64] The detailed connections with the Coulomb gas models are discussed in Refs. 82, 83 and 85. The phase diagram for models with $p \geq 5$ is similar except for a region of massless phase separating the two ordered phases from the disordered phase.[60]

Other interesting models of experimental relevance are obtained by including both the first and second harmonics of the $p$-fold anisotropy. The

XY model with both four- and eight-fold anisotropy terms

$$\frac{H}{k_B T} = -K \sum_{rr'} \cos(\theta(\mathbf{r}) - \theta(\mathbf{r}')) - h_4 \sum_{\mathbf{r}} \cos(4\theta(\mathbf{r}))$$

$$+ h_8 \sum_{\mathbf{r}} \cos(8\theta(\mathbf{r})) \qquad (1.79)$$

is interesting as a theoretical model but can also be used as a model describing the adsorption system H/W(100), hydrogen adsorbed on the (100) surface of tungsten. Changing the hydrogen coverage does not affect $h_8$, but drives the value of $h_4$ through zero, which corresponds to a structural transition of the tungsten surface.[87,88] When $K^{-1} \leq \pi/2$, Eq. (1.79) has two ordered phases with finite magnetization in one of the four directions $\theta(\mathbf{r}) = n\pi/2$ with $n = 0, 1, 2, 3$ when $h_4 > 0$, and $\theta(\mathbf{r}) = (2n+1)\pi/4$ when $h_4 < 0$. The two ordered phases are separated by a continuous transition in the universality class of the 2D XY model at $h_4 = 0$. The extra anisotropy term, $h_8 \sum \cos 8\theta(\mathbf{r})$, is irrelevant for $K^{-1} \geq \pi/8$ and relevant for $K^{-1} < \pi/8$. Thus, in the temperature range $\pi/8 \leq K^{-1} \leq \pi/2$, the ordered phases will be separated by a line of continuous transitions. However, at lower temperatures, $K^{-1} < \pi/8$, $h_8$ becomes relevant and has important effects. There are two possible situations, (i) $h_8 < 0$ when the orderings favored by $h_8$ are compatible with those favored by a finite $h_4$, independent of the sign of $h_4$. In this case, the transition at $h_4 = 0$ between the two ordered phases will be a discontinuous first order transition.[87,88] In case (ii), when $h_8 > 0$, the orderings favored by $h_4$ and $h_8$ are incompatible, resulting in an extra phase for small $|h_4|$ with the new phase separated from the the two ordered phases by continuous Ising transitions. The first scenario is compatible with the experimental data,[89,90] which seems to favor a first order transition at low temperature.

Another system, in which the first two harmonics of the local anisotropy potential have important effects, is tilted hexatic phases of liquid crystals. Experiments on the transitions between different tilted hexatic phases in thermotropic liquid crystals have observed a weak first order transition from the hexatic-$I$ to hexatic-$F$ phase.[91] Also, in an analogous layered lyotropic liquid crystal, the $L_{\beta I}$ and $L_{\beta F}$ phases, corresponding to the hexatic-$I$ and hexatic-$F$ phases, were shown to be separated by a third $L_{\beta L}$ phase with continuous Ising-like transitions between the phases.[92] These surprising observations can be explained by coupled XY models for the bond angle, $\theta(\mathbf{r})$, of the hexatic and the tilt-azimuthal angle, $\phi(\mathbf{r})$, of the director with

Hamiltonian,[93,94]

$$\beta H = \int d^2\mathbf{r} \left[\frac{1}{2}K_6(\nabla\theta)^2 + \frac{1}{2}K_1(\nabla\phi)^2 + g\nabla\theta\cdot\nabla\phi + V(\theta-\phi)\right], \quad (1.80)$$

$$V(\theta-\phi) = -h_6\cos(6(\theta-\phi)) - h_{12}\cos(12(\theta-\phi)) + \cdots,$$

which is similar to Eq. (1.79). The low temperature region is the one of interest, when one can write Eq. (1.80) as,

$$\beta H = \int d^2\mathbf{r} \left[\frac{1}{2}K_+(\nabla\theta_+)^2 + \frac{1}{2}K_-(\nabla\theta_-)^2 + V(\theta_-)\right],$$

$$\theta_+ = \frac{(K_6+g)\theta + (K_1+g)\phi}{K_6+K_1+2g}, \quad \theta_- = \theta - \phi, \quad (1.81)$$

$$K_+ = K_6 + K_1 + 2g, \quad K_- = \frac{K_1 K_6 - g^2}{K_+}.$$

This model system has been analyzed in detail,[94] with results in accord with experiment with the various scenarios determined by the sign of the coefficient $h_{12}$.

### 1.6.3. *Melting of a 2D crystal*

In our early paper[17] we proposed that a two-dimensional crystal melts by the unbinding of bound pairs of dislocations which are the appropriate topological defects which destroy the translational order in a periodic crystal. It is now apparent that these physical ideas are correct but incomplete.[42] A periodic crystal has two types of order (i) translational order and (ii) orientational order, and a fundamental question is: how are these orders destroyed as the temperature increases and what sort of transitions occur? The theory of 2D melting was worked out in 1978–79 in the seminal papers of Halperin, Nelson, and Young[42,43,95] where it was shown that, as $T$ is increased, there is first a dislocation unbinding transition from a low temperature crystal phase with power law decay of translational order to a fluid phase with exponentially decaying translational order, as we expected.[17] The breakthrough of Halperin, Nelson and Young was to recognize the importance of the angular terms in the dislocation–dislocation interaction. We ignored these angular terms on the grounds that these are subleading. By analyzing the system carefully, they realized that the dislocation unbinding transition at $T_m$ did not lead to an isotropic fluid, but to an intermediate anisotropic fluid phase with exponentially decaying translational order, but algebraic orientational order, due to remnants of the orientational order of the crystal. At a higher temperature, $T_I$, the orientational order is destroyed, and the resulting high temperature phase is an isotropic fluid with exponential decay of both translational and orientational orders. Some experimental investigations and computer simulations of melting followed, but seemed to favor a direct first order transition from a solid to an isotropic fluid, in disagreement with theory which predicts two successive continuous transitions, with an intermediate partially ordered phase. At present, to our knowledge, computer simulations are unable to reproduce the theoretical predictions. Surprisingly, experiments on melting in $2D$ are finally in agreement with theory, but these experiments[72] have been possible only over the last decade!

The notion of a crystal in $2D$ was somewhat controversial in 1972, when we began to think about the physics of melting because of the rigorous theorem that there is no crystalline order in two dimensions.[56] This theorem is correct, but somewhat misleading, as all it shows is that true long range translational order does not exist in $2D$, but it does allow for a finite elastic shear modulus in $2D$, as was realized by the author of the theorem.[56] We pictured a $2D$ elastic crystal as a periodic array of points (particles) on an elastic sheet. Then we asked: what happens to the periodic array when the sheet is elastically distorted without tearing? It is obvious that, locally, the lattice does not change much, but the relative separation of any pair of distant points can undergo a very large change. With this pictorial interpretation in mind, the idea of a $2d$ crystal became much clearer. We realized that the essential feature is not the translational order, but that the elastic moduli are finite. This picture is readily written in mathematical language by writing the local density, $\rho(\mathbf{r})$, as

$$\rho(\mathbf{r}) = \rho_0(\mathbf{r}) + \sum_{\mathbf{G}} \rho_{\mathbf{G}} e^{i\mathbf{G}\cdot\mathbf{u}(\mathbf{r})}, \qquad (1.82)$$

where $\mathbf{r}$ denotes the sites of the periodic reference lattice, $\mathbf{G}$ are the reciprocal lattice vectors obeying $\mathbf{G}\cdot\mathbf{r} = 2\pi n$ and $\mathbf{u}(\mathbf{r})$ is the displacement of the particle from its equilibrium position $\mathbf{r}$.

The most common crystal structure in $2D$ is a triangular lattice in which each particle has six neighbors. There is also a six-fold rotational symmetry as the crystal axes make angles $2\pi n/6$ relative to one another. Thus, the translational and orientational order parameters are[42]

$$\rho_{\mathbf{G}}(\mathbf{r}) = e^{i\mathbf{G}\cdot\mathbf{u}(\mathbf{r})} \quad \text{and} \quad \psi_6(\mathbf{r}) = e^{6i\theta(\mathbf{r})}, \qquad (1.83)$$

where $\theta(\mathbf{r})$ is the angle the local crystal axis makes with some arbitrary direction. Today, the natural way to proceed is to write down a Ginsburg–Landau–Wilson functional in terms of the order parameters $\rho_{\mathbf{G}}(\mathbf{r})$ and $\psi_6(\mathbf{r})$, which are invariant under the appropriate symmetry transformations. This was achieved in 1979 by Halperin, as reported by Nelson,[65] by constructing an appropriate free energy functional in terms of the first six translational order parameters $\rho_\alpha(\mathbf{r})$ where $\alpha = 1, 2, \ldots, 6$. Following the treatment by de Gennes[96] for a smectic-A–nematic transition, one obtains,

$$F[\rho_\alpha, \theta] = \sum_{\alpha=1}^{6} \int d^2\mathbf{r} \left( \frac{A}{2}|\mathbf{G}_\alpha \cdot \mathbf{D}_\alpha \rho_\alpha|^2 + \frac{B}{2}|\mathbf{G}_\alpha \times \mathbf{D}_\alpha \rho_\alpha|^2 + \frac{r(T)}{2}|\rho_\alpha|^2 \right)$$

$$+ \int d^2\mathbf{r} \left( w(\rho_1\rho_3\rho_5 + \rho_2\rho_4\rho_6) + \frac{K_A}{2}(\nabla\theta(\mathbf{r}))^2 \right), \quad (1.84)$$

$$\mathbf{D}_\alpha = \nabla - i(\hat{\mathbf{z}} \times \mathbf{G}_\alpha)\theta(\mathbf{r}),$$

where $\mathbf{G}_\alpha$, with $|\mathbf{G}_\alpha| = G_0$, are the six smallest reciprocal lattice vectors of the underlying lattice, and $\hat{\mathbf{z}}$ is a unit vector normal to the plane of the system. In the mean field approximation, the presence of third order terms in the free energy functional makes the transition first order. However, following the philosophy of the planar rotor model, we assume that the temperature, $T$, is well below the mean field transition temperature so that $r(T) \ll 0$ in Eq. (1.84). This means the $|\rho_\alpha|$ are a constant, $\rho_0$, so one can make the phase only approximation by writing $\rho_\alpha(\mathbf{r}) = \rho_0 e^{i\mathbf{G}_\alpha \cdot \mathbf{u}(\mathbf{r})}$. The free energy of Eq. (1.84) is minimized when[65]

$$\theta(\mathbf{r}) = \frac{1}{2}(\nabla \times \mathbf{u}(\mathbf{r})), \quad (1.85)$$

and the elastic free energy for a 2D triangular crystal is

$$F = \frac{1}{2} \int d^2\mathbf{r}[2\mu u_{ij}^2(\mathbf{r}) + \lambda u_{kk}^2(\mathbf{r})], \quad (1.86)$$

where the Lamé coefficients are, in terms of $A$ and $B$ of Eq. (1.84) are,[65]

$$\mu = \frac{3}{4}\rho_0^2(A+B)G_0^4, \qquad \lambda = \frac{3}{4}(A-B)G_0^4. \quad (1.87)$$

At quadratic order, the free energy is a functional of the linearized symmetric strain tensor,

$$u_{ij} = \frac{1}{2}\left(\frac{\partial u_i}{\partial r_j} + \frac{\partial u_j}{\partial r_i}\right). \quad (1.88)$$

The strain field can be decomposed as $u_{ij}(\mathbf{r}) = \phi_{ij}(\mathbf{r}) + u_{ij}^s(\mathbf{r})$ where $\phi_{ij}(\mathbf{r})$ is the smooth part of $u_{ij}$ and $u_{ij}^s$ is the singular part due to dislocations which can be characterized by the amount for which the integral of the displacement **u** round a closed contour fails to close,

$$\oint d\mathbf{u} = a_0 \mathbf{b}(\mathbf{r}) = a_0(n(\mathbf{r})\hat{\mathbf{e}}_1 + m(\mathbf{r})\hat{\mathbf{e}}_2). \tag{1.89}$$

Here, $a_0$ is the lattice spacing, $\mathbf{b}(\mathbf{r})$ is the dimensionless Burger's vector[44] of the singularity, $n(\mathbf{r}), m(\mathbf{r})$ are integers, and $\hat{\mathbf{e}}_{1,2}$ are the unit lattice vectors of the underlying triangular lattice.

Using results from linear elasticity theory,[44] one can solve for the strain $u_{ij}^s(\mathbf{r})$ due to a dislocation density, $\mathbf{b}(\mathbf{r}') = \sum_\alpha \mathbf{b}_\alpha \delta(\mathbf{r}' - \mathbf{r}_\alpha)$, to obtain

$$u_{ij}^s(\mathbf{r}) = \left( \frac{1}{2\mu} \epsilon_{ik}\epsilon_{jl} \frac{\partial^2}{\partial r_k \partial r_l} - \frac{\lambda}{4\mu(\lambda+\mu)} \delta_{ij} \nabla^2 \right) a_0 \sum_{\mathbf{r}'} b_m G_m(\mathbf{r},\mathbf{r}'). \tag{1.90}$$

The vector Green's function, $\mathbf{G}(\mathbf{r},\mathbf{r}')$, satisfies the biharmonic equation,

$$\nabla^4 G_m(\mathbf{r},\mathbf{r}') = -K_0 \epsilon_{mn} \frac{\partial}{\partial r_n} \delta(\mathbf{r} - \mathbf{r}'),$$

$$K_0 = \frac{4\mu(\mu+\lambda)}{(2\mu+\lambda)}. \tag{1.91}$$

To solve this differential equation for $G_m(\mathbf{r},\mathbf{r}')$, one can consider a crystal with free boundaries which requires von Neumann boundary conditions, $\nabla G_m(\mathbf{r},\mathbf{r}') = $ constant, for $\mathbf{r}$ on the boundary, so that

$$G_m(\mathbf{r},\mathbf{r}') = -\frac{K_0}{4\pi} \epsilon_{mn}(r_n - r'_n) \left[ \ln\left(\frac{|\mathbf{r}-\mathbf{r}'|}{a}\right) + C \right]. \tag{1.92}$$

Here, the constant $C > 0$ is a measure of the ratio of the core diameter to the cut-off $a$. This can be absorbed into an effective dislocation core energy, exactly as is done for the simpler vortex case of Sec. 1.5. The final result of these tedious calculations is that the effective elastic Hamiltonian, $H_E$, can be decomposed into a smooth part and a singular part due to dislocations as,[42]

$$H_E = H_0 + H_D,$$

$$\frac{H_0}{k_B T} = \frac{1}{2} \int \frac{d^2 \mathbf{r}}{a_0^2} (2\tilde{\mu}\phi_{ij}^2(\mathbf{r}) + \tilde{\lambda}\phi_{kk}^2(\mathbf{r})),$$

$$\frac{H_D}{k_B T} = -\frac{K}{8\pi} \sum_{\mathbf{r} \neq \mathbf{r}'} b_\alpha(\mathbf{r}) b_\beta(\mathbf{r}') \left( \ln\left[\frac{|\mathbf{r}-\mathbf{r}'|}{a}\right] \delta_{\alpha\beta} - \frac{(\mathbf{r}-\mathbf{r}')_\alpha (\mathbf{r}-\mathbf{r}')_\beta}{|\mathbf{r}-\mathbf{r}'|^2} \right) \tag{1.93}$$

$$+ \frac{E_c}{k_B T} \sum_{\mathbf{r}} |\mathbf{b}(\mathbf{r})|^2,$$

where $K = K_0 a_0^2/(k_B T)$, $\tilde{\mu} = \mu/(k_B T)$, and $\tilde{\lambda} = \lambda/(k_B T)$. The dislocation density is subject to the neutrality condition, $\sum_{\mathbf{r}} \mathbf{b}(\mathbf{r}) = 0$, because the energy of an isolated dislocation diverges as $\ln L$, where $L$ is the linear size of the system.

In analogy with the $2D$ superfluid, we immediately conclude that the low $T$ phase of our $2D$ crystal is one in which there are no isolated dislocations. There will be a finite concentration of bound pairs and triplets of dislocations, which renormalize the elastic constants downwards from their bare values. The crystal is described by the Gaussian Hamiltonian $H_0$ of Eq. (1.93), which gives the Debye–Waller factor $C_G(\mathbf{r})$ and the structure function $S(q)$:

$$C_G(\mathbf{r}) = \langle \rho_G(\mathbf{r}) \rho_G^*(0) \rangle^0 \sim r^{-\eta_G(T)},$$

$$\eta_G(T) = \frac{k_B T |G|^2}{4\pi} \frac{3\mu + \lambda}{\mu(2\mu + \lambda)}, \tag{1.94}$$

$$S(q) = \langle |\rho(q)|^2 \rangle = \sum_{\mathbf{r}} e^{i\mathbf{q}\cdot\mathbf{r}} \langle e^{i\mathbf{q}\cdot[\mathbf{u}(\mathbf{r}) - \mathbf{u}(0)]} \rangle \sim |\mathbf{q} - \mathbf{G}|^{-2+\eta_G(T)}.$$

This means that there is no long range translational order in a $2D$ crystal because $\langle \rho_G(\mathbf{r}) \rangle = 0$, in agreement with the Mermin–Wagner theorem,[28,56] but the harmonic crystal has a finite shear modulus $\tilde{\mu}(T) > 0$. Interestingly, the structure function $S(\mathbf{q})$ diverges at a set of small reciprocal lattice vectors $\mathbf{G}$ when $\eta_G(T) < 2$, and has finite cusp singularities when $\eta_G(T) > 2$.[65] In principle, this can be seen by X-ray or neutron diffraction, and the exponent $\eta_G(T)$ measured, which would be a rigorous experimental test of the theory. Unfortunately, this has not been possible because the required resolution in $\mathbf{q}$ has not been achieved to date.

Earlier,[16,17] we argued that, if the $2D$ crystal contains a finite concentration of free, unbound dislocations, then, under an arbitrarily small shear stress, these dislocations move to relax the stress to zero, so that the crystal has a fluid-like response to the stress. Thus, the system with free dislocations is melted and is a fluid at long length scales. Also, we were certain that the full dislocation Hamiltonian of Eq. (1.93) would yield the detailed behavior of the $2D$ system at the melting transition. However, around 1973, there was some controversy about the relative sign of the two terms in the dislocation part, $H_D$, of the Hamiltonian of Eq. (1.93). At the time, this apparently minor technical problem seemed unimportant, because the vital term seems to be the $\ln r$ term in the interaction, as this dominates the angular term at large separations $r$. Ignoring the angular term leads to a rather uninteresting (and incorrect) problem so, fortunately, we did no more with it. A few years later, Halperin, Nelson and Young solved the melting problem properly[42,43] by treating the angular term in $H_D$ correctly. The hexatic phase was discovered and the modern KTHNY theory of melting was born.

The natural way to go beyond the Gaussian theory, which uses only the smooth part of the lattice displacements in $H_0$ of Eq. (1.93), is to compute perturbative corrections to the bare elastic moduli due to dislocations as a power series in the dislocation fugacity $y = \exp(-E_c/k_B T)$. This can be done in an analogous way to computing the renormalized stiffness $\rho_R^s$ of Sec. 1.5 for a superfluid film. One must first identify the appropriate correlation function from which the renormalized elastic constants $\mu_R(T)$ and $\lambda_R(T)$ can be extracted. This has been done by Halperin and Nelson[42] by expressing the tensor of renormalized elastic constants in terms of a correlation function,

$$C_{R,ijkl} = \tilde{\mu}_R(T)(\delta_{ik}\delta_{jl} + \delta_{il}\delta_{jk}) + \tilde{\lambda}_R(T)\delta_{ij}\delta_{kl},$$

$$C^{-1}_{R;ijkl} = \frac{1}{\Omega a_0^2}\langle U_{ij}U_{kl}\rangle, \qquad (1.95)$$

$$U_{ij} = -\frac{1}{2}\oint_P dl(u_i n_j + u_j n_i),$$

where $\Omega$ is the area of the crystal, $\hat{n}$ is a unit vector normal to the edge of the system directed outwards, and the integration is over the perimeter $P$ of the crystal. Remembering that, in the presence of dislocations, the displacement field, $\mathbf{u}(\mathbf{r})$, is multivalued,[77] one can apply Green's theorem to $U_{ij}$ in Eq. (1.95) to obtain

$$U_{ij} = \int d^2\mathbf{r}\, u_{ij}(\mathbf{r}) + \frac{1}{2}a_0 \sum_{\mathbf{r}}(b_i(\mathbf{r})\epsilon_{jk}\mathbf{r}_k + b_j(\mathbf{r})\epsilon_{ik}\mathbf{r}_k). \qquad (1.96)$$

Now that we have identified the appropriate correlation function defining the renormalized elastic moduli, $\tilde{\mu}_R$ and $\tilde{\lambda}_R$, all that remains is some tedious but straightforward algebra to obtain the recursion relations for the running coupling constants under a rescaling of the short distance cut off, $a \to a e^{\delta l}$.[42]

$$\frac{d\tilde{\mu}^{-1}(l)}{dl} = 3\pi y^2(l) e^{K(l)/8\pi} I_0\left(\frac{K(l)}{8\pi}\right) + \mathcal{O}(y^3),$$

$$\frac{d(\tilde{\mu}(l) + \tilde{\lambda}(l))^{-1}}{dl} = 3\pi y^2(l) e^{K(l)/8\pi} \left[I_0\left(\frac{K(l)}{8\pi}\right) - I_1\left(\frac{K(l)}{8\pi}\right)\right] + \mathcal{O}(y^3),$$

$$\frac{dy(l)}{dl} = \left(2 - \frac{K(l)}{8\pi}\right) y(l) + 2\pi y^2(l) e^{K(l)/16\pi} I_0\left(\frac{K(l)}{8\pi}\right) + \mathcal{O}(y^3),$$

$$\frac{dK^{-1}(l)}{dl} = \frac{3}{4}\pi y^2(l) e^{K(l)/8\pi} \left[2I_0\left(\frac{K(l)}{8\pi}\right) - I_1\left(\frac{K(l)}{8\pi}\right)\right] + \mathcal{O}(y^3),$$

$$K(l) = \frac{4\tilde{\mu}(l)(\tilde{\mu}(l) + \tilde{\lambda}(l))}{2\tilde{\mu}(l) + \tilde{\lambda}(l)},$$

(1.97)

where $I_0(x)$ and $I_1(x)$ are modified Bessel functions.

The recursion relations of Eq. (1.97) have flows for the parameters, $\tilde{\mu}(l)$, $\tilde{\lambda}(l)$, and $y(l)$, very similar to Eq. (1.35) for the $XY$ model. Above a temperature, $T_m$, the dislocation fugacity, $y(l)$, increases as $l$ increases, implying that dislocations unbind and become free to move under any small stress, so that the crystal has melted. We can integrate these flow equations up to a length scale $\xi_+(T)$,[42] where

$$\xi_+(T) = e^{l^*} = \exp(b't^{-\nu}) \quad \text{with} \quad \nu = 0.3696\cdots. \quad (1.98)$$

The scale $\xi_+(T)$ is interpreted as the scale below which dislocations can be regarded as bound in pairs or triplets, while, on scales larger than $\xi_+(T)$, dislocations must be regarded as free to move under arbitrarily small stress, so that the system responds like a fluid and has melted. On the other hand, for $T \leq T_m$, the fugacity, $y(l)$, flows to the stable Gaussian fixed line $y(\infty) = 0$, and the renormalized elastic constants $\tilde{\mu}_R(T)$ and $\tilde{\lambda}_R(T)$ have finite values, in the same way as the renormalized superfluid density $\rho_R^s(T)$. Thus, the low temperature phase of this system is an elastic solid with finite elastic constants, neither of which are universal at $T_m^-$. For $T \leq T_m$, from Eq. (1.97), one obtains[42]

$$\tilde{\mu}_R(T) = \tilde{\mu}_R(T_m^-)(1 + b|t|^\nu),$$
$$\tilde{\lambda}_R(T) = \tilde{\lambda}_R(T_m^-)(1 + b|t|^\nu).$$

(1.99)

Although these elastic moduli are not individually universal at $T = T_m^-$, the combination $K_R(T_m^-)$ is universal. From Eq. (1.97), we have

$$K_R(T_m^-) = \frac{4\tilde{\mu}_R(\tilde{\mu}_R + \tilde{\lambda}_R)}{2\tilde{\mu}_R + \tilde{\lambda}_R} = \lim_{l \to \infty} K(l) = 16\pi. \quad (1.100)$$

The flow equations of Eq. (1.97) are very similar to Eq. (1.35) for the planar rotor model and the solutions also behave similarly. There is one major complication in the dislocation case for a solid with an underlying triangular symmetry. A bound pair of elementary dislocations can be equivalent to a third elementary dislocation, which leads to the $\mathcal{O}(y^2)$ contribution to the fugacity, $y(l)$, flow in Eq. (1.97). Although this does not make a major qualitative difference from the planar rotor vortex behavior, it does introduce some major technical complications, and gives a value of the exponent $\nu = 0.3696\cdots$ which is different from the planar rotor value of $\nu = 1/2$. At present, this difference in $\nu$ from $1/2$ is beyond experimental resolution.

One can also obtain the measured or renormalized Debye–Waller factor, which is closely related to the pair distribution function,

$$g(r) = \frac{\Omega}{N^2} \left\langle \sum_{ij} \delta(\mathbf{r} - \mathbf{r}_i + \mathbf{r}_j) \right\rangle, \quad (1.101)$$

and the Debye–Waller factor is

$$C_G(r) = \langle \rho_\mathbf{G}(\mathbf{r}) \rho_\mathbf{G}^*(0) \rangle \sim r^{-\eta_G(T)},$$

$$\eta_G(T) = \frac{|G|^2}{4\pi} \frac{3\tilde{\mu}_R + \tilde{\lambda}_R}{\tilde{\mu}_R(2\tilde{\mu}_R + \tilde{\lambda}_R)}. \quad (1.102)$$

Here, the $T$ dependence has been absorbed into the renormalized elastic moduli, $\tilde{\mu}_R(T)$ and $\tilde{\lambda}_R(T)$, of Eq. (1.99). The numerical value of $\eta_G(T)$ cannot be calculated exactly because the elastic moduli $\tilde{\mu}_R(T)$ and $\tilde{\lambda}_R(T)$ are not individually universal. However, for long ranged repulsive interactions, it is clear that $\tilde{\lambda}_R \gg \tilde{\mu}_R$ so that $\eta_G(T) \approx |G|^2/(4\pi\tilde{\mu}_R(T))$. The smallest value of $\eta_G$ gives the dominant behavior of $C_G(r)$ and, for a hexagonal crystal, $\eta_G(T_m^-) \geq \eta_{G_0}(T_m^-) = 1/3$.

At $T > T_m$, the dislocation fugacity is relevant, which implies that there is a finite density of free dislocations in the system. These can be regarded as bound in pairs up to a length scale $e^{l^*} = \xi_+(T)$ of Eq. (1.98), but are free, unbound at larger scales. The positional correlation function is expected to decay exponentially as

$$C_G(r) \sim e^{-r/\xi_+(T)} \text{ for } r \gg \xi_+(T). \quad (1.103)$$

Also, when $T > T_m$, the length dependent elastic moduli $\mu_R(l), \lambda_R(l) = 0$ for $e^l > \xi_+(T)$ but will be finite at length scales $e^l < \xi_+(T)$. This is an example of a system which has solid-like behavior at short length scales, $e^l < \xi_+(T)$ and is fluid-like at larger scales. This behavior is analogous to the behavior of the scale dependent stiffness constant $K(l)^{18}$ of the planar rotor model of Sec. 1.5.

We now discuss the behavior of the fluid phase when all dislocation pairs and triplets are unbound. This fluid phase is not the familiar isotropic fluid, but has remnants of the six-fold orientational order of the underlying triangular lattice of the low temperature solid phase. These possible orientational correlations in the fluid can be characterized by a finite renormalized Frank constant $K_A(T)$ of the anisotropic fluid. This can be expressed as a bond angle correlation function,[42]

$$K_A^{-1}(T) = \lim_{q \to 0} \frac{q^2}{\Omega} \langle \theta(q)\theta(-q) \rangle. \tag{1.104}$$

The orientational correlations in a fluid are characterized by $\infty > K_A^{-1}(T) > 0$, so a finite value of $K_A(T)$ implies that the fluid has orientational correlations. At temperatures, $T > T_m$, we expect exponential decay of translational order as $\exp(-r/\xi_+(T))$, but the system is elastic up to a length scale $e^l = \xi_+(T)$, and the dislocations are free at larger scales. One expects that the bond angles to be controlled by a free energy,[42]

$$\frac{F_A}{k_B T} = \frac{1}{2} K_A \int d^2\mathbf{r} (\nabla \theta)^2. \tag{1.105}$$

In the solid, the bond angle, $\theta_s(\mathbf{r})$, due to a set of dislocations is

$$\theta_s(\mathbf{r}) = -\frac{a_0}{2\pi} \sum_{\mathbf{r}'} \frac{\mathbf{b}(\mathbf{r}') \cdot (\mathbf{r} - \mathbf{r}')}{|\mathbf{r} - \mathbf{r}'|^2}, \tag{1.106}$$

from which one can express the stiffness constant as a correlation function,

$$K_A^{-1} = \lim_{q \to 0} \frac{a_0^2}{\Omega} \frac{q_i q_j}{q^2} \langle b_i(q) b_j(-q) \rangle. \tag{1.107}$$

The dislocation correlation function is calculated from the dislocation free energy of Eq. (1.93), written in Fourier space as,

$$\frac{H_D}{k_B T} = \frac{1}{2\Omega} \sum_q \left( \frac{K}{q^2} \left[ \delta_{ij} - \frac{q_i q_j}{q^2} \right] + \frac{2 E_c a^2}{k_B T} \delta_{ij} \right) b_i(q) b_j(-q). \tag{1.108}$$

At this point, we can see why the sign of the angular term in the dislocation interaction is vital and, also, why ignoring the angular term leads to incorrect results. The existence of a finite Frank constant, $K_A(T)$, depends on the transverse projection operator in the dislocation energy of Eq. (1.108), which arises from the form of the dislocation interaction of Eq. (1.93).

Here, one has assumed that the substrate potential can be written as $\sum_{\mathbf{K}} h_{\mathbf{K}} e^{i\mathbf{K}\cdot\mathbf{r}+\mathbf{u}(\mathbf{r})}$, where $\{\mathbf{K}\}$ are the reciprocal lattice vectors of the periodic substrate potential. These are incommensurate with the reciprocal lattice vectors $\{\mathbf{G}\}$ of the overlayer. Another assumption is that the adsorbate is in its low temperature floating solid phase, described by a harmonic free energy with renormalized elastic constants $\mu_R$ and $\lambda_R$. By redefining the displacement field,

$$u_i(\mathbf{q}) \to u_i'(\mathbf{q}) = u_i(\mathbf{q}) + i D_{ij}^{-1}(\mathbf{q}) \sum_{\mathbf{K}} h_{\mathbf{K}} K_j \Delta_{\mathbf{K},\mathbf{q}}$$

$$\Delta_{\mathbf{K},\mathbf{q}} = 1 \text{ if } \mathbf{K} = \mathbf{q} + \mathbf{G}$$

$$\frac{H}{k_B T} = \frac{1}{2k_B T} \int \frac{d^2 q}{(2\pi)^2} u_i'(\mathbf{q}) D_{ij}(\mathbf{q}) u_j'(-\mathbf{q}) \qquad (1.118)$$

$$- \frac{\Omega}{2} \sum_{\mathbf{K}} h_{\mathbf{K}}^2 K_i D_{ij}^{-1}(\mathbf{K}) K_j$$

where $\{\mathbf{G}\}$ is a reciprocal lattice vector of the adsorbed layer. One can write the second term in Eq. (1.118) as

$$C(\theta) = -\frac{\Omega}{2} \sum_{\mathbf{K}} \sum_{s=1}^{2} h_{\mathbf{K}}^2 \left( \frac{\mathbf{K}\cdot\hat{\epsilon}_s(\mathbf{K})}{\omega_s(\mathbf{K})} \right)^2, \qquad (1.119)$$

where $\hat{\epsilon}_s(\mathbf{K})$ is the $s^{\text{th}}$ polarization vector and $\omega_s(\mathbf{K})$ is the eigenfrequency of the matrix $D_{ij}(\mathbf{K})$.

Now, for a triangular adsorbate on a triangular substrate, $\theta = 0$ means perfect alignment, and the Fourier series is

$$C(\theta) = \Omega \sum_{k=0}^{\infty} c_k \cos(6k\theta) \qquad (1.120)$$

and, for a triangular adsorbate on a square substrate,

$$C(\theta) = \Omega \sum_{k=0}^{\infty} c_k \cos(12k\theta). \qquad (1.121)$$

Note that we have assumed that the adsorbate has a periodicity $\mathbf{G}$ which is incommensurate with the substrate periodicity $\mathbf{K}$, so that the only effect of the substrate potential is on the orientation of the crystal axes of the adsorbate. One can investigate the dislocation induced melting of the incommensurate adsorbate by writing the displacement as $\mathbf{u} = \phi + \mathbf{u}_s$, where $\mathbf{u}_s$ is the part due to the dislocations.

of a system which has solid-like behavior at short length scales, $e^l < \xi_+(T)$ and is fluid-like at larger scales. This behavior is analogous to the behavior of the scale dependent stiffness constant $K(l)$[18] of the planar rotor model of Sec. 1.5.

We now discuss the behavior of the fluid phase when all dislocation pairs and triplets are unbound. This fluid phase is not the familiar isotropic fluid, but has remnants of the six-fold orientational order of the underlying triangular lattice of the low temperature solid phase. These possible orientational correlations in the fluid can be characterized by a finite renormalized Frank constant $K_A(T)$ of the anisotropic fluid. This can be expressed as a bond angle correlation function,[42]

$$K_A^{-1}(T) = \lim_{q \to 0} \frac{q^2}{\Omega} \langle \theta(q) \theta(-q) \rangle. \quad (1.104)$$

The orientational correlations in a fluid are characterized by $\infty > K_A^{-1}(T) > 0$, so a finite value of $K_A(T)$ implies that the fluid has orientational correlations. At temperatures, $T > T_m$, we expect exponential decay of translational order as $\exp(-r/\xi_+(T))$, but the system is elastic up to a length scale $e^l = \xi_+(T)$, and the dislocations are free at larger scales. One expects that the bond angles to be controlled by a free energy,[42]

$$\frac{F_A}{k_B T} = \frac{1}{2} K_A \int d^2 \mathbf{r} (\nabla \theta)^2. \quad (1.105)$$

In the solid, the bond angle, $\theta_s(\mathbf{r})$, due to a set of dislocations is

$$\theta_s(\mathbf{r}) = -\frac{a_0}{2\pi} \sum_{\mathbf{r}'} \frac{\mathbf{b}(\mathbf{r}') \cdot (\mathbf{r} - \mathbf{r}')}{|\mathbf{r} - \mathbf{r}'|^2}, \quad (1.106)$$

from which one can express the stiffness constant as a correlation function,

$$K_A^{-1} = \lim_{q \to 0} \frac{a_0^2}{\Omega} \frac{q_i q_j}{q^2} \langle b_i(q) b_j(-q) \rangle. \quad (1.107)$$

The dislocation correlation function is calculated from the dislocation free energy of Eq. (1.93), written in Fourier space as,

$$\frac{H_D}{k_B T} = \frac{1}{2\Omega} \sum_q \left( \frac{K}{q^2} \left[ \delta_{ij} - \frac{q_i q_j}{q^2} \right] + \frac{2 E_c a^2}{k_B T} \delta_{ij} \right) b_i(q) b_j(-q). \quad (1.108)$$

At this point, we can see why the sign of the angular term in the dislocation interaction is vital and, also, why ignoring the angular term leads to incorrect results. The existence of a finite Frank constant, $K_A(T)$, depends on the transverse projection operator in the dislocation energy of Eq. (1.108), which arises from the form of the dislocation interaction of Eq. (1.93).

To proceed, we use the recursion relations of Eq. (1.97) up to some $l \leq l^* = \ln \xi_+(T)$, when they are still valid, to obtain,

$$K_A(T) = K_A(\tilde{\mu}, \tilde{\lambda}, y) = e^{2l} K_A(\tilde{\mu}(l), \tilde{\lambda}(l), y(l)), \qquad (1.109)$$

since there are two additional factors of $q$ in the definition of $K_A$ compared to $\tilde{\mu}, \tilde{\lambda}$. At the scale $e^{l^*} = \xi_+(T)$, we can assume that the density of free dislocations is large, and a reasonable approximation is to regard the field $\mathbf{b}(\mathbf{q})$ as continuous, ignoring the discrete nature of the $\mathbf{b}(\mathbf{r})$. This Debye–Hückel approximation allows the correlation function in Eq. (1.107) to be computed, because the transverse part in $H_D$ of Eq. (1.108) does not contribute. We obtain,

$$K_A^{-1}(T) \approx \frac{k_B T}{2 E_c a^2} > 0, \qquad (1.110)$$

which establishes that the fluid has orientational correlations.

We are left with a 2D XY problem with initial parameter $K_A(l^*)$, which we know how to solve. The topological defects which control the six-fold orientational order, when $T > T_m$, are $\pm\pi/3$ disclinations in the underlying hexagonal lattice. These are precisely analogous to the vortices in a 2D superfluid, and we can decompose the bond angle field $\theta(\mathbf{r})$ into a smooth part $\phi(\mathbf{r})$ and a singular part due to a set of disclinations to obtain,

$$\frac{H_D}{k_B T} = \frac{1}{2} \int d^2 \mathbf{r} (\nabla \phi(\mathbf{r}))^2 - \frac{\pi K_A}{36} \sum_{\mathbf{r} \neq \mathbf{r}'} m(\mathbf{r}) m(\mathbf{r}') \ln \left( \frac{|\mathbf{r} - \mathbf{r}'|}{a} \right)$$

$$+ \frac{E_c}{k_B T} \sum_{\mathbf{r}} m^2(\mathbf{r}). \qquad (1.111)$$

Here, $m(\mathbf{r}) = 0, \pm 1, \pm 2 \cdots$ is an integer measure of the strength of the disclination at $\mathbf{r}$, $a$ is the disclination core size, $E_c$ the disclination core energy, and the disclinations are subject to the neutrality condition, $\sum_\mathbf{r} m(\mathbf{r}) = 0$. We can immediately take over the results for the 2D superfluid of Sec. 1.5 to deduce that a disclination unbinding transition occurs at $T = T_I$, when

$$\frac{36}{2\pi K_A(T_m^-)} = \frac{1}{4}. \qquad (1.112)$$

The renormalized Frank constant, $K_A(T)$, behaves just like the stiffness constant, $K_R(T)$, of the 2D superfluid, and jumps discontinuously to zero at $T_I$,

$$K_A(T) = \frac{72}{\pi}(1 + b|t|^{1/2}), \quad \text{when } t = \frac{(T - T_I)}{T_I} \leq 0,$$

$$K_A(T) = 0, \quad \text{when } t > 0. \qquad (1.113)$$

When $T > T_I$, the orientational order is short ranged with

$$\langle \psi_6(r)\psi_6^*(0)\rangle \sim e^{-r/\xi_6(T)}, \tag{1.114}$$

where the orientational correlation length, $\xi_6(T)$, diverges exponentially,

$$\xi_6(T) = \exp\left(\frac{b}{(T/T_I - 1)^{1/2}}\right). \tag{1.115}$$

Assuming the initial value of $K_A(l)$ is larger than the critical value, it will flow to a value $K_A(\infty) \geq 72/\pi$. For $T$ just above the melting temperature, $T_m$, we expect that $K_A(T) \sim \xi_+^2(T)$, and, at the hexatic-isotropic liquid transition temperature, $T_I$, this theory predicts that $K_A(T_I^-) = 72/\pi$. Presumably, if the core energy $E_c$ is, for some reason, small so that the initial $K_A(l^*) < K_{0c}$, it will immediately flow to zero and there will be no hexatic phase. It is also possible that the hexatic phase exists over a very narrow temperature interval, $T_m < T \leq T_I$. Below the melting temperature, in the crystal phase, the correlation length $\xi_+(T) = \infty$, so that $K_A(T \leq T_m) = \infty$, and there is true long range orientational order. Alternatively, one can calculate $\langle \psi(\mathbf{r})\psi^*(0)\rangle$ at $r = \infty$ and find that it is finite,[42]

$$|\langle \psi(\mathbf{r})\rangle| \approx \exp\left(-\frac{9k_BT}{8\pi\mu_R(T)}\right) > 0. \tag{1.116}$$

This was noticed earlier by Mermin.[56]

### 1.6.4. Substrate effects on 2D melting

Most experiments on 2D melting are performed on a layer of adsorbed molecules on some substrate such as graphite, which forms relatively large well oriented domains of a periodic array of binding sites. It is therefore important to assess the effects of such a substrate. In their seminal paper,[42] Halperin and Nelson performed this investigation, summarized below. Both the graphite substrate and the adsorbate overlayer have the same six-fold orientational symmetry, allowing the relative orientations to vary slowly in space leading to an elastic free energy,

$$-\frac{H}{k_BT} = -\frac{1}{2k_BT}\int \frac{d^2q}{(2\pi)^2} u_i(\mathbf{q})D_{ij}(\mathbf{q})u_j(-\mathbf{q})$$
$$+ \sum_{\mathbf{r},\mathbf{K}} h_\mathbf{K} e^{i\mathbf{K}\cdot\mathbf{r}}(1 + i\mathbf{K}\cdot\mathbf{u}(\mathbf{r}) + \cdots) \tag{1.117}$$

$$D_{ij}(\mathbf{q}) = \mu_R q^2 \delta_{ij} + (\mu_R + \lambda_R)q_i q_j.$$

Here, one has assumed that the substrate potential can be written as $\sum_{\mathbf{K}} h_{\mathbf{K}} e^{i\mathbf{K}\cdot\mathbf{r}+\mathbf{u}(\mathbf{r})}$, where $\{\mathbf{K}\}$ are the reciprocal lattice vectors of the periodic substrate potential. These are incommensurate with the reciprocal lattice vectors $\{\mathbf{G}\}$ of the overlayer. Another assumption is that the adsorbate is in its low temperature floating solid phase, described by a harmonic free energy with renormalized elastic constants $\mu_R$ and $\lambda_R$. By redefining the displacement field,

$$u_i(\mathbf{q}) \to u'_i(\mathbf{q}) = u_i(\mathbf{q}) + i D_{ij}^{-1}(\mathbf{q}) \sum_{\mathbf{K}} h_{\mathbf{K}} K_j \Delta_{\mathbf{K},\mathbf{q}}$$

$$\Delta_{\mathbf{K},\mathbf{q}} = 1 \quad \text{if} \quad \mathbf{K} = \mathbf{q} + \mathbf{G}$$

$$\frac{H}{k_B T} = \frac{1}{2 k_B T} \int \frac{d^2 q}{(2\pi)^2} u'_i(\mathbf{q}) D_{ij}(\mathbf{q}) u'_j(-\mathbf{q}) \tag{1.118}$$

$$- \frac{\Omega}{2} \sum_{\mathbf{K}} h_{\mathbf{K}}^2 K_i D_{ij}^{-1}(\mathbf{K}) K_j$$

where $\{\mathbf{G}\}$ is a reciprocal lattice vector of the adsorbed layer. One can write the second term in Eq. (1.118) as

$$C(\theta) = -\frac{\Omega}{2} \sum_{\mathbf{K}} \sum_{s=1}^{2} h_{\mathbf{K}}^2 \left( \frac{\mathbf{K} \cdot \hat{\epsilon}_s(\mathbf{K})}{\omega_s(\mathbf{K})} \right)^2, \tag{1.119}$$

where $\hat{\epsilon}_s(\mathbf{K})$ is the $s^{\text{th}}$ polarization vector and $\omega_s(\mathbf{K})$ is the eigenfrequency of the matrix $D_{ij}(\mathbf{K})$.

Now, for a triangular adsorbate on a triangular substrate, $\theta = 0$ means perfect alignment, and the Fourier series is

$$C(\theta) = \Omega \sum_{k=0}^{\infty} c_k \cos(6k\theta) \tag{1.120}$$

and, for a triangular adsorbate on a square substrate,

$$C(\theta) = \Omega \sum_{k=0}^{\infty} c_k \cos(12k\theta). \tag{1.121}$$

Note that we have assumed that the adsorbate has a periodicity $\mathbf{G}$ which is incommensurate with the substrate periodicity $\mathbf{K}$, so that the only effect of the substrate potential is on the orientation of the crystal axes of the adsorbate. One can investigate the dislocation induced melting of the incommensurate adsorbate by writing the displacement as $\mathbf{u} = \phi + \mathbf{u}_s$, where $\mathbf{u}_s$ is the part due to the dislocations.

After some straightforward but tedious algebra, one finds that the elastic free energy can be written as,[42,43]

$$\frac{H_E}{k_B T} = \frac{1}{2} \int \frac{d^2 \mathbf{r}}{a_0^2} (2\tilde{\mu}\phi_{ij}^2 + \tilde{\lambda}\phi_{kk}^2 + \tilde{\gamma}(\partial_y \phi_x - \partial_x \phi_y)^2) + \frac{H_D}{k_B T},$$

$$\frac{H_D}{k_B T} = -\frac{1}{8\pi} \sum_{\mathbf{r} \neq \mathbf{r}'} b_i(\mathbf{r}) b_j(\mathbf{r}') \left( K_1 \ln \frac{|\mathbf{r} - \mathbf{r}'|}{a} \delta_{ij} - K_2 \frac{(\mathbf{r} - \mathbf{r}')_i (\mathbf{r} - \mathbf{r}')_j}{|\mathbf{r} - \mathbf{r}'|^2} \right)$$

$$+ \frac{E_c}{k_B T} \sum_{\mathbf{r}} |\mathbf{b}(\mathbf{r})|^2,$$

$$= \frac{1}{2} \int_{\mathbf{q}} \left[ \frac{K_1 + K_2}{2} \frac{1}{q^2} \left( \delta_{ij} - \frac{q_i q_j}{q^2} \right) \right.$$

$$\left. + \frac{K_1 - K_2}{2} \frac{q_i q_j}{q^4} + \frac{2 E_c a_0^2}{k_B T} \delta_{ij} \right] b_i(\mathbf{q}) b_j(-\mathbf{q}), \qquad (1.122)$$

$$K_1 = \frac{4\tilde{\mu}(\tilde{\mu} + \tilde{\lambda})}{2\tilde{\mu} + \tilde{\lambda}} + \frac{4\tilde{\mu}\tilde{\gamma}}{\tilde{\mu} + \tilde{\gamma}},$$

$$K_2 = \frac{4\tilde{\mu}(\tilde{\mu} + \tilde{\lambda})}{2\tilde{\mu} + \tilde{\lambda}} - \frac{4\tilde{\mu}\tilde{\gamma}}{\tilde{\mu} + \tilde{\gamma}}.$$

The renormalization group flows for the parameters $K_1(l)$, $K_2(l)$ and $y(l)$ were first worked out by Young[43] who showed that

$$\frac{dy}{dl} = \left( 2 - \frac{K_1}{8\pi} \right) y + 2\pi y^2 I_0 \left( \frac{K_2}{8\pi} \right) + \mathcal{O}(y^3). \qquad (1.123)$$

The same physics as for the isotropic elastic solid, when $K_2 = K_1$, holds in the ordered phase. Melting happens at $K_{1R}(T_m^-) = 16\pi$ and $K_{2R}(T_m^-)$ has some finite value depending on the orientational coupling strength, $\gamma_R(T_m^-)$. The relative values of the $K_i$ depend on the lattice structures of the substrate and the overlayer. The value of the exponent $\nu$ in Eqs. (1.98) and (1.99) depend on the value of $K_{2R}$.[42,43]

In the situation, when the substrate and overlayer reciprocal lattice vectors, $\{\mathbf{K}\}, \{\mathbf{G}\}$, have a set, $\{\mathbf{M}\}$, in common, there is the possibility that the overlayer is locked to the substrate, and so is a commensurate locked phase. We consider the smallest common vectors of length $M_0 = |\mathbf{M}|$ and write the Hamiltonian as,[42]

$$\frac{H}{k_B T} = \frac{H_E}{k_B T} + \sum_{|\mathbf{M}| = M_0} h_\mathbf{M} \sum_\mathbf{R} e^{i \mathbf{M} \cdot \mathbf{u}(\mathbf{R})}, \qquad (1.124)$$

where we assume that the effects of all other incommensurate terms have been incorporated into the elastic Hamiltonian, $H_E$. This is possible because these and commensurate terms in the substrate potential,

$$\frac{H_s}{k_B T} = \sum_{\mathbf{K}} h_{\mathbf{K}} \sum_{\mathbf{r}} e^{i\mathbf{K}\cdot\mathbf{u}(\mathbf{r})}, \qquad (1.125)$$

with $|\mathbf{M}| > M_0$ are less relevant than $|\mathbf{M}| = M_0$.

To assess the relevance of $h_{\mathbf{M}}$, we calculate the correlation function with a renormalized elastic Hamiltonian, $H_E$, with $h_{\mathbf{M}} = 0$,

$$\langle e^{i\mathbf{M}\cdot[\mathbf{u}(\mathbf{r})-\mathbf{u}(0)]}\rangle_{H_E} \sim r^{-\eta_{\mathbf{M}}},$$

$$\eta_{\mathbf{M}}(T) = \frac{M_0^2 k_B T}{4\pi} \frac{3\mu_R + \lambda_R + \gamma_R}{(\mu_R + \gamma_R)(2\mu_R + \lambda_R)}. \qquad (1.126)$$

From this, we can see immediately that the RG eigenvalue associated with $h_{M_0}(l) = h_{M_0}(0) e^{\lambda_{M_0} l}$ is

$$\lambda_{M_0}(T) = 2 - \frac{1}{2}\eta_{M_0}$$

$$= 2 - \frac{M_0^2 k_B T}{8\pi} \frac{3\mu_R + \lambda_R + \gamma_R}{(\mu_R + \gamma_R)(2\mu_R + \lambda_R)}. \qquad (1.127)$$

From Eq. (1.127), we see that, at low enough $T$, all the $h_{\mathbf{M}}$ are relevant, and the overlayer is locked to the substrate, implying that a lattice gas description is appropriate. The substrate mesh is sufficiently coarse, $\lambda_{M_0}(T_m) > 0$, which means that the overlayer is locked to the substrate up to the melting temperature, $T_m$, for $h_{\mathbf{M}} = 0$. The dislocation melting mechanism is not appropriate at any $T$ in this case. For large $M_0$, corresponding to a finer substrate mesh, $\lambda_{M_0}(T_m) < 0$, the overlayer will become unlocked from the substrate at some $T < T_m$, leading to a floating incommensurate solid, which may be accidentally commensurate with the substrate. One can also derive a criterion for the temperature, $T < T_m$, at which the overlayer unlocks from the substrate.[42]

The most dramatic effect of the substrate potential, even when it does not lock the overlayer into a commensurate phase, is on the fluid phases for $T > T_m$. We can perform a similar analysis for the Frank constant $K_A$ defined as a correlation function in Eq. (1.107), using a Debye–Hückel approximation wih the dislocation Hamiltonian, $H_D$, of Eq. (1.122). Since, as in Eq. (1.107), to compute $K_A$, we require the correlation function, in the

Debye–Hückel approximation,

$$K_A^{-1} = \lim_{\mathbf{q}\to 0} \frac{a_0^2}{\Omega} \frac{q_i q_j}{q^2} \langle b_i(\mathbf{q}) b_j(-\mathbf{q})\rangle ,$$

$$\langle b_i(\mathbf{q}) b_j(-\mathbf{q})\rangle = \frac{2\Omega q^2}{K_1 + K_2 + \frac{4E_c q^2 a^2}{k_B T}}\left(\delta_{ij} - \frac{q_i q_j}{q^2}\right) + \frac{2\Omega q^2}{K_1 - K_2 + \frac{4E_c q^2 a^2}{k_B T}} \frac{q_i q_j}{q^2},$$

$$K_A^{-1}(T) = \lim_{\mathbf{q}\to 0} \frac{2q^2}{K_1 - K_2 + \frac{4E_c a^2 q^2}{k_B T}} = 0, \text{ for } K_1 \neq K_2.$$

(1.128)

Thus, we recover the smooth substrate result that $K_A > 0$, while, for melting on a periodic substrate, $K_A(T) = \infty$. This means that there is no hexatic-isotropic fluid transition on any periodic triangular substrate, as the orientation of the hexatic fluid remains locked to the substrate orientation at all $T > T_m$.[42] For a triangular adsorbate on a substrate with a weak binding potential of square symmetry, the hexatic-isotropic transition will be in the 2D Ising universality class.

### 1.6.5. *Scaling in superconducting films*

We can write a scaling assumption for the resistivity $\mathcal{R}$, consistent with the renormalization group, by introducing all the relevant length scales, $\lambda_{\text{eff}}, \xi_+, \xi_-$, and $\xi_I$. The scales $\xi_+ = \xi_-^{2\pi}$ of Eq. (1.45) depend on $T$ only, while $\xi_I = I/I_0$ is the current length scale, and the penetration depth $\lambda_{\text{eff}}$ is an effective system size. We can make the scaling ansatz,[117]

$$\frac{V}{I} = e^{-zl}\mathcal{R}(\lambda_{\text{eff}} e^{-l}, \xi_+ e^{-l}, \xi_I e^{-l}, l/\ln \xi_-), \quad (1.129)$$

where the dynamical exponent $z = 2$, and $\mathcal{R}$ is an unknown scaling function, familiar from more conventional scaling functions, except for the combination $l/\ln \xi_-$, which arises from the marginally relevant and irrelevant scaling fields of the problem. The effect of this combination is to produce temperature and current dependent power laws, after making an appropriate choice for the length scale $e^l$ such that the scaling function $\mathcal{R}$ is known or can be computed. In this system, there are four potentially divergent length scales, $\xi_+, \xi_-, \xi_I$, and $\lambda_{\text{eff}}$, so the behavior of the scaling function and the $IV$ characteristics are very complex and difficult to obtain.

However, when the current $I$ is not too small so that $\xi_I < \xi_+, \lambda_{\text{eff}}$, one can choose the arbitrary scale $e^l = \xi_I$. At this scale, all vortex pairs are unbound by the current, and the free vortices may be regarded as moving independently, driven by the current in a viscous medium, so that

$$V/I = I^2 \mathcal{R}(\lambda_{\text{eff}}/\xi_I, \xi_+/\xi_I, 1, \ln \xi_I / \ln \xi_-). \tag{1.130}$$

For $T \leq T_{KT}$ and $\lambda_{\text{eff}} > \xi_I > \xi_-$, we expect that

$$\mathcal{R}(\infty, \infty, \ln \xi_I / \ln \xi_-) \sim I^{1/\ln \xi_-(T)} \sim I^{x(T)}, \tag{1.131}$$

where $\xi_-(T) = \exp[1/x(T)]$ with $x(T) = b|T - T_{KT}|^{1/2}$, which leads to the usual nonlinear $IV$ relation $V \sim I^{\pi K_R(T)+1}$.[55] When $\ln \xi_I < \ln \xi_-$, we expect that $V \sim I^3$ which is the $IV$ characteristic at $T_{KT}$ when $\xi_-(T) \to \infty$. Finite size dominated behavior is also contained in Eq. (1.129) by taking $\lambda_{\text{eff}} < \xi_I$ and $e^l = \lambda_{\text{eff}}$ to obtain

$$V/I = \lambda_{\text{eff}}^{-2} \mathcal{R}(1, \infty, \xi_I / \lambda_{\text{eff}}, \ln \lambda_{\text{eff}} / \ln \xi_-) \sim \lambda_{\text{eff}}^{-2}. \tag{1.132}$$

Identical arguments lead to a linear $IV$ relation $V/I \sim \lambda_{\text{eff}}^{-\pi K_R(T)}$ when $\lambda_{\text{eff}} > \xi_-$ and $V/I \sim \lambda_{\text{eff}}^{-2}$ when $\lambda_{\text{eff}} < \xi_-$. In principle, in the scaling regime of large correlation lengths one should be able to obtain analytically the scaling function $\mathcal{R}$, but this has not yet been achieved. Some simulations have been performed on lattice models of coupled Josephson junctions in a square array[117] which agree with the analytic predictions of Refs. 67 and 55, while other simulations[116] disagree obtaining $V \sim I^{2\pi K_R(T)-1}$, which agrees with the AHNS theory[67] only at $T = T_{KT}$, but not for any other temperature.

Experimental measurements of the $IV$ relation are, in general, consistent with the power law relation $V \sim I^{a(T)}$, with $a(T) = \pi K_R(T) + 1$ at intermediate values of the current $I$, but there is no discontinuity of the power $a(T)$ where $a(T) \geq 3$ for $T \leq T_{KT}$ and $a(T) = 1$ for $T > T_{KT}$.[118] Some measurements find evidence for a KT transition as these use a system with a very large penetration depth.[119–121] However, theory[17] predicts that there should be no true superconducting phase at any $T > 0$ because there would always be a finite concentration of free unbound vortices $n_f$ and $V/I \sim n_f > 0$ for sufficiently small $I$. In dirty superconducting films, the penetration depth $\lambda_{\text{eff}}$ is sufficiently large so that $n_f \ll 1$ and the linear resistivity is swamped by finite current effects. However, measurements on small arrays of Josephson junctions,[122] and on films of $YBCO$ one unit cell thick,[123] find indications of a linear $IV$ relation at very small currents at low $T < T_{KT}$, indicating a finite concentration of free vortices and no $KT$ transition, as predicted in an early paper.[17] The conclusion one seems to be forced to adopt is that, in superconducting films, the $I \to 0$ predictions are not accessible because of finite current effects, and a full crossover analysis is needed to fit the experimental data.

## 1.7. Experiments and Simulation

We have already given, in Sec. 1.5, a brief discussion of the key experiments[21,66,69] that have established the linear relation between the minimum superfluid density at which superfluidity can exist and the transition temperature at which this occurs. In this section we give a fuller discussion of these experiments. We also discuss experiments on other systems which can have transitions controlled by defects, and simulations of such systems, both simulations based on computer calculations, and those based on fabricated analog systems.

### 1.7.1. Measurements on superfluid films

The first thing to notice is that the superfluid density is, in fact, the $q, \omega \to 0$ limit of a response function. The Bishop–Reppy torsional oscillator experiment[66] measures the resonant frequency and the drive needed to overcome the damping due both to the torsion fiber and, in the neighborhood of the transition temperature, to the mutual friction between the normal fluid and the superfluid, the complex frequency dependent behavior of $\rho_s^R(q=0,\omega)$ where $\omega$ is the oscillation frequency of the torsional oscillator. The force exerted by the film on the oscillating substrate leads to a shift $\Delta P$ in the period $P$ of the oscillator, and the power dissipated in the film leads to a finite $Q$ factor[67] where

$$\frac{\Delta P}{P} = \frac{A}{2M} \operatorname{Re}[\rho_s^R(\omega)],$$

$$Q^{-1} = -\frac{A}{M} \operatorname{Im}[\rho_s^R(\omega)]. \tag{1.133}$$

Here, $A$ is the area of the substrate, $M$ is the mass of the oscillating substrate and $\omega$ is the frequency of oscillation. The extension of the Kosterlitz–Thouless static theory to finite frequency was essential for the comparison to experiment because, by necessity, the superfluid density is measured in a dynamical experiment at finite $\omega$ and finite superfluid velocity $u_s$. The conventional interpretation of the superfluid density is at $u_s = \omega = q = 0$ which is not accessible to experiment. The BKT static theory predicts only the $\omega = q = 0$ component of $\rho_s^R(q,\omega)$. The AHNS theory[67] for $\rho_s^R(q,\omega,v_n)$ is based on the static theory and yields an excellent fit to the torsional oscillator experiments and to third sound measurements[69,70] which also indicated that

$\rho_s^R(T_c^-) > 0$. This is in contrast to superfluids in $3D$ where the superfluid density vanishes at $T_c$.

In the torsional oscillator experiments at higher temperatures, where the helium is a normal liquid, the fluid moves with the substrate, and so the position of the resonance is determined in the usual way by the stiffness of the torsion thread, and by the total classical moment of inertia of the combination of the Mylar roll and the liquid film adhering to it. Most of the damping of the oscillator comes from dissipative effects in the torsion thread, with some small contribution from the viscosity of the helium gas surrounding the rotating cylinder. As the transition temperature is approached from above there begin to be regions which are superfluid, presumably predominantly in regions where the helium film is thickest, and such superfluid regions need to move with potential flow, $\nabla \times \mathbf{v} = 0$. Such regions of superfluidity reduce the moment of inertia of the film and increase the damping of the oscillator, so the onset of superfluidity is marked by an increase in the frequency of the torsional oscillator and a reduction in its amplitude. As the temperature is lowered further, the whole film becomes superfluid, with a superfluid density that tends to a limit lower than the total density of the film (in particular, the layer in contact with the solid never shares in the superfluidity), while the damping has a sharp maximum close to the superfluid transition, and is significant, but well below the maximum.

In the paper by Bishop and Reppy,[66] a figure to display the linear relation between transition temperature and the superfluid density (per unit area) at the transition is combined with experimental results based on measurements of third sound speed and attenuation. Third sound is a surface wave in a superfluid, which involves a wave in the superfluid density, balanced by a wave motion of the entropy, so that the normal fluid atoms, locked to the substrate, do not have to move. The analysis of this system is more complicated than the analysis of the torsion oscillator. We have already mentioned that Rudnick's group published an analysis showing a discontinuity of $\rho_s$ in 1969,[21] but published a reanalysis a year later[22] which no longer claimed this discontinuity of the superfluid density at the phase transition.

### 1.7.2. *Experimental measurements on 2D melting*

Around 1980, all these interesting theoretical predictions for a solid/hexatic/fluid system in $2D$ were known, in particular that a $2D$ crystal melts, at temperature $T_m$, to an anisotropic hexatic fluid by a continuous transition. At temperature $T_I > T_m$, the hexatic fluid transitions to an isotropic fluid via another continuous transition. This is in contrast to measurements of melting in 3D, which seems always to be a discontinuous first order transition. Also, prior to this, melting in 2D was believed by almost all physicists to be a single first order transition directly to an isotropic fluid. The status of the 2D melting problem in 1988 is reviewed in depth by Strandburg.[97] The status of 2D solids and fluids was already under fairly intense scrutiny, but with little success, although apparently continuous melting transitions were observed in several systems, and a continuous hexatic to isotropic fluid transition in a few cases. The natural sequence of experimental studies of melting in 2D would be to study first the simplest system of particles confined to a plane by a smooth, featureless potential to eliminate any extra complications. Unfortunately, this simplicity was not realized experimentally until 1999.[98]

The experimental challenge is to find convincing evidence for or against the appealing theoretical picture of a two-stage melting process, with an intervening anisotropic hexatic fluid phase between the low temperature triangular crystal phase and the high temperature isotropic fluid. An experiment to test the dislocation theory of melting must satisfy at least two conflicting criteria, a large, flat, well characterized substrate to support the system, and, preferably, featureless to avoid as many complications as possible. Ideally, a solid state substrate should be atomically flat, with no steps, over distances of several correlation lengths, $\xi_\pm(T)$, which diverge exponentially near the transitions. A typical domain on a flat substrate is less than $10^4$ atomic spacings, which can be exceeded by $\xi_\pm(T)$, implying that measurements tend to be finite size limited. Another major problem with solid state substrates is that they tend to be a crystal surface. Graphite cleaves into fairly large, atomically flat regions, which seem ideal until one remembers that a graphite surface tends to be a set of binding sites which form a triangular lattice of a certain spacing which may not be commensurate with the natural spacing of the adsorbate lattice. The most obvious effect of a graphite substrate is to impose a hexatic ordering field which induces long range hexatic order in the adsorbed layer, so that the fluid phase is a hexatic fluid at all $T > T_m$. Experiments on xenon adsorbed on graphite,[99] and on the (111) face of silver[100] are consistent with this. An extensive survey of experiments with different combinations of adsorbates and substrates is in Strandburg's review article.[97] As shown in Sec. 1.6.4, the solid still melts due to dislocations and the transition is characterized by slightly different singularities than melting on a smooth substrate, but the differences are beyond present experimental resolution.

The first attempt to investigate the melting of a $2D$ periodic crystal, without the extra complications of a periodic substrate, was done by Grimes and Adams.[101] They trapped a layer of electrons of density, $n_s = 10^8 - 10^9 \text{cm}^{-2}$, approximately 100 Å above the surface of superfluid $^4$He. In this range of density, the electrons behave classically, with a repulsive $1/r$ Coulomb interaction, characterized by a dimensionless parameter, $\Gamma = e^2\sqrt{\pi n_s}/k_B T$, where $n_s$ is the number density. For $\Gamma > \Gamma_c = 137 \pm 15$, the electrons form a hexagonal lattice, whose longitudinal vibrations are coupled to surface waves of the $^4$He.[102] This coupled system has measurable response functions, which allow one to draw some conclusions about the electron lattice. There is no discontinuity in the frequency or amplitude of the resonances as $\Gamma$ is varied, but there is a pronounced maximum in the amplitude at $\Gamma \approx 137$, and no hysteresis in the response. From this, one can conclude that the lattice melts at $\Gamma = \Gamma_m = 137 \pm 15$ by a continuous transition, consistent with KTHNY theory.[102] Thouless attempted to calculate $\Gamma_m$[103] using the $T = 0$ values, $\lambda = \infty$ and $\mu = 0.245 e^2 n^{3/2}$, in $k_B T_m = a_0^2 \mu$ to obtain $\Gamma_m = (\pi n)^{1/2} e^2/(k_B T_m) = 78.71$, which differs from the experimental value of $\Gamma_m = 137 \pm 15$.[101] A value $\Gamma_m \approx 95 \pm 2$ was obtained by Hockney and Brown using molecular dynamics.[104] A more sophisticated calculation by Morf,[105] taking into account the downward renormalization of $\mu(T)$ from its $T = 0$ value, agrees with the measured value of $\Gamma_m$.[101] Unfortunately, with this system, the question of the existence of a hexatic fluid phase could not be addressed, as there are no signals of a hexatic fluid in this system.

In 1987, Murray and Van Winkle[106] examined colloidal suspensions of charged polystyrene spheres of diameter approximately 0.30 μm. When confined between glass plates, separated by $1 - 4$ μm, in a solution of the correct salt concentration, these spheres form a hexagonal lattice, which can be seen with a microscope, the images digitized, and the structure function $S(\mathbf{q})$ calculated. At the widely spaced end of the wedge, the spheres form a well defined hexagonal solid, and, at the thin end, the diffraction pattern is a uniform ring of an isotropic $2D$ fluid. In the intermediate regime, the diffraction pattern has a hexagonal modulation, consistent with a hexatic phase. The visual data of this experiment is consistent with the KTHNY scenario, but is not definitive.

In 1995, an ingenious refinement was attempted[107] by confining an uncharged colloidal suspension between two glass plates. A static electrostatic field between them induces an electric dipole moment on the neutral polystyrene spheres, which repel each other. The electric dipole moment, $\mathbf{p}_s$, induced on a dielectric sphere of radius $r_0$, and dielectric constant $\epsilon_s$, by an electric field **E** in a medium of dielectric constant $\epsilon_m$, is[108]

$$\mathbf{p}_s = -\frac{\epsilon_m(\epsilon_m - \epsilon_s)}{2\epsilon_m + \epsilon_s} r_0^3 \mathbf{E}. \tag{1.134}$$

Assuming that the dipoles are all oriented in the same direction, normal to the confining glass plates, leads to a repulsive dipole–dipole energy,

$$V(r) = \epsilon_m \left(\frac{\epsilon_m - \epsilon_s}{2\epsilon_m + \epsilon_s}\right)^2 \left(\frac{r_0}{r}\right)^3 r_0^3 E^2. \tag{1.135}$$

The effective interaction strength, $V(r)$, between particles can be controlled by varying the applied electric field $E$ normal to the glass plates. Data on the particle configurations were collected by viewing with a CCD camera, and recording the signal for analysis.

The results of this investigation are consistent with the KTHNY scenario, since melting is observed to occur by a two-stage process, with an intermediate hexatic phase. The exponent, $\eta_{G_0}(T)$, is estimated to be very close to the predicted value of $1/3$ at melting of the solid to hexatic and, at the hexatic to isotropic transition, the data is consistent with the predicted value, $\eta_6 = 1/4$.[107] However, this investigation suffered from difficulties of equilibration, and these measurements were actually done as the system was evolving towards equilibrium, but are in agreement with previous studies.[106,109] A more serious difficulty is the confining glass plates, which, inevitably, have random trapped electric charges and dipoles. The resulting random pinning potential has all sorts of undetermined effects on the trapped colloid monolayer. However, the three predicted phases, periodic solid, hexatic, and isotropic fluid, are observed, which do agree with theory.

More recent experiments, carried out by Maret and coworkers,[72] from 1999–2006, on colloidal films trapped at the water–air interface of an inverted droplet, have provided data which agree, in almost every respect, with the predictions of the theory of $2D$ melting, which seem to confirm the KTHNY dislocation theory of melting in $2D$. The electric dipoles induced on the polystyrene spheres[107] are replaced by doping the colloidal particles with superparamagnetic $Fe_2O_3$, and the external electric field, $E$, is replaced by an applied magnetic field, $H$, normal to the colloidal film. The doped colloidal spheres are denser than water and are trapped at the water–air interface, which is extremely flat and contains no trapped immobile magnetic moments. This substrate is well characterized and very uniform, so this system is a very good realization of an ideal smooth flat substrate, leading to an experimental realization of an isolated two-dimensional system of interacting spherically symmetric particles, that is very close to the theoretical conditions. Surprisingly, despite the apparently ideal substrate, which has no pinning centers, equilibrating this very small 2D system proves to be surprisingly difficult, and time consuming.[110] Equilibration is vital because the experiments are designed to check the analytic theory, which is based on equilibrium statistical mechanics. The experiment must obey the theoretical conditions for the comparison to have meaning! Several days, combined with oscillating the magnetic field, are needed. The positions of the colloidal particles are tracked by a CCD camera, recorded and analyzed in detail. In these experiments, the interaction strength $\Gamma$ is given by[72]

$$\Gamma = \frac{\mu_0}{4\pi} \frac{\chi^2 H^2 (\pi\rho)^{3/2}}{k_B T} \propto \frac{1}{T_{\text{eff}}}, \qquad (1.136)$$

where $H$ is the magnetic field normal to the plane of the film, $\chi$ the susceptibility per particle, and $\rho$ the area number density of particles. The particle–particle interaction,

$$\frac{V(r)}{k_B T} = \frac{\Gamma}{(r/r_0)^3} \quad \text{where} \quad r_0 = (\pi\rho)^{-1/2}, \qquad (1.137)$$

is controlled by the magnetic field, $H$. The whole film contains up to $3 \times 10^5$ particles of which, typically, $3 \times 10^3$ are in the field of view. The water–air interface is kept flat to within 250 nm by controlling the amount of water in the inverted droplet.[72]

Overall, the results from these experiments agree quantitatively with the predictions of the KTHNY theory of melting, finding a power law decay of the positional correlation function for $\Gamma \geq \Gamma_m$,

$$C_{G_0}(r) \sim r^{-\eta_{G_0}(T)}, \qquad (1.138)$$

with the exponent $\eta_{G_0}(T) = 1/3$ at $T_m$, and decreasing as $T$ decreases, as expected. Also, for $T > T_m$, the positional correlation function is observed to decay faster than any power as $r$ increases, consistent with the predicted exponential behavior of Eq. (1.103).[98] Also, the elastic moduli, $\mu_R(T)$, $\lambda_R(T)$, and the Young's modulus, $K_R(T)$, are measured,[111] and found to agree remarkably well, for $T \leq T_m$, with the theoretical predictions of Nelson and Halperin.[42]

Knowing the individual positions of every particle in the field of view, enables one to study, quantitatively, the orientational order by calculating the orientational correlation function,[112] which is just the two-point correlation function of the planar rotor model, Eq. (1.55),

$$G_6(r) = \langle \psi_6(\mathbf{r})\psi_6^*(0) \rangle,$$

$$\psi_6(\mathbf{r}) = \psi_k = \frac{1}{N_k} \sum_j e^{6i\theta_{jk}}, \qquad (1.139)$$

where $\mathbf{r}$ is the position of the particle $k$, and $\theta_{jk}$ is the angle between a fixed reference axis and the bond between particles $k$ and $j$. The average is over all nearest neighbor particle pairs for each image, but also over many independent images. Theory predicts that

$$\begin{aligned} G_6(r) &\neq 0, & r \to \infty, & \quad T \leq T_m, \\ G_6(r) &\sim r^{-\eta_6(T)}, & & \quad T_m < T \leq T_I, \\ G_6(r) &\sim e^{-r/\xi_6(T)}, & & \quad T > T_I. \end{aligned} \qquad (1.140)$$

The orientational correlation length, $\xi_6(T) = \exp\bigl(b|\Gamma^{-1} - \Gamma_I^{-1}|^{-1/2}\bigr)$, and the exponent, $\eta_6(\Gamma)$, is related to the Frank constant, $K_A(\Gamma)$, as,

$$\eta_6(\Gamma) = \frac{18 k_B T}{\pi K_A(\Gamma)}. \qquad (1.141)$$

As we have seen, $\eta_6(\Gamma_I) = 1/4$, and, just above the melting temperature $T_m$,

$$K_A(\Gamma) \sim \xi_6^2 \sim \exp\left(\frac{2b}{|\Gamma^{-1} - \Gamma_m^{-1}|^\nu}\right), \qquad (1.142)$$

where $\nu = 0.3696\cdots$. The data fit these theoretical expressions up to $r/a = 30$ rather well, and can be considered as very strong evidence for the dislocation unbinding mechanism of melting, followed by a disclination unbinding transition between an intermediate hexatic fluid, with finite orientational stiffness and algebraic order, to an isotropic fluid at $T_I > T_m$.

Despite the rather good fits of the experimental data to all the theoretical predictions of the dislocation theory, one cannot exclude the possibility that weakly first order transitions preempt the continuous transitions of theory. For example, if the discontinuity in the Young's modulus at the melting temperature, $T_m$, turns out to be just slightly larger than the universal value,

$$\lim_{T \to T_m^-} K_R(T) = 16\pi, \qquad (1.143)$$

the theoretical continuous transition must be preempted by a first order melting transition. A similar discussion holds for the hexatic–isotropic transition. The available experimental data, although very good, is not sufficiently detailed, nor of sufficient resolution to *exclude* a weak first order transition. Perhaps, in the future, a smoking gun will be discovered which is capable of distinguishing between a weak first order melting transition and the continuous melting transition of KTHNY theory. At least, in the paramagnetic colloid system of Maret and coworkers,[72] the data agree quantitatively, to within experimental resolution, with all theoretical predictions of the KTHNY topological defect theory, and is completely consistent with two continuous transitions.

There have also been some experimental studies of anisotropic colloidal crystals in $2D$,[110,113] by tilting the magnetic field away from the normal to the plane of the colloidal system. This tilt modifies the interaction between particles from an isotropic $1/r^3$ interaction, when the magnetic field, $H$, normal to the droplet interface to an anisotropic dipole–dipole interaction,

$$V(\mathbf{r}) = \frac{\mu_0(\chi H)^2}{8\pi r^3}(1 - 3(\hat{\mathbf{H}} \cdot \hat{\mathbf{r}})^2), \quad (1.144)$$

where $\hat{\mathbf{H}}$ and $\hat{\mathbf{r}}$ are unit vectors in the direction of the field $\mathbf{H}$ and the separation $\mathbf{r}$ respectively. For small angles of tilt, a two-step melting process of a distorted hexagonal lattice is observed. For larger angles of tilt, the equilibrium crystal phase is a very anisotropic centered rectangular lattice, which melts into a $2D$ smectic-like phase.[113] This is similar to theoretical expectations for dislocation mediated melting of anisotropic layers.[114,115]

### 1.7.3. *Simulations of 2D melting*

There have been a number of numerical simulations of $2D$ melting on a smooth substrate. Alder and Wainwright's study of hard discs in 1962,[124] on the basis of a van der Waals loop in the pressure, claimed that the melting of this system is of first order. As bigger, better, and faster computers became available, larger and larger arrays of discs were simulated, but the first order/continuous melting controversy escalated, instead of settling to an agreed answer.[97] Most early simulations on hard discs, or on discs with a $1/r^3$ repulsion, failed to identify a hexatic phase, and concluded that melting was a direct first order transition, with supercooling and superheating, from a solid phase to an isotropic fluid.[125–129] There was also no consensus about the mechanism of melting in the hard disc system.[130–132] Bagchi et al.[133] did a finite size analysis of bond orientational order in a system of particles interacting with a repulsive $1/r^{12}$ interaction and found agreement with KTHNY theory. However, conclusive evidence for a hexatic phase was not found. Jaster, in a series of large scale simulations of the hard disc system,[134–136] obtained good agreement with theory for the melting transition, and some evidence for the existence of a hexatic phase, but the expected algebraic decay of the orientational correlation function was missed. Simulations of an electron system with a $1/r$ interaction did find algebraic decay of the correlation function,[137] successfully demonstrating the existence of the hexatic phase. Finally, Lin et al.[138] performed a finite size scaling analysis of a system of up to $N = 16384$ particles, with a dipole–dipole interaction, succeeded in fitting their data to KTHNY theory for both positional and orientational correlations. At about the same time, Mak[139] performed a simulation of up to $N = 2048^2$ hard discs, and demonstrated, by a finite size scaling analysis, that the van der Waals loops at the two transitions seem to scale away. This is consistent with both the solid–hexatic and the hexatic–isotropic transitions being continuous, and a van der Waals loop being a finite size artifact. However, there seems to still be some uncertainty about the isotropic–hexatic transition in the $2D$ hard disc system.[139]

An important contribution to simulations of $2D$ melting was made by Saito,[140] who performed Monte Carlo simulations on the dislocation Hamiltonian part, $H_D$, of Eq. (1.93), allowing the core energy, $E_c$, to be an independent parameter. The positions of the Burgers vector variables, $\mathbf{b}(\mathbf{r})$ lie on the sites, $\mathbf{r}$, of a triangular mesh of size up to $N = 1762$ points. The simulation is carried out in a rectangular box of size $\Omega = L_x L_y$, with periodic boundary conditions. The mesh site $\mathbf{r} = (x, y) = a(l, \sqrt{3}m)$, with $0 \leq l < L = L_x/a$ and $0 \leq m < L_y/a\sqrt{3}$, with $l + m$ even. The Burgers vectors are restricted to their six smallest values, $\mathbf{b} = (\pm a_0, 0), (\pm a_0/2, \pm\sqrt{3}a_0/2)$. The simulations, with a large core energy, $E_c$, are consistent with a continuous KTHNY melting transition.[140] On the other hand, the simulation data for small core energy, $E_c$, is more consistent with a first order transition. Interestingly, a study of the distribution of dislocations indicates that they are organized into grain boundaries, for small $E_c$, which is consistent with the grain boundary melting picture of Chui.[141] These simulations are unable to study the hexatic phase or the hexatic/isotropic transition, as the essential disclinations are excluded in the dislocation Hamiltonian of Eq. (1.93).

The first order versus continuous transition controversy led to the introduction, in 1982, of the Laplacian roughening model.[142] This model, defined on a triangular lattice, is dual to a melting model of interacting *disclinations*. The variables are a set of integer valued heights, $-\infty < h(\mathbf{r}) < +\infty$, with Hamiltonian,

$$H = -\frac{J}{2} \sum_{\mathbf{r}} \left( \frac{2}{3} \sum_{i=1}^{6} (h(\mathbf{r} + \delta_i) - h(\mathbf{r})) \right)^2, \quad (1.145)$$

where $\delta_i$ is an elementary lattice vector. By using the Poisson summation formula,[143]

$$\sum_{h=-\infty}^{+\infty} f(h) = \sum_{s=-\infty}^{+\infty} \int_{-\infty}^{+\infty} dh f(h) e^{2\pi i s h}, \qquad (1.146)$$

in the partition function, one obtains,

$$Z = \prod_{\mathbf{r}} \sum_{h(\mathbf{r})} e^{-\beta H[h]} = \prod_{\mathbf{r}} \sum_{s(\mathbf{r})=-\infty}^{+\infty} \int_{-\infty}^{+\infty} dh(\mathbf{r}) \exp\left(-\beta H(h) + 2\pi i \sum_{\mathbf{r}} s(\mathbf{r}) h(\mathbf{r})\right),$$

$$\propto \sum_{\{s(\mathbf{r})\}} \exp\left(-\frac{2\pi^2 k_B T}{J} \sum_{\mathbf{r} \neq \mathbf{r}'} V(\mathbf{r} - \mathbf{r}') s(\mathbf{r}) s(\mathbf{r}')\right),$$

$$\propto \sum_{\{s(\mathbf{r})\}} \exp\left(\frac{K}{16\pi} \sum_{\mathbf{r} \neq \mathbf{r}'} |\mathbf{r} - \mathbf{r}'|^2 \ln|\mathbf{r} - \mathbf{r}'| s(\mathbf{r}) s(\mathbf{r}') + E_c \sum_{\mathbf{r}} s^2(\mathbf{r})\right),$$

$$V(\mathbf{r}) \approx \frac{1}{8\pi}(r^2 \ln r + Ar^2 - B), \quad K = \frac{4\pi^2 k_B T}{J}, \quad E_c = \frac{BK}{16\pi}. \qquad (1.147)$$

In the dual representation, the integer variable, $s(\mathbf{r})$, is the strength of a disclination at site $\mathbf{r}$. These are subject to neutrality conditions, $\sum_{\mathbf{r}} s(\mathbf{r}) = 0 = \sum_{\mathbf{r}} \mathbf{r} s(\mathbf{r})$. In the disclination representation, $E_c$ represents the core energy of a disclination, which also determines the core energy of a dislocation, as a dislocation can be regarded as a disclination dipole.[142] One can change $E_c$ from its value obtained from the pure nearest neighbor roughening model of Eq. (1.145) by adding some further neighbor interactions to this Hamiltonian.[144,145] Strandburg[145] made a series of simulations on a modified Laplacian roughening model, designed to test the effect of the size of the dislocation and disclination core energies on the order of the 2D melting transitions. There have been some numerical studies on the effects of lowering the vortex core energy on the transition in the planar rotor model[146–148] with the conclusion that the continuous KT transition of Sec. 1.5 becomes first order when the defect core energy is sufficiently small. However, another study of the roughening representation of a 2D planar rotor model coupled to an Ising degree of freedom, where the two transitions could merge to become a single first order transition, showed that the first order transition did not occur, but the two second order lines came very close to each other.[149] One might speculate that reducing the defect core energy might make the hexatic phase disappear resulting in a direct first order transition from a periodic solid to an isotropic fluid. The simulations by Strandburg[144,145] are designed to test this hypothesis. The simulations were performed on a Laplacian roughening model on a triangular lattice described by a Hamiltonian,

$$\frac{H[h]}{k_B T^*} = \sum_{\mathbf{q}} [G^{-2}(\mathbf{q}) - \gamma G^{-4}(\mathbf{q})] h(\mathbf{q}) h(-\mathbf{q}), \qquad (1.148)$$

where $\gamma = \frac{\sqrt{3}\delta E_c}{32\pi}$ and $\delta E_c$ is the change in the dislocation core energy.

The results of these simulations are, for the disclination core energy $E_c > E_c^* \approx 2.7$, the duals of the three phases predicted by theory[42] are clearly visible, with quantitative agreement with theory. For the core energy, $E_c < E_c^*$, the two transitions seem to merge into a single first order transition.[145] These simulations are in accord with expectations, with the caveat that it is possible that the separation between the two transition lines becomes so small that they look like a single first order transition.

# References

1. L. Onsager, *Nuovo Cimento* **6**, Suppl. 2, 249 (1949).
2. R. P. Feynman, in *Progress in Low-Temperature Physics* **1**, pp. 17–53, ed. C. J. Gorter (North-Holland, Amsterdam, 1955).
3. P. W. Anderson and G. Yuval, *Phys. Rev. Lett.* **23**, 89 (1969).
4. D. R. Hamann, *Phys. Rev. Lett.* **23**, 95 (1969).
5. G. Yuval and P. W. Anderson, *Phys. Rev. B* **1**, 1522 (1970).
6. P. W. Anderson, G. Yuval and D. R. Hamann, *Phys. Rev. B* **1**, 4464 (1970).
7. D. Ruelle, *Commun. Math. Phys.* **9**, 267 (1968).
8. F. J. Dyson, *Commun. Math. Phys.* **12**, 91 (1969).
9. L. D. Landau and E. M. Lifshitz, *Statistical Physics* (Pergamon Press, Ltd., London), p. 482 (1958).
10. D. J. Thouless, *Phys. Rev.* **187**, 732 (1969).
11. F. J. Dyson, *Commun. Math. Phys.* **21**, 269 (1971).
12. J. Fröhlich and T. Spencer, *Commun. Math. Phys.* **84**, 87 (1982).
13. M. Aizenman, J. M. Chayes, L. Chayes and C. M. Newman, *J. Stat. Phys.* **50**, 1 (1988).
14. H. E. Hall and W. F. Vinen, *Proc. Roy. Soc. (London)* **238**, 204 and 215 (1956).
15. H. Lamb, *Hydrodynamics*, 6th ed. reprinted (Dover, New York, 1945), pp. 615–617.
16. J. M. Kosterlitz and D. J. Thouless, *J. Phys. C: Solid St. Phys.* **5**, L124 (1972).
17. J. M. Kosterlitz and D. J. Thouless, *J. Phys. C: Solid St. Phys.* **6**, 1181 (1973).
18. D. R. Nelson and J. M. Kosterlitz, *Phys. Rev. Lett.* **39**, 1201 (1977).

19. V. L. Berezinskii, *Zhur. Eksp. Teor. Fiz.* **59**, 907 (1970); *Sov. Phys. JETP* **32**, 493 (1970).
20. V. L. Berezinskii, *Zhur. Eksp. Teor. Fiz.* **61**, 1144 (1971); *Sov. Phys. JETP* **34**, 610 (1971).
21. R. S. Kagiwada, J. C. Fraser, I. Rudnick and D. Bergman, *Phys. Rev. Lett.* **22**, 338 (1969).
22. I. Rudnick and J. C. Fraser, *J. Low Temp. Phys.* **3**, 225 (1970).
23. M. E. Fisher, *Rep. Progr. Phys.* **30**, 731 (1967).
24. K. G. Wilson, *Phys. Rev. B* **4**, 3174, 3184 (1971).
25. K. G. Wilson and M. E. Fisher, *Phys. Rev. Lett.* **28**, 240 (1972).
26. R. E. Peierls, *Helv. Phys. Acta* **7**, Suppl. II, 81 (1934).
27. R. E. Peierls, *Ann. Inst. Henri Poincaré* **5**, 177 (1935).
28. N. D. Mermin and H. Wagner, *Phys. Rev. Lett.* **17**, 1133 (1966).
29. P. C. Hohenberg, *Phys. Rev.* **158**, 383 (1967).
30. K. G. Wilson and J. Kogut, *Phys. Rep. C* **12**, 76 (1974).
31. H. E. Stanley and T. A. Kaplan, *Phys. Rev. Lett.* **17**, 916 (1966).
32. H. E. Stanley, *Phys. Rev. Lett.* **20**, 150 (1968).
33. M. A. Moore, *Phys. Rev. Lett.* **23**, 861 (1969).
34. A. A. Belyavin and A. M. Polyakov, *Zhur. Eksp. Teor. Fiz. Pisma* **22**, 503 (1975) [translation in JETP Lett. **22**, 245].
35. T. H. R. Skyrme, *Proc. Roy. Soc. (London)* **262**, 237 (1961).
36. S. L. Sondhi, A. Karlhede, S. A. Kivelson and E. H. Rezayi, *Phys. Rev. B* **47**, 16419 (1993).
37. H. A. Fertig, L. Brey, R. Cote and A. H. MacDonald, *Phys. Rev. B* **50**, 11018 (1994).
38. A. Schmeller, J. P. Eisenstein, L. N. Pfeiffer and K. W. West, *Phys. Rev. Lett.* **75**, 4290 (1995).
39. Z. F. Ezawa and G. Tsitsishvili, *Repts. Progr. Phys.* **72**, 086502 (2009).
40. P. Debye and E. Hückel, *Physik Z.* **24**, 185 (1923).
41. J. M. Kosterlitz, *J. Phys. C: Solid St. Phys.* **7**, 1046 (1974).
42. B. I. Halperin and D. R. Nelson, *Phys. Rev. Lett.* **41**, 121 (1978); *Phys. Rev. B* **19**, 2457 (1979).
43. A. P. Young, *Phys. Rev. B* **19**, 1855 (1979).
44. F. R. N. Nabarro, *Theory of Crystal Dislocations* (Pergamon Press, Oxford, 1967), pp. 688–690.
45. S. Timoshenko, *Theory of Elasticity* (MacGraw-Hill, New York, 1951).
46. Y. Imry and L. Gunther, *Phys. Lett.* **29A**, 483 (1969).
47. E. Orowan, *Zeits. f. Phys.* **89**, 605 (1934).
48. M. Polanyi, *Zeits. f. Phys.* **89**, 660 (1934).
49. G. I. Taylor, *Proc. Roy. Soc. (London)* **145**, 362 (1934); *Proc. Roy. Soc. (London)* **145**, 405 (1934).
50. M. J. Bowick and A. Travesset, *Phys. Rept.* **344**, 255 (2001).
51. R. L. Elgin and D. L. Goodstein, in *Monolayer and Submonolayer Films*, ed. J. G. Daunt and E. Lerner (Plenum Press, New York, 1973), pp. 35–52.
52. M. Tinkham, *Introduction to Superconductivity* (McGraw-Hill, New York, 1975).

53. J. Pearl, *Appl. Phys. Lett.* **5**, 65 (1964).
54. D. R. Nelson, *Defects and Geometry in Condensed Matter Physics* (Cambridge University Press, Cambridge, New York, Port Melbourne, Madrid, Cape Town, 2002).
55. B. I. Halperin and D. R. Nelson, *J. Low Temp. Phys.* **36**, 599 (1979).
56. N. D. Mermin, *Phys. Rev.* **176**, 250 (1968).
57. L. D. Landau, *Phys. Z. Sowjetunion* **11**, 26 (1937).
58. Y. Imry and L. Gunther, *Phys. Rev. B* **3**, 3939 (1971).
59. A. P. Young, *J. Phys. C: Solid State Phys.* **11**, L453 (1978).
60. J. José, L. P. Kadanoff, S. Kirkpatrick and D. R. Nelson, *Phys. Rev. B* **16**, 1217 (1977): Errata in *Phys. Rev. B* **17**, 1477 (1978).
61. P. C. Hohenberg and P. C. Martin, *Ann. Phys.* **34**, 291 (1965).
62. B. D. Josephson, *Phys. Lett.* **21**, 608 (1966).
63. E. Lieb, *Phys. Rev. Lett.* **18**, 1046 (1967).
64. E. Lieb and F. Y. Wu, in *Phase Transitions and Critical Phenomena*, eds. C. Domb and M. S. Green (Academic Press, London), Vol. 1, (1972).
65. D. R. Nelson, in *Phase Transitions and Critical Phenomena* Vol. 7, eds. C. Domb and M. S. Green (Academic Press, London, 1983).
66. D. J. Bishop and J. D. Reppy, *Phys. Rev. Lett.* **40**, 1727 (1978).
67. V. Ambegaokar, B. I. Halperin, D. R. Nelson and E. D. Siggia, *Phys. Rev. Lett.* **40**, 783 (1978); *Phys. Rev. B* **21**, 1806 (1980).
68. I. Rudnick, *Phys. Rev. Lett.* **40**, 1454 (1978).
69. K. R. Atkins and I. Rudnick, *Prog. Low Temp. Phys.* **6**, 37 (1970).
70. K. K. Telschow, D. T. Ekholm and R. B. Hallock, in *Quantum Solids and Fluids*, eds. S. B. Trickey *et al.* (Plenum, New York, 1977), p. 421.
71. C. Rottman and M. Wortis, *Phys. Rept.* **103**, 59 (1984).
72. H. H. von Grünberg, P. Keim and G. Maret in *Soft Matter, Vol. 3: Colloidal Order from Entropic and Surface Force*, eds. G. Gompper and M. Schick (Wiley-VCH Verlag GmbH & Co. KGaA, Weinheim, 2007), p. 40.
73. P. E. Wolf, S. Balibar and F. Gallet, *Phys. Rev. Lett.* **51**, 1366 (1983).
74. S. T. Chui and J. D. Weeks, *Phys. Rev. B* **14**, 4978 (1976).
75. S. T. Chui and J. D. Weeks, *Phys. Rev. Lett.* **40**, 733 (1978).
76. C. Jayaprakash, W. F. Saam and S. Teitel, *Phys. Rev. Lett.* **50**, 2017 (1983).
77. P. M. Chaikin and T. C. Lubensky, *Principles of Condensed Matter Physics* (Cambridge University Press, Cambridge, New York, Melbourne, Madrid, Cape Town, Singapore, São Paulo, 1995).
78. K. O. Keshishev, A. Ya. Parshin and A. V. Babkin, *Sov. Phys. JETP Lett.* **30**, 56 (1979): *JETP* **53**, 362 (1981).
79. C. A. Young, R. Pindak, N. A. Clark and R. B. Meyer, *Phys. Rev. Lett.* **40**, 773 (1978).
80. R. Pindak, C. A. Young, R. B. Meyer and N. A. Clark, *Phys. Rev. Lett.* **45**, 1193 (1980).
81. X. Du, I. Skachko, A. Barker and E. Andrei, *Nature Nanotechnol.* **3**, 491 (2008).
82. L. P. Kadanoff, *Phys. Rev. Lett.* **39**, 903 (1977).
83. L. P. Kadanoff and A. C. Brown, *Ann. Phys. (New York)* **121**, 318 (1979).

84. M. P. M. den Nijs, *J. Phys. A* **12**, 1857 (1979).
85. H. J. F. Knops, in *Fundamental Problems in Statistical Mechanics*, Vol. V, ed. E. G. D. Cohen (North Holland, New York, 1980).
86. J. Villain, *J. Phys. (Paris)* **36**, 581 (1975).
87. K. Kankaala, T. Ala-Nissila and S.-C. Ying, *Phys. Rev. B* **47**, 2333 (1993).
88. T. Ala-Nissila, E. Granato, K. Kankaala, J. M. Kosterlitz and S.-C. Ying, *Phys. Rev. B* **50**, 12692 (1994).
89. K. Griffiths, D. A. King and G. Thomas, *Vacuum* **31**, 671 (1981).
90. J. J. Arrecis, Y. J. Chabal and S. B. Christman, *Phys. Rev. B* **33**, 7906 (1986).
91. S. B. Dierker and R. Pindak, *Phys. Rev. Lett.* **59**, 1002 (1987).
92. G. S. Smith, E. B. Sirota, C. R. Safinya and N. A. Clark, *Phys. Rev. Lett.* **60**, 813 (1988).
93. David R. Nelson and B. I. Halperin, *Phys. Rev. B* **21**, 5312 (1980).
94. J. V. Selinger and D. R. Nelson, *Phys. Rev. Lett.* **61**, 416 (1988); *Phys. Rev. A* **39**, 3135 (1989).
95. D. R. Nelson, *Phys. Rev. B* **18**, 2318 (1978)
96. P. G. de Gennes, *Mol. Cryst. Liq. Cryst.* **21**, 49 (1973).
97. K. J. Strandburg, *Rev. Mod. Phys.* **60**, 161 (1988).
98. K. Zahn, R. Lenke and G. Maret, *Phys. Rev. Lett.* **82**, 2721 (1999).
99. P. A. Heiney, P. W. Stephens, R. J. Birgenau, P. M. Horn and D. E. Moncton, *Phys. Rev. B* **28**, 6416 (1983).
100. N. Greiser, G. A. Held, R. Frahm, R. L. Greene, P. M. Horn and R. M. Suter, *Phys. Rev. Lett.* **59**, 1706 (1987).
101. C. C. Grimes and G. Adams, *Phys. Rev. Lett.* **42**, 795 (1979).
102. D. S. Fisher, B. I. Halperin and P. M. Platzman, *Phys. Rev. Lett.* **42**, 798 (1979).
103. D. J. Thouless, *J. Phys. C* **11**, L189 (1978).
104. R. W. Hockney and T. R. Brown, *J. Phys. C: Solid St. Phys.* **8**, 1813 (1975).
105. R. H. Morf, *Phys. Rev. Lett.* **43**, 931 (1979).
106. C. A. Murray and D. H. van Winkle, *Phys. Rev. Lett.* **58**, 1200 (1987).
107. R. E. Kusner, J. A. Mann and A. J. Dahm, *Phys. Rev. B* **51**, 5746 (1995).
108. J. D. Jackson, *Classical Electrodynamics* (Wiley, New York, Chichester, Weinheim, Brisbane, Singapore, Toronto, 1998).
109. Y. Tang, A. J. Armstrong, R. C. Mockler and W. J. O'Sullivan, *Phys. Rev. Lett.* **62**, 2401 (1989). C. A. Murray and D. H. van Winkle, *Phys. Rev. Lett.* **58**, 1200 (1987).
110. C. Eisenmann, P. Keim, U. Gasser and G. Maret, *J. Phys.: Condens. Matter* **16**, 4095 (2004).
111. H. H. von Grünberg, P. Keim, K. Zahn and G. Maret, *Phys. Rev. Lett.* **93**, 255703 (2004).
112. P. Keim, G. Maret and H. H. von Grünberg, *Phys. Rev. E* **75**, 031402 (2007).
113. C. Eisenmann, U. Gasser, P. Keim and G. Maret, *Phys. Rev. Lett.* **93**, 105702 (2004).
114. S. Ostlund and B. I. Halperin, *Phys. Rev. B* **23**, 335 (1981).
115. J. Toner and D. R. Nelson, *Phys. Rev. B* **21**, 316 (1981).

116. P. Minnhagen, O. Westman, A. Jonsson and P. Olsson, *Phys. Rev. Lett.* **74**, 3672 (1994).
117. M. V. Simkin and J. M. Kosterlitz, *Phys. Rev. B* **55**, 11646 (1997).
118. M. R. Beasley, J. E. Mooij and T. P. Orlando, *Phys. Rev. Lett.* **42**, 1165 (1979).
119. A. F. Hebard and A. T. Fiory, *Phys. Rev. Lett.* **50**, 1603 (1983).
120. P. A. Bancel and K. E. Gray, *Phys. Rev. Lett.* **46**, 148 (1981).
121. S. A. Wolf, D. U. Gubser, W. W. Fuller, J. C. Garland and R. S. Newrock, *Phys. Rev. Lett.* **47**, 1071 (1981).
122. H. S. J. van der Zant, H. A. Rijken and J. E. Mooij, *J. Low Temp. Phys.* **79**, 289 (1990).
123. J. M. Repaci, C. Kwon, Qi Li, Xiuguang Jiang, T. Venkatessan, R. E. Glover III, C. J. Lobb and R. S. Newrock, *Phys. Rev. B* **54**, R9674 (1996).
124. B. J. Alder and T. E. Wainwright, *Phys. Rev.* **127**, 359 (1962).
125. R. K. Kalia and P. Vashista, *J. Phys. C* **14**, L643 (1981).
126. V. M. Bedanov, G. V. Gadiyak and Yu. E. Lozovik, *Phys. Lett. A* **92**, 400 (1982).
127. S. Toxvaerd, *Phys. Rev. A* **24**, 2735 (1982).
128. H. Weber, D. Marx and K. Binder, *Phys. Rev. B* **51**, 14636 (1995).
129. P. S. Branicio, J. P. Rino and N. Studart, *Phys. Rev. B* **64**, 193413 (2001).
130. J. A. Zollweg, G. V. Chester and P. W. Leung, *Phys. Rev. B* **39**, 9518 (1989).
131. J. A. Zollweg and G. V. Chester, *Phys. Rev. B* **46**, 11186 (1992).
132. J. F. Fernández, J. J. Alonso and J. Stankiewicz, *Phys. Rev. Lett.* **75**, 3477 (1995); *Phys. Rev. E* **55**, 750 (1997).
133. K. Bagchi, H. C. Andersen and W. Swope, *Phys. Rev. Lett.* **76**, 255 (1996).
134. A. Jaster, *Europhys. Lett.* **42**, 277 (1998).
135. A. Jaster, *Phys. Rev. E* **59**, 2594 (1999).
136. A. Jaster, *Phys. Lett. A* **330**, 120 (2004).
137. W. J. He, T. Cui, Y. M. Ma, Z. M. Liu and G. T. Zou, *Phys. Rev. B* **68**, 195104 (2003).
138. S. Z. Lin, B. Zheng and S. Trimper, *Phys. Rev. E* **73**, 066106 (2006).
139. C. H. Mak, *Phys. Rev. E* **73**, 065104 (2006).
140. Y. Saito, *Phys. Rev. Lett.* **48**, 1114 (1982); *Phys. Rev. B* **26**, 6239 (1982).
141. S. T. Chui, *Phys. Rev. Lett.* **48**, 1114 (1982).
142. D. R. Nelson, *Phys. Rev. B* **26**, 269 (1982).
143. I. M. Gel'fand and G. E. Shilov, *Generalized Functions* (Academic Press, New York, 1964), Vol. I, 332.
144. K. J. Strandburg, S. A. Solla and G. V. Chester, *Phys. Rev. B* **28**, 2717 (1983).
145. K. J. Strandburg, *Phys. Rev. B* **34**, 3536 (1989).
146. R. H. Swendsen, *Phys. Rev. Lett.* **49**, 1302 (1982).
147. E. Domany, M. Schick and R. H. Swendsen, *Phys. Rev. Lett.* **52**, 1535 (1984).
148. J. E. van Himbergen, *Phys. Rev. Lett.* **53**, 5 (1984).
149. S. Lee, K.-C. Lee and J. M. Kosterlitz, *Phys. Rev. B* **56**, 340 (1997).

# Chapter 2

# Duality, Gauge Symmetries, Renormalization Groups and the BKT Transition

Jorge V. José[*]

*Department of Physics, Indiana University, Carmichael Center, 530 E. Kirkwood Avenue, Bloomington, IN 47408, USA*

In this chapter, I will briefly review, from my own perspective, the situation within theoretical physics at the beginning of the 1970s, and the advances that played an important role in providing a solid theoretical and experimental foundation for the Berezinskii–Kosterlitz–Thouless theory (BKT). Over this period, it became clear that the Abelian gauge symmetry of the 2D-XY model had to be preserved to get the right phase structure of the model. In previous analyses, this symmetry was broken when using low order calculational approximations. Duality transformations at that time for two-dimensional models with compact gauge symmetries were introduced by José, Kadanoff, Nelson and Kirkpatrick (JKKN). Their goal was to analyze the phase structure and excitations of XY and related models, including symmetry breaking fields which are experimentally important. In a separate context, Migdal had earlier developed an approximate Renormalization Group (RG) algorithm to implement Wilson's RG for lattice gauge theories. Although Migdal's RG approach, later extended by Kadanoff, did not produce a true phase transition for the XY model, it almost did asymptotically in terms of a non-perturbative expansion in the coupling constant with an essential singularity. Using these advances, including work done on instantons (vortices), JKKN analyzed the behavior of the spin–spin correlation functions of the 2D XY-model in terms of an expansion in temperature and vortex-pair fugacity. Their analysis led to a perturbative derivation of RG equations for the XY model which are the same as those first derived by Kosterlitz for the two-dimensional Coulomb gas. JKKN's results gave a theoretical formulation foundation and justification for BKT's sound physical assumptions and for the validity of their calculational approximations that were, in principle, strictly valid only at very low temperatures, away from the critical $T_{\text{BKT}}$ temperature. The theoretical predictions were soon

---

[*]Also at Cellular & Integrative Physiology, Indiana University School of Medicine.

tested successfully against experimental results on superfluid helium films. The success of the BKT theory also gave one of the first quantitative proofs of the validity of the RG theory.

## 2.1. Introduction

An important problem in condensed matter physics has been to understand the possible ordered phases in systems with restricted dimensionalities. Back in the 1930s, Peierls[1] and Landau and Lifshitz,[2] among others, had already addressed this question. They provided heuristic arguments showing that no stable phase was possible in one- and two-dimensional systems with short-range interactions. However, in 1944, Onsager exactly solved the discrete $\mathbb{Z}_2$ symmetry Ising model in two dimensions, proving that this model has a finite temperature second-order phase transition with power law singularities.[3,4] For systems with continuous symmetries, however, Mermin and Wagner, and others rigorously proved that there cannot be a broken symmetry, stable, thermodynamic ordered phase in one- and two-dimensional systems.[5] Using Bogoliubov's inequality, they explicitly showed that there was no *Long-Range Order*[6] (LRO) in this type of model.

More explicitly, in the case of spin systems, LRO can be defined in terms of the two-point spin–spin correlation function:

$$\langle \vec{S}_{\vec{X}} \bullet \vec{S}_{\vec{X}'} \rangle = \frac{1}{Z} \text{Tr}_{\{\vec{S}\}} e^{-H[\vec{S}]/k_B T} \vec{S}_{\vec{X}} \bullet \vec{S}_{\vec{X}'}. \tag{2.1}$$

Here $Z = \text{Tr}_{\{\vec{S}\}} e^{-H[\vec{S}]/k_B T}$ is the partition function; $k_B$ Boltzmann's constant and $T$ the temperature. We shall take $k_B = 1$ henceforth. The trace, $\text{Tr}_{\{\vec{S}\}}$, is evaluated over the set of spin degrees of freedom $\{\vec{S}_{\vec{X}}\}$, with $\vec{X}$ a lattice site vector.

There is LRO in a spin model system if there is a critical temperature $T_c$ with $0 < T \leq T_c < \infty$ such that

$$\lim_{|\vec{X}-\vec{X}'|\to\infty} \langle \vec{S}_{\vec{X}} \bullet \vec{S}_{\vec{X}'} \rangle_{T<T_c} \to \langle \vec{S}_{\vec{X}} \rangle^2_{T<T_c} \to M^2 \neq 0. \tag{2.2}$$

Above $T_c$, the correlation function decays exponentially:

$$\lim_{|\vec{X}-\vec{X}'|\to\infty} \langle \vec{S}_{\vec{X}} \bullet \vec{S}_{\vec{X}'} \rangle_{T>T_c} \sim e^{-m(T)|\vec{X}-\vec{X}'|}, \tag{2.3}$$

where $m(T)(>0)$ is the inverse correlation length, which diverges at $T_c$. At $T_c$, the spin–spin correlation function decays algebraically,

$$\lim_{|\vec{X}-\vec{X}'|\to\infty} \langle \vec{S}_{\vec{X}} \bullet \vec{S}_{\vec{X}'} \rangle_{T=T_c} \sim \frac{A}{|\vec{X}-\vec{X}'|^\eta}, \tag{2.4}$$

with $\eta$ a constant critical exponent. Equations (2.2)–(2.4) characterize a system with a standard second order phase transition. The results presented above are correct both for a quantum or a classical system.

There was an important idea developed in the 1970s known as the *Universality hypothesis*.[7,8] It states that only two parameters are needed to classify the long-range order properties of a model system exhibiting a second-order phase transition: the dimensionality of geometric space, "$d$", and the dimensionality of the symmetry of the order parameter "$n$". Given these parameters, one can, in principle, then study the simplest model with the right $(d, n)$ combinations to describe the critical properties of all systems within the same $(d, n)$ since they will belong to the same "*universality class*". The $n = 1$ case corresponds to the Ising model; $n = 2$ to the XY or planar model and $n = 3$ to the Heisenberg model. The Ising case has discrete $\mathbb{Z}_2$ symmetry; the XY model has $O(2)$ planar rotational symmetry and the $n = 3$ case has $O(3)$ rotational symmetry.

Mermin and Wagner showed that when $d = 2$, $n = 2$,

$$|M| < \frac{\text{Const}}{T^{1/2}} \frac{1}{\sqrt{\ln |h|}}, \qquad (2.5)$$

where $h$ is a constant external magnetic field, showing that there is no LRO in zero field in this case.

High temperature series expansion calculations were carried out in the 1960s for $d = 2$ and $n = 2, 3$ to find out what actually happens in these cases. The results showed that indeed there was no LRO, but instead evidence was found for a finite critical temperature, $T_c$, below which the magnetic susceptibility $\chi = \infty$.[9,10] The numerical evidence for this behavior was stronger in the $n = 2$ case than for $n = 3$.[9,10]

In the mid-1970s, Polyakov,[11] Migdal,[12,13] and Brezin and Zinn-Justin,[14] did an $\varepsilon = d - 2$ RG expansion in the nonlinear $\sigma$ model to analyze the critical properties of the $n$-vector model. They found that there is a critical temperature given to leading order in $\varepsilon$ by[11–14]

$$T_c = \frac{2\pi\varepsilon}{n - 2} + O(\varepsilon^2). \qquad (2.6)$$

This clearly indicated that there was no phase transition for $n \geq 3$, with $d = 2$ being the lower critical dimension for the $n$-vector model to have LRO. However, their results indicate that the $n = 2$ case is special.

To address that issue, we next consider the classical two-dimensional XY or Planar model ($n = 2$). In this case, we can take as statistical variable the normalized ($|\vec{S}_{\vec{X}}| = 1$) classical spin vector,

$$\vec{S}_{\vec{X}} = (\cos\theta(\vec{X}), \sin\theta(\vec{X})), \qquad (2.7)$$

with the orientation angle $0 < \theta(\vec{X}) \leq 2\pi$ defined on the plane; $\vec{X}$ is the two-dimensional lattice vector. The corresponding Heisenberg Hamiltonian reads,

$$H_\Lambda = -\sum_{\langle \vec{X},\vec{X}'\rangle} K(\vec{X}-\vec{X}')\vec{S}_{\vec{X}}\bullet\vec{S}_{\vec{X}'} = K[1-\cos(\theta(\vec{X})-\theta(\vec{X}'))], \qquad (2.8)$$

where $\Lambda$ is the volume of the lattice; Here $K(\vec{X}-\vec{X}') = J(\vec{X}-\vec{X}')/T$ with $J(\vec{X}-\vec{X}')$ is the exchange interaction, which we will take as the constant $J$ from now on. The corresponding partition function is:

$$Z_\Lambda = \int_{|\vec{S}|=1} \prod_{\vec{X}\in\Lambda} d\vec{S}_{\vec{X}}\, e^{-H_\Lambda} = \int_0^{2\pi} D[\theta(\vec{X})] \exp{-K\sum_{\langle\vec{X},\vec{X}'\rangle}[1-\cos[\theta(\vec{X})-\theta(\vec{X}')]]}. \qquad (2.9)$$

Here $K = J/T$. At very low temperatures ($T \ll 1$) the angular orientation difference between nearest-neighbor spins is very small. Several authors thus approximated the Hamiltonian by its leading quadratic term:

$$H_\Lambda = -K\sum_{\vec{\mu}}(\theta(\vec{X})-\theta(\vec{X}+\vec{\mu}))^2, \qquad (2.10)$$

with $\vec{\mu}$ a unitary vector in the lattice. Within this quadratic approximation, the two-point spin-spin correlation function gives:[15,16]

$$\lim_{|\vec{X}-\vec{X}'|\to\infty} \langle\vec{S}_{\vec{X}}\bullet\vec{S}_{\vec{X}'}\rangle_{T\ll 1} = \lim_{|\vec{X}-\vec{X}'|\to\infty} \langle\cos[\theta(\vec{X})-\theta(\vec{X}')]\rangle_{T\ll 1} \cong \frac{1}{|\vec{X}-\vec{X}'|^{\eta(T)}}, \qquad (2.11)$$

with the "critical exponent"

$$\eta(T) = T/2\pi J. \qquad (2.12)$$

This result shows that at long distances, and at very low temperatures, there is a line of critical points with divergent susceptibility at each $T$. This result is consistent with the Mermin–Wagner theorem since,

$$\lim_{|\vec{X}-\vec{X}'|\to\infty} \langle\vec{S}_{\vec{X}}\bullet\vec{S}_{\vec{X}'}\rangle_{T\ll 1} = M^2 = 0. \qquad (2.13)$$

The question then is what happens at higher temperatures, since we expect that eventually the correlation function will decay exponentially with distance. Will this power law behavior continue before becoming exponential or will something else happen in between that would modify it in an important way? There were indeed some theoretical papers suggesting that this behavior would be modified before $T_c$.[16,23,24] Furthermore, experimental results with Helium films clearly indicated signatures of having a superfluid behavior in these systems.[17] Although there was no true LRO at finite temperatures, there was then something else happening. As mentioned above, high temperature series expansion results had indicated a possible phase transition to an infinite susceptibility phase below some critical temperature.[9,10]

In 1970, Berezinskii considered this problem and noted that the important angular periodicity of the original Eq. (2.8) Hamiltonian had been broken when keeping only the leading quadratic approximation.[18] To preserve the angular symmetry, he considered instead the approximation:

$$H_\Lambda = -K \sum_{\vec{\mu}} [\theta(\vec{X}) - \theta(\vec{X} + \vec{\mu}) - 2\pi n(\vec{X}, \vec{X} + \vec{\mu})]^2, \qquad (2.14)$$

where $\vec{\mu}$ is a unitary vector between two nearest neighbor lattice sites and with the integer bonds $n(\vec{X}, \vec{X} + \vec{\mu}) \in \mathbb{Z}$. He noticed this Hamiltonian is invariant under the discrete transformations

$$\begin{aligned}\theta(\vec{X}) &\to \theta(\vec{X}) + 2\pi m(\vec{X}) \\ n(\vec{X}, \vec{X}') &= n(\vec{X}, \vec{X}') + m(\vec{X}) - m(\vec{X}'),\end{aligned} \qquad (2.15)$$

with $m(\vec{X})$ an integer field. He did not state that this is a local gauge transformation – a fact we will discuss in more detail later in this chapter. He moved on to recognize that these integers were related to the *vorticity* of the vortex excitations associated with the angular phase variables in the model. He realized that the energy associated with one vortex excitation diverges logarithmically with distance and thus it is a thermodynamic unstable excitation in the system. He then concluded that to be thermodynamically stable, vortices must appear as bound $(+, -)$ pairs. The vortex-pair logarithmic interaction strength is inversely proportional to temperature and thus it decreases in strength as the temperature increases. This would lead to a phase transition at a critical temperature, above which the vortex pairs unbind.

Totally independently, Kosterlitz and Thouless had been considering this problem by way of a different route. The history of how they developed their ideas and their work is beautifully discussed in full detail in the first Chapter of this volume. Here, I would just like to highlight some of the main elements in their seminal ideas, and the reasons some authors expressed concerns about the validity of their physical intuition and computational analysis.[19,20] Kosterlitz and Thouless's first paper began by suggesting that melting occurs in a two-dimensional lattice via the unbinding of bound dislocations having Burger's vectors of opposite polarity. They also considered the 2D superfluid problem under similar assumptions of having slow fluctuations of the phase of the order parameter, and vortex excitations due to the singularities in the solution to Laplace's equation in two dimensions. The solution to this equation for the Coulomb interactions between charges is logarithmic in 2D. The energy associated with one charge diverges logarithmically for large distances. Based on previous qualitative arguments used by Thouless to determine the critical temperature of the one-dimensional Ising model with $1/r^2$ interaction, KT concluded that only pairs of positive and negative bound charges could be thermodynamically stable. By analyzing the free energy of one set of bound charged pairs, they estimated a critical temperature to have a pair of charges unbinding, i.e. going from an insulating phase at low temperatures to a conductive screened phase at high temperatures. The KT paper gave a qualitative estimation of $T_c$ but also used a quantitative iterative Mean Field Theory analysis supporting their qualitative seminal idea of having a topological phase transition with vortex pairs as topological excitations. As mentioned in the KT paper in this volume, a paper by Anderson and Yuval[21] played an important role in their thinking and analysis: that paper studied the one-dimensional Ising model with an $1/r^2$ interaction. There Anderson and Yuval did a RG calculation describing the properties of that 1D Ising model. Kosterlitz used a parallel approach to study the two-dimensional Coulomb gas problem. He derived the RG equations with important consequences[22] which we will describe in more detail after we rederive those equations following a different approach.

What was of concern to some authors about the KT theory was that they assumed a complete decoupling between spin-waves, or slow phase fluctuations, from the 2D Coulomb gas charged or vortex excitations, leaving out higher order interactions between those two types of excitations in the Hamiltonian. Their approximation was correct at very low temperatures, but close to $T_c$, the higher order corrections in the cosine expansion of the Hamiltonian, having oscillatory signs "+" and "–", were harder to justify within the BKT approximation.[16,23]

There was an interesting paper written by Jacques Villain[24] after the BKT papers were published, which we will mention in more detail later in this chapter. He noted that the approximation given in Eq. (2.10) broke the angular symmetry of the Hamiltonian in Eq. (2.8). He instead studied a similar type of Hamiltonian as was considered by Berezinskii, by using a Self Consistent Harmonic Approximation (SCHA) analysis. He did recognize that his approximation was also difficult to justify against higher corrections in the Cosine Hamiltonian. A paper by Zittartz[16] and another one by Luther and Scalapino[23] provided further analysis for the 2D XY model using alternative approaches. Both analyses concluded that the spin-wave vortex decoupling was hard to justify close to $T_c$, and in fact, doubted the general validity of the BKT approximations and conclusions. This was the situation in the mid-1970s. As we will see, different developments in theoretical physics that happened just around that time provided further proof that BKT approximations were indeed correct in the asymptotic renormalization group sense and that the physical motivation and quantitative results from BKT could, in fact, be fully mathematically justified.

## 2.2. Duality Transformations in the 2D XY Model

Duality transformations were first introduced by Kramers and Wannier in 1941 in their study of the two-dimensional Ising model in zero magnetic field.[25] They showed that there was a mathematical transformation between the high and low temperature regimes of the partition function expressions so that the equations in both asymptotic limits are identical to each other. The duality transformation definition for the 2D Ising model can be written as:

$$Z(K)_L = Z^*(K^*)_{L^*}. \qquad (2.16)$$

Here $L$ and $L^*$ define the geometry of the original and the dual lattices, respectively. The coupling constants are related to each other by the functional transformations:

$$K = K(K^*)$$
$$K^* = K^*(K).$$

The 2D Ising model in the square lattice is self-dual, i.e. $L = L^*$. The Ising $\overset{D}{\Leftrightarrow}$ Ising coupling constants duality transformation found by Kramers and Wannier are given by the expressions,

$$\text{Tanh}(K) = e^{-2K^*}$$
$$\text{Tanh}(K^*) = e^{-2K}. \qquad (2.17)$$

The duality transformation relates an Ising model in the disordered phase to an Ising model in the ordered phase. This is the power of a duality transformation: if we know the behavior of the system in either regime, we can deduce the behavior in the other one by using the duality transformation. By assuming that there was only one critical temperature in the 2D Ising model, and by equating the two expressions for the partition functions of the model, KW were able to determine the critical temperature of the 2D Ising model exactly by setting $K = K^* = K_c(T_c = 2.2692J)$.

Three years later, in 1944, using the KW duality transformations, Onsager derived the exact solution for 2D Ising model in zero magnetic field in the thermodynamic limit.[3,4] The resulting partition function showed singularities in the second order derivatives of the corresponding free energy. Onsager's Ising model solution is considered one of the towering landmarks of exact mathematical solutions in theoretical physics. Nowadays, duality transformations are used widely in many areas of theoretical physics.[26]

As mentioned in the introduction, the classical XY model in two dimensions can be used to describe the critical properties of all systems, quantum or classical, with $d = 2$ and $n = 2$. Here, we will show that there is also a duality transformation in this case, first shown by JKKN, which was very helpful in understanding the phase transition properties of the 2D XY model.[27]

JKKN considered the partition function:[27]

$$Z_\Lambda = \int_0^{2\pi} D[\theta(\vec{X})] \exp\left\{-K \sum_{\langle \vec{X},\vec{X}'\rangle} [1 - \cos[\theta(\vec{X}) - \theta(\vec{X}')]] + \sum_{\langle \vec{X},p\rangle} h_p \cos(p\theta(\vec{X}))\right\}. \tag{2.18}$$

This partition function includes the experimentally important symmetry breaking fields, $h_p = B_p/T$, characterized by the "$p$" symmetry breaking parameter: $p = 1$ (magnetic field), $p = 2$ (Ising model or quadratic symmetry), $p = 3$ (cubic symmetry), $p = 4$ (quartic symmetry), $p = 6$ (hexagonal symmetry), etc.

We will analyze first the $h_p = 0$ case. Note that $H$ can be expressed in terms of the periodic potential, $V[\theta(\vec{X}) - \theta(\vec{X}')]$,

$$H[\theta] = -\sum_{\langle \vec{X},\vec{X}'\rangle} K\{1 - \cos[\theta(\vec{X}) - \theta(\vec{X}')]\} = \sum_{\langle \vec{X},\vec{X}'\rangle} V[\theta(\vec{X}) - \theta(\vec{X}')]. \tag{2.19}$$

Because of the $V$ periodicity (i.e. $U(1)$ compact symmetry) the partition function can be written as:

$$Z_\Lambda = \int_0^{2\pi} D[\theta(\vec{X})] e^{V[\theta(\vec{X}) - \theta(\vec{X}')]} = \int_0^{2\pi} D[\theta(\vec{X})] e^{V[\theta(\vec{X}) - \theta(\vec{X}') + 2\pi\nu(\vec{X},\vec{X}')]}. \tag{2.20}$$

We can next use the Peter–Weyl Fourier transform group theorem[28]

$$e^{V(g)} = \sum_{\nu=-\infty}^{\infty} \chi_\nu(g) e^{\tilde{V}(\nu)},$$

for each periodic term in the integrand of Eq. (2.20). Here $\chi_\nu(g)$ are the characters of the group $G$ (i.e. the complete set of inequivalent irreducible representations of $G$). In the compact Abelian $G = U(1)$ case the characters are given by:

$$\chi_\nu = e^{i\theta\nu}; \nu \in \mathbb{Z}. \tag{2.21}$$

So that the discrete Fourier expansion for each term is,

$$e^{V(\theta)} = \sum_{\nu=-\infty}^{\infty} e^{i\nu\theta + \tilde{V}(\nu)}. \tag{2.22}$$

The partition function in Eq. (2.20) can then be written as:

$$Z = \int D[\theta] \prod_{\langle \vec{X}.\vec{X}'\rangle} \sum_{\{\nu(\vec{X},\vec{X}')\}=-\infty}^{\infty} e^{i\nu(\vec{X},\vec{X}')[\theta(\vec{X})-\theta(\vec{X}')]+\tilde{V}[\nu(\vec{X},\vec{X}')]}. \tag{2.23}$$

The angular integrals can be done immediately giving,

$$Z_\Lambda = \prod_{\langle \vec{X}.\vec{X}'\rangle} \sum_{\{\nu(\vec{X},\vec{X}')\}=-\infty}^{\infty} \delta(\nu(\vec{X},\vec{X}')) e^{\tilde{V}(\nu(\vec{X},\vec{X}'))}. \tag{2.24}$$

The set of constraints $\{\delta(\nu(\vec{X},\vec{X}'))\}$ can be solved by recognizing that they represent a "curl" of the vector number field $\vec{N}(\vec{R})$, i.e. $\nabla \Lambda \vec{N}(\vec{R}) = 0$, with a finite difference nearest neighbor lattice gradient solution:

$$\nabla N(\vec{R}) \equiv N(\vec{R}) - N(\vec{R}'). \tag{2.25}$$

The $\vec{R}$ vectors are defined at the center of each plaquette in the squares of the $\vec{X}$-lattice. Replacing this solution in Eq. (2.24) we get,

$$Z^*_{L^*}[N] = \sum_{\{N(\vec{R})\}=-\infty}^{\infty} e^{\sum_{\langle \vec{R},\vec{R}'\rangle} \tilde{V}(N(\vec{R})-N(\vec{R}'))}. \tag{2.26}$$

Here, we have defined the $\vec{R}$ dual lattice geometry as $L^*$. We can immediately write the duality transformation between the angular $\theta$-representation and the integer $N$-representation as:

$$Z_L[\theta] \overset{D}{\Leftrightarrow} Z^*_{L^*}[N]. \tag{2.27}$$

Notice that this XY-duality transformation for the planar model is exact, not self-dual, and in contrast to the Ising model, it maps a partition function in terms of angular variables to a partition function in terms of integer field variables and in the dual lattice. By preserving the periodicity of the Hamiltonian in Eq. (2.8), we have derived a dual representation of the partition function in terms of a set of topological interacting integer field excitations. The partition functions still have to be evaluated but we can do so in the different parameter regimes of interest using the appropriate dual representation of the partition function.

To calculate this type of integer sums, we use the Poisson summation formula. Defining the Dirac "comb" or "Cha" harmonic analysis function as the sum

$$\prod(x) = \sum_{n=-\infty}^{\infty} \delta(x-n), \qquad (2.28)$$

we can determine any continuous function $f(x)$ at a particular set of integers "$n$" by using the inner product

$$\left\langle \prod(x), f(x) \right\rangle = \sum_{n=-\infty}^{\infty} f(n). \qquad (2.29)$$

Defining the Fourier transform of a function as

$$[\hat{\bullet}] \equiv \int_{-\infty}^{\infty} d\phi [\bullet] e^{-2\pi i m \phi}, \qquad (2.30)$$

we have,

$$\left\langle \prod(x), f(x) \right\rangle = \left\langle \prod(t), \hat{f}(t) \right\rangle.$$

It follows that

$$\sum_{n=-\infty}^{\infty} f(n) = \sum_{m=-\infty}^{\infty} \hat{f}(m). \qquad (2.31)$$

Applying these results to Eq. (2.22) we get

$$e^{V(\theta)} = \sum_{n=-\infty}^{\infty} e^{in\theta + \tilde{V}(n)} = \sum_{m=-\infty}^{\infty} \int_{-\infty}^{\infty} d\phi e^{i\phi(n-2\pi m) + \tilde{V}(n)} = \sum_{m-\infty}^{\infty} e^{V'(\theta - 2\pi m)}, \qquad (2.32)$$

with $V'$ defined by the integral,

$$e^{V'(\theta)} = \int_{-\infty}^{\infty} d\phi e^{\tilde{V}(\phi) + i\theta\phi}.$$

We can use these results to rewrite the partition function in Eq. (2.20) as

$$Z_\Lambda = \int_0^{2\pi} D[\theta(\vec{X})] e^{V[\theta(\vec{X}) - \theta(\vec{X}')]}$$

$$= \sum_{\{\nu(\vec{X},\vec{X}')\}=-\infty}^{\infty} \int_0^{2\pi} D[\theta(\vec{X})] e^{V'[\theta(\vec{X}) - \theta(\vec{X}') - 2\pi\nu(\vec{X},\vec{X}')]}. \qquad (2.33)$$

Using Eq. (2.32) backwards we get exactly

$$Z_{\Lambda^*}[M(\vec{R})] = \sum_{\{M(\vec{R})\}=-\infty}^{\infty} \prod_{\vec{R}} \int_{-\infty}^{\infty} d\phi[(\vec{R})] e^{\sum_{\langle \vec{R}, \vec{R}' \rangle} \tilde{V}(\phi(R)-\phi(R')) + \sum_{R} 2\pi i M(\vec{R})\phi(\vec{R})}.$$

(2.34)

Here, we see the appearance of the plaquette integer field $M(\vec{R})$, which will play the role of the quantized vortex excitations of the model. To control the total number of possible vortex-pairs excitations JJKN introduced a "fugacity" parameter "$y_0$" to the Hamiltonian to control the number of vortex excitations in the model as:

$$Z_{\Lambda^*}[M(\vec{R})]$$
$$= \sum_{\{M(\vec{R})\}=-\infty}^{\infty} \prod_{\vec{R}} \int_{-\infty}^{\infty} d\phi[(\vec{R})] e^{\sum_{\langle \vec{R}, \vec{R}' \rangle} \tilde{V}(\phi(R)-\phi(R')) + \sum_{R} 2\pi i M(\vec{R})\phi(\vec{R}) + \ln y_0 \sum_{R} M(\vec{R})^2}.$$

(2.35)

We can now carry out each integrals over "$\theta$" in Eq. (2.18) ($h_p = 0$) getting:

$$e^{\tilde{V}[N]} = e^{-K} I_N(K);$$

Here $I_N(K)$ is the modified Bessel function of the second kind of index "$N$". The two asymptotic expressions for $I_N(K)$ as a function of $K$ are:

$$\tilde{V}(N) = -N^2/2K - \frac{1}{2}\ln(2\pi K); \lim K \to \infty$$
$$\tilde{V} = \ln\left\{\left(\frac{1}{N!}\right)\left(\frac{1}{2}K\right)^N\right\}; \lim K \to 0.$$

(2.36)

From these expressions we can analyze the weak and strong coupling limits of the partition function. By using first the top equation in (36), the low temperature approximation for the partition function reads:

$$Z_{\Lambda^*}[M(\vec{R})] = \sum_{\{M(\vec{R})\}} \left[\prod_{\vec{R}} \int_{-\infty}^{\infty} \frac{d\phi(\vec{R})}{2\pi K}\right] e^{-\frac{1}{2K} \sum_{\langle \vec{R}, \vec{R}' \rangle} (\phi(\vec{R})-\phi(\vec{R}'))^2 + \sum_{R} 2\pi i M(\vec{R})\phi(\vec{R})}.$$

(2.37)

This is a Gaussian model in $\phi(\vec{R})$ with $1/K$ coupling constant. It corresponds to the spin-wave approximation for the partition function. We can carry out the Gaussian integrals over $\phi(\vec{R})$ in Eq. (2.37), and using (2.32), we get:

$$Z_\Lambda = \sum_{\{\nu(\vec{X},\vec{X}')\}=-\infty}^{\infty} \int_0^{2\pi} D[\theta(\vec{X})] e^{-\frac{K}{2} \sum_{\langle \vec{X},\vec{X}' \rangle} [\theta(\vec{X})-\theta(\vec{X}')-2\pi\nu(\vec{X},\vec{X}')]^2}, \quad (2.38)$$

which is the Villain model approximation for the original Hamiltonian. This model was studied by Villain in 1975. He pointed out that previous low temperature approximations had broken the periodic symmetry of the cosine Hamiltonian.[24] He indicated that his approximation was quite good for very low temperatures but not as good for higher temperatures. In Villain's analysis, Eq. (2.38) arose from a low temperature expansion of the cosine interaction but here it came from the low temperature asymptotic limit approximation of the exact duality transformation for the XY model given in Eq. (2.27). By following the duality transformation, we see that the vortices appear as a result of preserving the gauge invariance of the original model and they are the integer topological excitations in the model.

We note that in the periodic Gaussian approximation given in Eq. (2.38), spin waves and vortices do not interact. In the cosine planar model however, there are, in principle, higher order nonlinearities in the cosine expansion that could have an effect on the critical properties and critical exponents of the model. These concerns, as mentioned before, were expressed by several authors who questioned the general validity of completely decoupling spin waves from vortex excitations. As we shall see in the next section, this problem can be resolved to some extent after doing a Renormalization Group analysis of the full Cosine Hamiltonian.

## 2.3. Migdal–Kadanoff RG Approximation of the Two-Dimensional XY Model

Once we have exact dual representations of the two-dimensional XY model, we can use either limiting representation to calculate the thermodynamic critical properties of the model. Migdal introduced a Renormalization Group approximation that allowed the study of lattice gauge problems, in particular the XY model.[13] Kadanoff extended Migdal's approach in various ways and noted that the RG process was close in spirit to a one-dimensional "decimation" RG transformation, which had been used before to study other problems.[29] The MK renormalization group equation for our model can then be written as a one-dimensional decimation transformation:

$$Z'(\phi) = \sum_{\phi'} Z^2(\phi') Z^2(\phi - \phi'). \quad (2.39)$$

Here $Z'(\phi)$ is the "renormalized" partition function. The "2" comes from the dimensionality of the model. Depending on the dual representation of the partition function used, the trace over the statistical variables is defined as,

$$\sum_\theta \equiv \int_0^{2\pi} d[\theta] . \tag{2.40}$$

Or, if in the dual representation, the normalized sum is given by:

$$\sum_{\phi'} \equiv \frac{1}{(2\pi K)^{1/2}} \int_{-\infty}^{\infty} d\phi' \sum_{\{m'\}=-\infty}^{\infty} e^{2\pi i m' \phi'} . \tag{2.41}$$

Migdal and JKKN carried out the RG transformation given in Eq. (2.39) starting with the initial bond partition function expression:

$$\ln Z(\theta) = K \cos(\theta) . \tag{2.42}$$

They carried out a numerical evaluation of the recursion relation given in Eq. (2.39) and found that after a few initial iterations, the numerical results for the partition function preserved the angular periodicity and tended asymptotically to the periodic Gaussian form

$$\ln Z^s(\theta) \to -K^s(\theta - 2\pi n)^2 . \tag{2.43}$$

Here $s$ denotes the $s$ iteration in the RG recursion relation, with the renormalized coupling constants $K^s$. This result is important since it indicates that the periodic Gaussian approximation tends to be a fixed point of the MK–RG transformation, and that the higher order corrections in the cosine interaction become asymptotically irrelevant, close to the critical region.[27]

To get an idea analytically of what is happening about the asymptotic critical properties of the model, we can carry out the MK–RG transformation instead in the dual representation given in Eq. (2.39), using the trace definition given in Eq. (2.41). In this case, we take the initial partition function as:

$$\ln Z(\phi) = -\frac{\phi^2}{2K} + X(\phi) . \tag{2.44}$$

We have added the function $X(\phi)$ which represents all the other corrections we are leaving out from the Gaussian approximation, but which actually will be generated by the RG transformations. Note that the periodicity and presence of vortices is already contained in the definition of the trace. We can write the recursion formula from Eq. (2.39) as,

$$Z' = e^{-\frac{\phi^2}{2K'} + X'(\phi)} = \sum_{\phi'} e^{-\frac{\phi'^2}{K} + 2X(\phi') + 2X(\phi - \phi')} . \tag{2.45}$$

To evaluate Eq. (2.45) we start by using the change of variables $\phi' \to \psi + \frac{1}{2}\phi$ to write:

$$Z' = Z_0 \sum_{n=-\infty}^{\infty} \int \frac{d\psi}{(\pi K)^{1/2}} e^{i\pi n\phi - \frac{2\psi^2}{K} + 2X\left(\psi + \frac{1}{2}\phi\right) + 2X\left(\psi - \frac{1}{2}\phi\right)}. \quad (2.46)$$

Here $Z_0$ is the original partition function with $K$ coupling constant and $\phi$ is the new statistical variable. Defining the average of a function of $\psi$ as:

$$\langle F(\psi)\rangle = \int \frac{d\psi}{(\pi K)^{1/2}} e^{-\frac{2\psi^2}{K}} F(\psi), \quad (2.47)$$

we have:

$$\langle e^{i2\pi n\psi}\rangle = e^{-2\pi^2 n^2 K}. \quad (2.48)$$

Considering first the case $X = 0$ in Eq. (2.46), we have:

$$Z'(\phi) = Z_0(\phi) \sum_{m=-\infty}^{\infty} e^{im\pi\phi - \frac{1}{2}\pi^2 m^2 K}$$

$$\simeq Z_0[1 + 2e^{-\frac{\pi^2}{2}K}\cos(\pi\phi) + 2e^{-2\pi^2 K}\cos(2\pi\phi) + \cdots]. \quad (2.49)$$

Here, we have expanded the integer sum keeping only the first three terms. This result allows us to make a low temperature ansatz for the function $X$ in terms of a series expansion of the form:

$$X(\phi) = \sum_{n=1}^{\infty} a(n)\cos(2\pi\lambda^n \phi). \quad (2.50)$$

Equation (2.49) gives the $n = 1$, and $\lambda = 1/2$ correction to $Z'(\phi)$, with renormalized coupling constant $a'(n = 1) = \exp\left(-\frac{\pi^2}{2}K\right)$. Using this result in Eq. (2.50) to get the $n = 2$ contribution to $Z'_0(\phi)$ in Eq. (2.45) we can write

$$Z' \cong Z_0[f'_0 + f'_1 + \cdots], \quad (2.51)$$

where $f'_0$ is the term in parenthesis in Eq. (2.49). The approximate recursion relation for $a(n)$ then becomes:

$$a'(n) = 4^n a(n)[a(n=1)]^{\lambda^{2n}}. \quad (2.52)$$

We are, however, interested in getting the recursion relations for $K$ so we look at the coefficients of the renormalized coupling constant and taking their difference we have:

$$\frac{1}{K'} - \frac{1}{K} = 8\pi^2 a_1(n)\{1 - a_1(n-1)[a_1(n=1)]^{\lambda^{2n}}\}. \quad (2.53)$$

Taking now the $n \to \infty$ limit, and using the equality $\sum_{n=1}^{\infty} \lambda^{2n} = \frac{4}{3}$, we can write the corresponding continuous $\beta$-function RG equation:

$$\beta(K) = \frac{d}{dl}\frac{1}{K} = 8\pi^2 e^{-\frac{2}{3}\pi^2 K}. \qquad (2.54)$$

This is quite an interesting result. First it shows that the beta-function does not have a zero at any finite temperature. It shows that we have an essential singularity in $K$. It almost also gives a phase transition exponentially close in $K$, but not a true phase transition.

Both results, the ones obtained numerically leading to a fixed point type periodic Gaussian renormalized Hamiltonian, and the analytic result providing evidence for a singular perturbative term in the beta function, show that standard coupling constant expansions would not be able to exhibit these type of singular term contributions, which are in fact related to the presence of vortex topological excitations in the system.

## 2.4. Correlation Function Calculations and Renormalization Group Equations

Based on the duality transformations and the MK–RG results we could confidently proceed to the evaluation of the spin–spin correlation functions in either dual representation of the partition function. We do that here within the generalized Villain model approximation. The spin–spin correlation function of order $p$ is defined by:

$$G_p(\vec{r},\vec{r}') = \langle e^{ip(\theta(\vec{r})-\theta(\vec{r}'))}\rangle = \frac{1}{Z}\mathrm{Tr}\, e^{ip(\theta(\vec{r})-\theta(\vec{r}'))} e^{-H}. \qquad (2.55)$$

The generalized Villain model dual partition function reads:

$$Z_{\Lambda^*}[M(\vec{R})] = \sum_{\{M(\vec{R})\}=-\infty}^{\infty} \prod_{\vec{R}} \int_{-\infty}^{\infty} d\phi[(\vec{R})]$$

$$\times\, e^{\sum_{\langle \vec{R},\vec{R}'\rangle} -\frac{1}{K}(\phi(R)-\phi(R'))^2 + \sum_{\vec{R}} 2\pi i M(\vec{R})\phi(\vec{R}) + \ln y_0 \sum_{\vec{R}} M(\vec{R})^2}. \qquad (2.56)$$

Carrying out the Gaussian integrals over the $\phi(\vec{R})$ field, we get:

$$Z_{GV} = \sum_{\{M(\vec{R})\}}{}' e^{-H_{GV}[M(\vec{R})]}. \qquad (2.57)$$

Here "′" stands for the neutrality constraint $\sum_{\vec{R}} M(\vec{R}) = 0$, of having bound vortex pairs with opposite polarity. The Hamiltonianis $H_{GV}[M(\vec{R})]$ is:

$$H_{GV} = 2\pi K \sum_{\langle \vec{R}, \vec{R}' \rangle} M(\vec{R}) \alpha(\vec{R} - \vec{R}') M(\vec{R}') + \ln y \sum_{\vec{R}} M^2(\vec{R}). \quad (2.58)$$

The Coulomb Green function $\alpha(\vec{R})$ in two dimensions is asymptotically given by[30]

$$\alpha(\vec{R}) \cong \ln \left| \frac{\vec{R}}{a} \right| + \frac{1}{2} \pi. \quad (2.59)$$

With $a$ the unit lattice spacing and $y$ the new vortex fugacity:

$$y = y_0 e^{-\frac{\pi^2}{2} K}. \quad (2.60)$$

Neglecting first the vortex contributions (setting $y = 0$), within the Gaussian approximation, we recover the spin-wave correlation function result:

$$G_p^{SW}(\vec{r}, \vec{r}') = \frac{1}{|\vec{r} - \vec{r}'|^{\frac{p^2}{2\pi K}}}. \quad (2.61)$$

As mentioned before, this result was derived in different contexts and conditions by different authors.[15,16,18] This result is consistent with the Mermin–Wagner theorem result of not having a broken symmetry in the model. It also indicates that there is a continuous line of fixed points at very low temperatures, and that the long distance behavior is dominated by the Gaussian free field theory. Next, we include the vortex contributions.[27] Including the lowest order vortex correction leads to a multiplicative form of the correlation function, but with exactly the same long distance behavior as given in Eq. (2.61), except that we have instead $K \to K_R$, with $K_R$ a modified coupling constant i.e.

$$\lim_{|\vec{r} - \vec{r}'| \to \infty} G_p(\vec{r}, \vec{r}') \cong G_p^{SW}(\vec{r}, \vec{r}') G_p^V(\vec{r}, \vec{r}') \to \frac{1}{|\vec{r} - \vec{r}'|^{\frac{p^2}{2\pi K_R}}}. \quad (2.62)$$

This means that the modified correlation function that includes vortices to lowest order is still given by the free field theory Gaussian-type behavior. The modified coupling constant to leading order in $y$ is given by[27]:

$$K_R^{-1} = K^{-1} - \frac{1}{2} \pi^2 \sum_{\vec{R}} \left| \frac{\vec{R}}{a} \right|^2 \langle M(\vec{R}), M(\vec{0}) \rangle_{GV}. \quad (2.63)$$

The second term in this equation is the leading contribution to the polarizability of the vortex gas, written in terms of the two-point vortex correlation function evaluated with the Hamiltonian given in Eq. (2.58). The vortex BKT unbinding phase transition will occur when the polarizability diverges at long distances, i.e. when going from the insulating vortex bound pairs regime to the vortex unbounded, screened, high temperature phase.

After evaluating the vortex correlation function to lowest order, we get the self-consistent equation

$$K_R^{-1}(r) = K^{-1} + 2\pi^3 y_0^2 \int_a^\infty \frac{dr}{a} \left(\frac{r}{a}\right)^{3-2\pi K_R(r)}. \qquad (2.64)$$

The integral will diverge when $K_R < \frac{2}{\pi}$, i.e. when there is no longer a self-consistent solution to Eq. (2.64). By the small $\delta$ rescaling transformation $\int_a^\infty = \int_a^{a(1+\delta l)} + \int_{a(1+\delta l)}^\infty$, we obtain the leading order RG recursion equations[27]:

$$\frac{dK^{-1}(l)}{dl} = 2\pi^3 y^2(l) + O(y^4)$$

$$\frac{dy(l)}{dl} = (2 - \pi K(l))y(l) + O(y^3). \qquad (2.65)$$

These equations to lowest order in $y^2$ were first derived by Kosterlitz[22] for the two-dimensional Coulomb gas problem following the Anderson–Yuval technique.[21]

Pruisken and Kadanoff showed that the Kosterlitz RG equations are stable against higher order correction in an expansion in $y_0$ up to order $y_0^4$. They also showed that many two-dimensional theories, like the 8-vertex model, the Ashkin–Teller model, the F-model are in the same Gaussian universality class having continuously varying critical exponents.[31]

Two of the main conclusions from the analysis of the RG equations were that the correlation length diverges with a square root essential singularity:

$$\xi(T) \sim e^{b/|T-T_{\text{BKT}}|^{1/2}}. \qquad (2.66)$$

And that the correlation function exponent at criticality is:

$$\eta = \frac{1}{4}. \qquad (2.67)$$

This is of the same value as the 2D Ising model at criticality. We will see below that these results have important experimental consequences.

## 2.5. Symmetric Breaking Fields, Duality and RG Equations

Based on the results listed above, we can get back to the discussion on the contributions due to the symmetry breaking fields given in the Hamiltonian in Eq. (2.18). First we note that we can write[27]

$$e^{h_p \cos p\theta} \to \sum_{n=-\infty}^{\infty} e^{ipn\theta + \ln y_p n^2}, \qquad (2.68)$$

where $y_p \to 1$ is like $h_p \to \infty$, which would be the limit for the $p$-state vector clock model with $\theta(r) = \left(\frac{2\pi}{p}\right) q(r); q(r) = 0, 1, 2, \ldots, p-1$. Including this approximate contribution to the generalized Villain model Hamiltonian given in Eq. (2.58), an integer field "$n(\vec{r})$" is added that represents the symmetry breaking excitations. By using this approximation we now have two sets of integer field excitations. The "electric" "$n(\vec{r})$" charges and the "magnetic" "$M(\vec{R})$" dual lattice vortex charges — they each have a logarithmic interaction with an added $n(\vec{r}) \Leftrightarrow M(\vec{R})$ imaginary coupling interaction in the Hamiltonian. The leading approximation to the Hamiltonian, keeping just the quadratic terms, yields the partition function[27,32]:

$$Z[2\pi K, y_0, y_p] = \sum_{\{M(\vec{R}), n(\vec{r})\}=-\infty}^{\infty} e^{-H_C[M,n]}, \qquad (2.69)$$

with

$$H_C[M,n] = \pi K \sum_{\vec{R} \neq \vec{R}'} M(\vec{R}) \alpha(|\vec{R}-\vec{R}'|) M(\vec{R}')$$

$$+ \ln y \sum_{\vec{R}} M^2(\vec{R}) + \sum_{\vec{r} \neq \vec{r}'} \frac{p^2}{4\pi K} n(\vec{r}) \alpha(|\vec{r}-\vec{r}'|) n(\vec{r}')$$

$$+ \ln y_p \sum_{r} n^2(\vec{r}) + ip \sum_{\vec{R} \neq \vec{r}} M(\vec{R}) n(\vec{r}) \tan^{-1}\left(\frac{X-x}{Y-y}\right). \qquad (2.70)$$

We can immediately derive a duality transformation between the "$n(\vec{r})$" and "$M(\vec{R})$" vortices given by[27]:

$$Z(2\pi K, y_0, y_p) = Z\left(\frac{p^2}{2\pi K}, y_p, y_0\right) \left(\frac{p}{2\pi K}\right)^N. \qquad (2.71)$$

Here $N$ is the total number of lattice sites. We can write the corresponding RG recursion equations using the previously derived RG equations with $h_p = 0$,

$$\frac{dK^{-1}}{dl} = 2\pi^3 y^2 - \pi p^2 K^{-2} y_p^2, \quad \frac{dy}{dl} = (2-\pi K)y, \quad \frac{dy_p}{dl} = \left(2 - \frac{p^2}{4\pi K}\right) y_p. \qquad (2.72)$$

Note that the XY-stability region is restricted by the recursion relations to:

$$\frac{2}{\pi} \leq K_R(T) \leq \frac{p^2}{8\pi}. \tag{2.73}$$

For $p = 4$ the inequality reduces itself to a point. This is a multi-critical point which is related to the F-model, 8-vertex model, Ashkin–Teller model.[31,32] For $p \geq 5$, there is a temperature region with XY-like power low decaying spin–spin correlation functions and continuously varying critical exponents. Note that the duality transformation given in Eq. (2.71) is only asymptotically exact, in contrast to the duality transformation of Eq. (2.27). An alternative view of the analysis of the remaining problems of possible phases in the model defined by Eq. (2.18) is given in Chapter 4 of this book in a paper by Ortiz *et al.*

## 2.6. An Early Experimental Confirmation of the BKT Theory

Soon after the work discussed within the JKKN paper was done, Nelson and Kosterlitz considered the 2D superfluid Helium problem from the KT, Kosterlitz and JJKN theories' point of view.[33] They used a hydrodynamic approach to analyze the superfluid properties of $^4$He films. KT had already analyzed this problem in their original paper. They obtained a relationship between the superfluid density per unit area $\rho_s$ at $T_{\mathrm{BKT}}$,[20]

$$\rho_s(T_{\mathrm{BKT}}^-) = \frac{2m^2 T_{\mathrm{BKT}}^- k_B}{\pi \hbar^2}. \tag{2.74}$$

After looking at the published papers at the time, they concluded that it did not agree with the experimental results. Kosterlitz rederived this result within his RG analysis and pointed out that Eq. (2.74) implies that the superfluid jump is nonzero at onset.

Because of Universality, scaling and the universal RG results, in particular, the connection between Eq. (2.67) and $\rho_s$, Nelson and Kosterlitz concluded that there should be a universal jump in the superfluid density independent of the thickness of the superfluid film. They found that the jump was given in terms of fundamental physical constants by[33]:

$$\lim_{T \to T_{\mathrm{BKT}}^-} \frac{\rho_s(T_{\mathrm{BKT}}^-)}{T_{\mathrm{BKT}}^-} = \frac{2m^2}{\pi \hbar k_B T_{\mathrm{BKT}}^-} = 3.491 \times 10^{-9}\, \mathrm{gcm}^{-2}\, \mathrm{K}^{-1}. \tag{2.75}$$

As reviewed by Dash,[17] there were a number of papers prior to 1975 that found evidence of superfluidy in $^4$He films, but the analysis of the experimental results was inconclusive. Soon after, the Nelson–Kosterlitz paper, experimentalists went back to their old results and tried to reanalyze their data to check if the theoretical predictions were right. Unfortunately, the BKT theory is static and most of the experiments involved dynamical measurements. The static theory had to be extended to the dynamic regime. This, again, was done relatively soon by Ambegaokar, Halperin, Nelson, and Siggia[34] and by Huberman, Myerson and Doniach.[35] Using the dynamic BKT results, Bishop and Reppy[36] were able to analyze their torsional oscillator results that led to a quantitative confirmation of the BKT prediction given in Eq. (2.75). There were also third sound experiments by Rudnick[37] that further confirmed the BKT theory predictions.

## 2.3. Conclusions

In this chapter, I intended to give a historical perspective about the theoretical and experimental situation prior to and after the theoretical and experimental confirmation of the seminal BKT topological phase transition theory. As I mentioned in the introduction, this had been a problem of interest both theoretically and experimentally for more than 50 years before the publication of the Berezinskii[18] and Kosterlitz Thouless papers.[19,20,22] As one can read from their review paper in Chapter One, there were a number of previous advances that led BKT to make their proposal for these systems and models to have a *topological phase transition*. Their sound physical and mathematical analysis was based on a number of intuitive approximations that left some researchers wondering about the regions of validity of the BKT theory. A number of advances in statistical physics and in theoretical physics in general after the publication of their papers helped solidify the soundness and correctness of the BKT results. As described in the KT Chapter, when they wrote their first paper, they had not yet grasped the recent advances at the time from the Renormalization Group theory. Just after their two publications, Kosterlitz was able to do an RG calculation in the spirit of the Anderson–Yuval RG calculation of the one-dimensional Ising model with long range interactions, within the context of the 2D Coulomb gas approximation. But in the KT approach, there was an explicit decoupling of the spin-wave from vortex excitations that left out the possible relevance and possible correction to their picture due to higher order corrections. In my view, it was thanks to the exact duality transformations satisfied by the 2D XY model. The development of RG algorithms for problems with gauge symmetries by Migdal, and its extensions by Kadanoff, showed that approximations which preserved the gauge symmetry of the model led to fixed point type Hamiltonian after a few iterations of the MK–RG calculation. It was not enough to prove that the corrections due to the vortex excitation renormalized the critical exponent in the power law decay of the correlation function below the BKT critical temperature. Many further developments were made after the BKT theory was published that gave further support to the validity of the theory. Other contemporary contributors also helped extend the theory, providing extensions of the BKT theory to other fields of physics, some of which are presented in this volume marking the 40th Anniversary of the BKT theory.[38] Evidence of the importance of their work are the many thousands of papers published, citing BKT ideas, which is clear proof of the impact they have had on theoretical and experimental physics.

Suffice it to say, as Dirac emphasized many times, that when a theory makes physical sense and it is beautiful, it has to be correct.

## Acknowledgments

Many of the results and ideas I have described here are based on the joint work I did with Leo Kadanoff, David Nelson and Scott Kirkpatrick in the JKKN paper. The material here builds upon JKKN, and indeed, a lifetime of reading and reflection, and my own publications on the ideas of BKT. I have tried to cite all the sources that have kept me informed over the years, and am grateful to all of those whose work has contributed to this effort. I also take this opportunity to thank the NSF who funded my work related to the subject matter of this chapter and its extensions continuously to other problems for over two decades.

## References

1. R. Peierls, *Ann. Inst. Henri Poincare* **5**, 177 (1935), *Helv. Phys. Acta* **7**, Suppl. II, **81** (1936).
2. L. D. Landau, *Zh. Eksp. Teor. Fiz.* **7**, 627 (1937) (Sb. trudov (Collected Works), 1 (Nauka, 1969), No. 29); L. D. Landau and E. M. Lifshitz, *Statisticheskaya fizika* (Statistical Physics), (Nauka, 1964), Sec. 123 (English Transl., Addison-Wesley, 2nd ed., 1969).
3. L. Onsager, *Phys. Rev.* **65**, 117 (1944).
4. See also B. Kaufman and L. Onsager, *Phys. Rev.* **76**, 1244 (1949).
5. N. D. Mermin and H. Wagner, *Phys. Rev. Lett.* **17**, 1133 (1966); H. Wagner, *Z. Physik* **195**, 273 (1966); P. C. Hohenberg, *Phys. Rev.* **158**, 383 (1967); N. D. Mermin, *J. Math. Phys.* **8**, 1061 (1967), *Phys. Rev.* **176**, 250 (1968).
6. O. Penrose and L. Onsager, *Phys. Rev.* **104**, 576 (1956).
7. L. P. Kadanoff, *Critical Phenomena, Proc. Intl. School of Physics, "Enrico Fermi"*, Course LI, ed. M. S. Green (Academic Press, New York, 1971), p. 101.
8. R. B. Griffiths, *Phys. Rev. Lett.* **24**, 1479 (1970).

9. H. E. Stanley and T. A. Kaplan, *Phys. Rev. Lett.* **17**, 916 (1966).
10. H. E. Stanley, *Phys. Rev. Lett.* **20**, 150 (1968); M. A. Moore, *Phys. Rev. Lett.* **23**, 861 (1969).
11. M. Polyakov, *Phys. Lett. B* **59**, 79 (1975).
12. A. A. Migdal, *Zh. Eksp. Teoret. Fiz.* **69**, 810 (1975), *Sov. Phys. JETP* **42**, 433 (1976).
13. A. A. Migdal, *Zh. Eksp. Teoret. Fiz.* **69**, 1457 (1975), *Sov. Phys. JETP* **42**, 743 (1976).
14. E. Brezin and J. Zinn-Justin, *PRL* **36**, 691 (1976).
15. F. Wegner, *Z. Phys.* **206**, 465 (1967); J. W. Kane and L. P. Kadanoff, *Phys. Rev.* **155**, 80 (1967).
16. J. Zittartz, *Z. Phys. B* **22**, 63 (1976).
17. See, for example, the review book by J. G. Dash, *Films in Solid Surfaces* (Academic Press, New York, 1975).
18. V. L. Berezinskii, *Zhur. Eksp. Teor. Fiz.* **59**, 907 (1970), *Sov. Phys. JETP* **32**, 493 (1970); V. L. Berezinskii, *Zhur. Eksp. Teor. Fiz.* **61**, 1144 (1971), *Sov. Phys. JETP* **34**, 610 (1971).
19. J. M. Kosterlitz and D. J. Thouless, *J. Phys. C: Solid St. Phys.* **5**, L124 (1972).
20. J. M. Kosterlitz and D. J. Thouless, *J. Phys. C: Solid St. Phys.* **6**, 1181 (1973). See also their review in *Prog. Low. Temp. Phys.* **78**, 371 (1978), and their discussion in the first Chapter of this book.
21. P. W. Anderson and G. Yuval, *Phys. Rev. Lett.* **23**, 89 (1969), ibid. *J. Phys. C* **4**, 607 (1971).
22. J. M. Kosterlitz, *J. Phys. C: Solid St. Phys.* **7**, 1046 (1974).
23. A. Luther and D. J. Scalapino, *Phys. Rev. B* **16**, 1153 (1977).
24. J. Villain, *J. Phys. (Paris)* **36**, 581 (1975).
25. H. A. Kramers and G. H. Wannier, *Phys. Rev.* **60**, 252 (1941), ibid. *Phys. Rev.* **60**, 263 (1941).
26. See, for example, recent special issue where many applications are discussed, eds. A. J. Guttmann and J. L. Jacobsen, *J. Phys. A: Math. Theor.* **45**, 490301 (2012).
27. J. V. José, L. P. Kadanoff, S. Kirkpatrick and D. R. Nelson, *Phys. Rev. B* **16**, 1217 (1977); Errata in *Phys. Rev. B* **17**, 1477 (1978).
28. B. Doubrovine, S. Novikov and A. Fomenko, *Geometried Contempraine: Methods et Applications* (Editions MIR, Moscow, 1982). M. Sepanski, *Compact Lie Groups* (Springer-Verlag, 2006).
29. L. P. Kadanoff, *Ann. Phys.* **100**, 359 (1976).
30. F. Spitzer, *Principles of Random Walk* (Van Nostrand, Princeton, 1964), pp. 148–151.
31. A. M. M. Pruisken and L. P. Kadanoff, *Phys. Rev. B* **22**, 5154 (1980).
32. L. P. Kadanoff, *Phys. Rev. Lett.* **39**, 903 (1977), *J. Phys. C* **11**, 1399 (1978); *Ann. Phys.* **120**, 39 (1979).
33. D. R. Nelson and J. M. Kosterlitz, *Phys. Rev. Lett.* **39**, 1201 (1977); See also, D. R. Nelson, *Defects and Geometry in Condensed Matter Physics* (Cambridge University Press, Cambridge, New York, Port Melbourne, Madrid, Cape Town, 2002).

34. V. Ambegaokar, B. I. Halperin, D. R. Nelson and E. D. Siggia, *ibid.* **40**, 576 (1978).
35. B. A. Huberman, R. J. Myerson and S. Doniach, *Phys. Rev. Lett.* **40**, 780 (1978).
36. D. J. Bishop and J. D. Reppy, *Phys. Rev. Lett.* **40**, 1727 (1978).
37. I. Rudnick, *Phys. Rev. Lett.* **40**, 1454 (1978).
38. See review by R. Savit, *Rev. Mod. Phys.* **52**, 453 (1982).

# Chapter 3

# Berezinskii–Kosterlitz–Thouless Transition Through the Eyes of Duality

G. Ortiz* and E. Cobanera

*Department of Physics, Indiana University, Bloomington, IN 47405, USA*
*ortizg@indiana.edu

Z. Nussinov

*Department of Physics, Washington University, St. Louis, MO 63160, USA*

> A new "bond-algebraic" approach to duality transformations provides a very powerful technique to analyze elementary excitations in the classical two-dimensional XY and $p$-clock models. By combining duality and Peierls arguments we establish the non-Abelian symmetries, phase structure and transitions of these models, unveil the nature of their topological excitations, and explicitly show the continuous U(1) symmetry that emerges when $p \geq 5$. The latter is associated with the appearance of discrete vortices and Berezinskii–Kosterlitz–Thouless-type transitions.

## 3.1. Introduction

In this chapter we investigate, via *dualities*, classical two-dimensional ($D=2$) systems, such as the XY and clock models,[1,2] displaying Berezinskii–Kosterlitz–Thouless (BKT)-type transitions.[3–6] BKT transitions, notably characterized by essential singularities in the free energy, emerge in many physical situations including screening in Coulomb gases,[7] surface roughening,[8] melting in $D=2$ solids,[4,9] and many other classical and quantum problems, such as deconfinement in $D=3+1$ lattice gauge theories.[10–13] We study these models by invoking a *bond-algebraic* approach that we have recently developed.[12–15] Within our duality-based method, one relates singularities in the free energy at one temperature (or coupling constants) to those at a dual temperature (or dual coupling constants).

Specifically, we investigate exact dualities of the $D = 2$ XY and $p$-clock models, and exploit those dualities to unravel their phase structures. Those transformations are exact even for finite systems after appropriate boundary terms are included. It is noteworthy that unlike nearly all of the analytical work done to date, our dualities do not rely on the approximation scheme of Villain,[10,16–18] yet they can be related to the exact dualities of the Villain model in appropriate limits. Also, our analysis leads to exact dualities for general $p$-clock models and yields a better understanding of the appearance of two transitions in systems with $p \geq 5$ states (the XY model with only one transition is recovered in the $p \to \infty$ limit). By fusing our duality results with the *Peierls argument*, we will be able to (1) prove that $p \geq 5$ clock systems can be made to be self-dual; (2) prove by a Peierls argument that there exists a lower ordering temperature $T^{(1)} \sim 1/p^2$, associated to domain-wall excitations, below which the global $\mathbb{Z}_p$ symmetry is broken; (3) demonstrate that a second transition occurs at a temperature $T^{(2)} \sim \mathcal{O}(1)$ when $p \geq 5$ (in the self-dual case, it follows that if $T^{(1)}$ is not the self-dual temperature $T^*$, then there must necessarily be a second phase transition at $T^{(2)}$ of an identical character); and (4) characterize the nature of the topological excitations, and further explain that at $p = 5$ a new type of topological excitation, with an associated discrete winding number, appears. Our considerations suggest that these *discrete vortices* may proliferate above the temperature $T^{(2)}$. We will also (5) determine an analytic expression for the self-dual temperature $T^*$, which relates and clarifies temperature scales discussed in Ref. 19, and most importantly, (6) establish the non-Abelian polyhedral symmetry group $P(2,2,p)$ of the $p$-clock and related models, and explicitly unveil the U(1) continuous symmetry that *emerges* when $p \geq 5$. Indeed, the latter is intimately tied to the existence of the BKT transition.

Despite several analytic[10,20,21] and numerical calculations,[19,22,23] the precise nature of the two phase transitions ($p \geq 5$) is not completely understood. It was proven[21] that for large enough $p$, clock models exhibit a BKT-type transition (actually, it has only been proved that there exists an intermediate critical phase with power-law correlations). The question whether that still holds when $p = 5$ remains open.[22] By relying on exact results, we shed light on the character of the BKT transitions in these systems. We will relate the BKT transition to a continuous U(1) emergent symmetry of an usual type. Although BKT transitions are often discussed in terms of specific anomalous exponents and jumps in the helicity modulus, we will not address such non-universal issues.

Our treatment of classical dualities is based on a new approach developed in Refs. 13 and 14 that relies on the transfer matrix or operator formalism.[11] In statistical mechanics, two models $a$ and $b$ are dual if their partition functions $\mathcal{Z}_a = \text{tr}[T_a^N]$ and $\mathcal{Z}_b = \text{tr}[T_b^N]$ are related as ($N$ is the linear size of the system in $D = 2$)

$$\mathcal{Z}_a[K] = A(K, K^*)\mathcal{Z}_b[K^*],  \qquad (3.1)$$

with $A$ some analytic function of the set of couplings $K$ of model $a$, and dual couplings $K^*$. In principle, Eq. (3.1) establishes an extremely broad relationship that could be achieved through many transformation schemes, including the standard one based on taking the Fourier transform of individual Boltzmann weights.[17,24] However, it was discovered in Refs. 13 and 14 that low-temperature (strong coupling)/high-temperature (weak coupling) dualities correspond to a unitary equivalence of transfer matrices or operators, $T_a$ and $T_b$,

$$T_b = \mathcal{U}_\mathsf{d} T_a \mathcal{U}_\mathsf{d}^\dagger, \qquad (3.2)$$

with $\mathcal{U}_\mathsf{d}$ a unitary operator. This observation is extremely insightful because there is a simple and systematic way to look for unitary equivalences between physical operators, based on the notion of *bond algebra*.[13,14]

The outline of this chapter is as follows. In Sec. 3.2 we define the classical XY model, and then in Sec. 3.2.1, *establish its transfer operator*. In Sec. 3.2.2 we discuss the form of the exact one-dimensional quantum analogue whose partition function is that of the $D = 2$ classical XY model with coupling constants $K_1$ and $K_2$. In the limit of large coupling $K_2$ along columns, the quantum model is the O(2) quantum rotor model. In Sec. 3.2.3, we establish the duality of the $D = 2$ XY model to a solid-on-solid-like model and also to a lattice Coulomb gas-like and, moreover, determine the disorder variables. These dualities do not rely on the Villain approximation scheme but are *exact dualities* obtained by our bond-algebraic method.[13]

We next proceed to analyze in Sec. 3.3 the $p$-clock model.[1] This model provides a particular controlled limit to the XY model (the $p \to \infty$ limit). We replicate the same steps undertaken in the analysis of the XY model of Sec. 3.2, but now the Weyl algebra[25] and the theory of circulant matrices[26] play a key role. We construct in Sec. 3.3.1 its transfer matrix, and proceed in Sec. 3.3.2 to establish the corresponding one-dimensional quantum Hamiltonian, that is not self-dual for $p \geq 5$. We study the dualities of these systems in Sec. 3.3.3. The system is exactly self-dual for $p = 2, 3, 4$, and becomes approximately self-dual for $K_2 \gg K_1$ when $p \geq 5$. In Sec. 3.3.4, we introduce a variant of the classical $p$-clock model that is exactly self-dual for all $p$. We examine, in Sec. 3.3.5, the exact and emergent symmetries of these systems and, notably, unveil the U(1) symmetry that emerges when $p \geq 5$. This continuous emergent symmetry is responsible for the existence of the intermediate critical (massless) phase.

Finally, in Sec. 3.4 we utilize our previous findings to better understand the phase diagram of the $p$-clock model. Here we present an analytic expression for the self-dual temperature $T^*$, an important scale in the problem, and advance a Peierls argument. We also introduce a topological invariant, that we call the discrete winding number $k$, to unravel the nature of the topological excitations. Starting at $p \geq 5$, a new type of topological excitation appears with a non-zero value of $k$ that one may call discrete vortex, and which is responsible for the phase transition to a disordered state. Domain-wall excitations are key at low-temperatures and their energy cost depends on $p$, and on the relative spin configurations (except for $2 \leq p \leq 4$). By using both duality and energy-versus-entropy balance considerations we show that the transition from the critical to the disordered phase scales as $T^{(2)} \sim \mathcal{O}(1)$, while the one from the broken $\mathbb{Z}_p$ symmetry to the critical phase goes as $T^{(1)} \sim 1/p^2$ for large $p$. The appendices provide technical developments including a duality of the XY model to $q$-deformed bosons, illustrating the key physical difference between compact and non-compact degrees of freedom.

## 3.2. The XY Model: A Paradigm of BKT Phenomenology

The $D = 2$ classical XY model is the paradigmatic example of a system displaying a BKT transition at a finite temperature $T_{\mathsf{BKT}}^{(2)} > 0$. This model, also known as planar rotator or planar O(2), consists of an $N \times N$ array of classical two-component spins $\boldsymbol{S_r}$ located at the vertices $\boldsymbol{r} = i\,\boldsymbol{e_1} + j\,\boldsymbol{e_2}$ ($i, j$ being integers) of a square lattice with unit vectors $\boldsymbol{e}_\mu$, $\mu = 1, 2$, as indicated in Fig. 3.1. Its partition function is

$$\mathcal{Z}_{\mathsf{XY}}[K_\mu, \boldsymbol{h}] = \sum_{\{\boldsymbol{S_r}\}} \exp\left[\sum_{\boldsymbol{r}} \left(\sum_{\mu=1,2} K_\mu \, \boldsymbol{S_{r+e_\mu}} \cdot \boldsymbol{S_r} + \boldsymbol{h} \cdot \boldsymbol{S_r}\right)\right], \quad (3.3)$$

where the spin $\boldsymbol{S_r} = S_{\boldsymbol{r}}^x \, \boldsymbol{e_1} + S_{\boldsymbol{r}}^y \, \boldsymbol{e_2}$, the coupling $K_\mu = \beta J_\mu$ is the product of the inverse temperature $\beta = 1/k_B T$ and the exchange coupling $J_\mu$, and $\boldsymbol{h}$ is a temperature-rescaled external magnetic field. Fixing the magnitude of

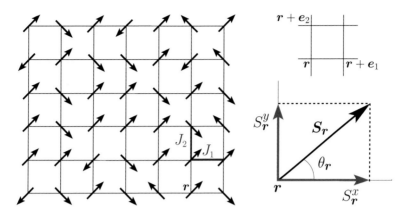

Fig. 3.1. The two-dimensional classical XY model. On each vertex $\boldsymbol{r} = i\,\boldsymbol{e}_1 + j\,\boldsymbol{e}_2$ of the square lattice there is a classical spin $\boldsymbol{S_r} = S_r^x\,\boldsymbol{e}_1 + S_r^y\,\boldsymbol{e}_2$ of magnitude 1, i.e. $\boldsymbol{S_r} \cdot \boldsymbol{S_r} = 1$, and $S_r^x = \cos\theta_r$, $S_r^y = \sin\theta_r$ with $\theta_r \in [0, 2\pi)$. Nearest neighbor spins interact with an exchange constant value $J_1$ or $J_2$ depending on the spatial direction.

the spin variable $\boldsymbol{S}_r^2$ to one allows us to rewrite the partition function as

$$\mathcal{Z}_{\mathsf{XY}}[K_\mu, h] = \sum_{\{\theta_r\}} \exp\left[\sum_{r}\left(\sum_{\mu=1,2} K_\mu \cos(\theta_{r+e_\mu} - \theta_r) + h\cos\theta_r\right)\right], \tag{3.4}$$

where the continuous angle variables $\theta_r$ take values in the interval $\theta_r = \theta_{i,j} \in [0, 2\pi)$, i.e. it is a compact variable. We assume, without loss of generality, that $\boldsymbol{h} = h\,\boldsymbol{e}_1$. The sum over configurations represents an integral

$$\sum_{\{\theta_r\}} = \int_0^{2\pi} \prod_r d\theta_r\,. \tag{3.5}$$

In the remainder of the chapter, we will concentrate on the case with zero external magnetic field, i.e. $h = 0$. The XY model displays a (global) continuous U(1) symmetry, which amounts to the invariance of the model under a simultaneous rotation of every spin in the lattice by the same angle. In low dimensional systems with continuous symmetries, such as the XY model above, long-range order is more fragile. Thermal fluctuations may induce instabilities with the end result that long-range order is actually non-existent in $D \leq 2$ dimensions. Spin-wave excitations are responsible for destroying such an order. The *Mermin–Wagner theorem* formalizes this qualitative picture. In the context of the $D = 2$ XY model, the theorem states that this system does not display spontaneous magnetization at finite temperatures. A common physical mechanism behind the formal proofs for both classical and quantum versions of the XY model, can be found in Ref. 11.

A phase transition is said to occur whenever a thermodynamic function of the system under study displays a non-analyticity. The latter may occur even when the ground state is unique, so that there is *no* spontaneous symmetry breakdown. This is the case for the $D = 2$ XY model, that is known to have a special phase transition at a finite, non-zero temperature $T_{\text{BKT}}^{(2)}$. This BKT *transition* is characterized by an essential singularity in the free energy and correlation length at $T_{\text{BKT}}^{(2)}$. If $T > T_{\text{BKT}}^{(2)}$, the correlators of the XY model decay exponentially with distance, as is typical of a disordered, paramagnetic phase. In the low-temperature phase, $T < T_{\text{BKT}}^{(2)}$, the correlators decay algebraically with distance, just as if every temperature below $T_{\text{BKT}}^{(2)}$ represented an ordinary critical point. The fact that this power-law behavior extends over the finite temperature range $0 < T < T_{\text{BKT}}^{(2)}$, without long-range order, is known as *quasi-long-range order*.

### 3.2.1. *A transfer operator for the XY model*

In this section we set up a transfer operator for the XY model, in preparation for the detailed study of its duality properties and symmetries. We assume open boundary conditions in the $e_1$-direction and periodic ones in the $e_2$-direction.

Since we are considering the XY model on a square lattice of size $N \times N$ we will need the operators

$$L_{z,i} = -i\frac{\partial}{\partial \theta_i}, \quad \text{and} \quad e^{\pm i\hat{\theta}_i}, \quad i = 1, 2, \ldots, N, \qquad (3.6)$$

satisfying the following commutation relations

$$[L_{z,i}, \, e^{\pm i\hat{\theta}_j}] = \pm \delta_{i,j} e^{\pm i\hat{\theta}_j}. \qquad (3.7)$$

The eigenstates of the unitary operators $e^{\pm i\hat{\theta}_i}$,

$$e^{\pm i\hat{\theta}_i} |\theta_i\rangle = e^{\pm i\theta_i} |\theta_i\rangle, \quad \theta_i \in [0, 2\pi), \qquad (3.8)$$

satisfy

$$\langle \theta_i' | \theta_i \rangle = \delta(\theta_i' - \theta_i), \quad \int_0^{2\pi} d\theta_i \, |\theta_i\rangle\langle\theta_i| = \mathbb{1}. \qquad (3.9)$$

The plane wave eigenstates $|n_i\rangle$ of $L_{z,i}$ form an orthonormal basis of the Hilbert space of square integrable functions on the circle $\mathcal{L}^2(S_1) = \mathcal{H}_i$, and are related to the states $|\theta_i\rangle$ via $\langle \theta_i | n_i \rangle = e^{i\theta_i n_i}/\sqrt{2\pi}$.

On the one hand, $e^{\pm i\hat{\theta}_i}$ represent position operators for the spin at site $i$, since their simultaneous eigenstates $|\theta_i\rangle$ specify one, and only one point on the unit circle. On the other hand, $L_{z,i}$ represents their canonically conjugate momentum, the infinitesimal generator of translations

$$e^{-i\delta L_{z,i}}|\theta_i\rangle = |\theta_i + \delta\rangle. \tag{3.10}$$

This last equation follows from Eq. (3.7), since

$$e^{i\delta L_{z,i}} e^{\pm i\hat{\theta}_i} e^{-i\delta L_{z,i}} = e^{\pm i(\hat{\theta}_i + \delta)}. \tag{3.11}$$

The product states,

$$|\theta\rangle = \bigotimes_i |\theta_i\rangle, \tag{3.12}$$

that are simultaneous eigenstates of all the position operators $e^{\pm i\hat{\theta}_i}$, form an orthonormal basis of the total Hilbert space $\mathcal{H} = \bigotimes_i \mathcal{H}_i$.

We have now all the ingredients needed to write down a transfer operator for the XY model. Consider for concreteness the following row-to-row ($j$ to $j+1$) matrix elements of the transfer operator

$$\langle \theta' | T_2 | \theta \rangle = \exp\left[\sum_{i=1}^{N} K_2 \cos(\theta_{i,j+1} - \theta_{i,j})\right], \tag{3.13}$$

and the diagonal matrix

$$T_1 |\theta\rangle = \exp\left[\sum_{i=1}^{N-1} K_1 \cos(\theta_{i+1,j} - \theta_{i,j})\right] |\theta\rangle. \tag{3.14}$$

Both matrices are defined in the basis of states introduced in Eq. (3.12). It is straightforward to check that if we set $T_{\mathsf{XY}} \equiv T_2 T_1$, then

$$\mathrm{tr}[T_{\mathsf{XY}}^N] = \mathcal{Z}_{\mathsf{XY}}[K_\mu, h=0], \tag{3.15}$$

recovers the partition function for the XY model of Eq. (3.3), provided we set the external magnetic field $h$ to zero.

Next, we rewrite the operators $T_1$, $T_2$ in terms of the operators introduced in Eq. (3.7). The result reads

$$T_1 = \prod_{i=1}^{N-1} e^{K_1 \cos(\hat{\theta}_{i+1} - \hat{\theta}_i)}, \quad T_2 = \prod_{i=1}^{N} \int_0^{2\pi} d\theta \, e^{K_2 \cos\theta} e^{-i\theta L_{z,i}}, \tag{3.16}$$

as can be checked by taking matrix elements of $T_2 T_1$ in the basis of Eq. (3.12). Notice that $T_1$ factors into a product of two-body operators that involves only nearest neighbors, while $T_2$ factors into a product of one-body operators. This important simplification is a direct reflection of locality.

The relevant symmetries of the classical XY model translate into unitary transformations that commute with $T_{\mathsf{XY}} \equiv T_2 T_1$. Besides the obvious geometrical symmetries of the lattice, $T_{\mathsf{XY}}$ commute with two operators that represent internal, global symmetries. The continuous global U(1) symmetry under global rotations of the classical spin direction $\theta_r \to \theta_r + \alpha$, $\forall r$, guarantees that $[L_z, T_{\mathsf{XY}}] = 0$, where $L_z = \sum_{i=1}^{N} L_{z,i}$ is the total angular momentum. There is also a discrete symmetry $\mathcal{C}_0 = \prod_{i=1}^{N} \mathcal{C}_{0i}$ that is alluded to, somewhat inaccurately, as "charge conjugation".[27] The operator $\mathcal{C}_{0i}$ acts on position eigenstates as $\mathcal{C}_{0i}|\theta_i\rangle = |2\pi - \theta_i\rangle$, and on angular momentum eigenstates as $\mathcal{C}_{0i}|n_i\rangle = |-n_i\rangle$. Thus, $\mathcal{C}_{0i}^2 = \mathbb{1}$ and $\mathcal{C}_{0i}^\dagger = \mathcal{C}_{0i}$. Since $\mathcal{C}_0$ does not commute but rather anticommutes with $L_z$, we notice that the full group of internal symmetries of the XY model is non-Abelian.

### 3.2.2. Hamiltonian form of the XY model

Models that can be written in terms of an Hermitian transfer matrix ($T$) or operator, such as the XY model, can be translated into quantum-mechanical problems[11] by simply defining a quantum Hamiltonian according to

$$H_{\mathsf{XY}} = -\ln(T_{\mathsf{XY}}), \quad \text{or equivalently}, \quad T_{\mathsf{XY}} = e^{-H_{\mathsf{XY}}}. \tag{3.17}$$

While this is a powerful tool, often used in the literature, its actual value is diminished by the technical problem of computing $\ln(T_{\mathsf{XY}})$ and, perhaps more importantly, because $H_{\mathsf{XY}}$ turns out to be a highly non-local operator. The standard way out of this difficulty is to make approximations that solve both of these problems. The qualitative picture that emerges is intuitively appealing but problematic if the approximations are not reasonably controlled.

We will not determine $H_{\mathsf{XY}}$ in closed form, but rather we will compute

$$H_\mu = -\ln T_\mu, \quad \mu = 1, 2. \tag{3.18}$$

in closed form, and then exploit the Baker–Campbell–Haussdorf (BCH) formula to obtain the expansion

$$H \equiv -\ln(e^{-H_2} e^{-H_1}) = H_1 + H_2 + [H_1, H_2]/2 + \cdots. \tag{3.19}$$

Since $T_{\mathsf{XY}} = T_2 T_1$, as defined in the previous section, is not Hermitian, we have to set $H_{\mathsf{XY}} = (H + H^\dagger)/2$. This only affects terms quadratic and higher order in the commutators. We will also study the conditions for the non-diagonal (kinetic) part of the Hamiltonian $H_2$ to reduce to the intuitively appealing form $H_2 \propto \sum_i L_{z,i}^2/2$.

Referring back to the operator form of $T_1$ and $T_2$, Eq. (3.16), we see that it is straightforward to compute $H_1$. Since $T_1$ is already in diagonal form in the $\theta_i$ basis, we obtain

$$H_1 = -\sum_{i=1}^{N-1} K_1 \cos(\hat{\theta}_{i+1} - \hat{\theta}_i). \qquad (3.20)$$

On the other hand, $T_2$ is not diagonal. Computing $H_2$ is further simplified by the fact that $T_2$ factors into the product of $N$ commuting one-body operators. It follows that $H_2 = \sum_{i=1}^{N} H_{2,i}$, with

$$e^{-H_{2,i}} = \int_0^{2\pi} d\theta\, e^{K_2 \cos\theta} e^{-i\theta L_{z,i}}. \qquad (3.21)$$

Next we notice that, since $e^{-i\theta_1 L_{z,i}} e^{-i\theta_2 L_{z,i}} = e^{-i(\theta_1+\theta_2) L_{z,i}}$, $H_{2,i}$ should be of the form (see Appendix A)

$$H_{2,i} = -\int_0^{2\pi} d\theta\, a_{K_2}(\theta) e^{-i\theta L_{z,i}}. \qquad (3.22)$$

By combining this expression with Eq. (3.21), we get an equality between functions of $L_{z,i}$ that can be evaluated in that operator's diagonal basis. The result is an infinite set of equations

$$\int_0^{2\pi} d\theta\, a_{K_2}(\theta) e^{-i\theta n} = \ln(2\pi I_n(K_2)), \quad n \in \mathbb{Z}, \qquad (3.23)$$

that one can use to determine $a_{K_2}(\theta)$ by taking the Fourier transform of the left-hand side. The modified Bessel functions of the first kind and of integer order $n$, $I_n(K_2)$, satisfy[28]

$$e^{K_2 \cos\theta} = \sum_{n \in \mathbb{Z}} I_n(K_2) e^{i\theta n} = I_0(K_2) + 2 \sum_{n=1}^{\infty} I_n(K_2) \cos(\theta n), \qquad (3.24)$$

after using the relation $I_{-n}(K) = I_n(K)$.[28]

It follows from Eq. (3.23) that $a_{K_2}(2\pi - \theta) = a_{K_2}(\theta)$, so that we can make the substitution $e^{-i\theta L_{z,i}} \to \cos(\theta L_{z,i})$ in Eq. (3.22). Then, we can Taylor-expand the cosine function to get

$$H_{2,i} = -\sum_{m=0}^{\infty} a_m(K_2) L_{z,i}^{2m}, \qquad (3.25)$$

with $a_m(K_2) = (-1)^m \int_0^{2\pi} d\theta\, \theta^{2m} a_{K_2}(\theta)/(2m)!$. This equation provides a very convenient representation of $H_{2,i}$, especially if we are allowed to discard terms beyond $m = 1$. To investigate this possibility, consider Eqs. (3.21) and (3.25), and evaluate these in the diagonal basis of $L_{z,i}$ to get

$$\sum_{m=0}^{\infty} a_m(K_2) n^{2m} = \ln(2\pi I_n(K_2)). \tag{3.26}$$

For large $K_2$, the functions $I_n(K_2)$ have the following asymptotic expansion[28]

$$I_n(K_2) \sim \frac{e^{K_2}}{\sqrt{2\pi K_2}} \sum_{m=0}^{\infty} (-1)^m \frac{c_m(n)}{K_2^m}, \tag{3.27}$$

with

$$c_0(n) = 1, \quad c_m(n) = \frac{(4n^2 - 1)(4n^2 - 3^2) \cdots (4n^2 - (2m-1)^2)}{m! \, 8^m}, \quad m \geq 1. \tag{3.28}$$

Notice that this can be rearranged into an expansion in $n^2$ that can be compared to the left-hand side of Eq. (3.26). Keeping, for each $m$, only the leading order in $1/K_2$, and expanding the logarithm accordingly ($\ln(1+x) \sim x$), we obtain

$$\ln(2\pi I_n(K_2)) \sim \ln\left(\sqrt{\frac{2\pi}{K_2}} e^{K_2}\right) + \sum_{m=1}^{\infty} (-1)^m \frac{n^{2m}}{2^m m! K_2^m}. \tag{3.29}$$

Comparing with Eq. (3.26),

$$a_0(K_2) \sim \ln\left(\sqrt{\frac{2\pi}{K_2}} e^{K_2}\right), \quad a_m(K_2) \sim \frac{(-1)^m}{2^m m! K_2^m}, \quad m \geq 1, \tag{3.30}$$

so that $a_{m+1}/a_m \sim -1/(2(m+1)K_2)$. In summary, in the large $K_2$ limit,

$$H_{\mathsf{XY}} \approx -N a_0(K_2) + \sum_{i=1}^{N} \frac{1}{2K_2} L_{z,i}^2 - \sum_{i=1}^{N-1} K_1 \cos(\hat{\theta}_{i+1} - \hat{\theta}_i), \tag{3.31}$$

which is the one-dimensional O(2) quantum rotor model.

### 3.2.3. *Duality of the XY model without the Villain approximation*

We now exploit the detailed understanding we have gained on the exact operator structure of the XY model to look for its dual representations. The standard approach to the dualities of the XY model starts by replacing it with the Villain model (see Appendix C), then mapping the latter to the solid-on-solid (SoS) model, and finally mapping the SoS model to a lattice Coulomb gas.[11] In contrast, in this section we establish directly *exact* dual representations of the XY model. In Appendix B, we establish a duality to a

$q$-deformed boson Hamiltonian which illustrates the fact that non-canonical bosons need to emerge because of the compact nature of the degrees of freedom of the XY model.

The starting point are the transfer operators $T_1$, $T_2$ introduced in Eq. (3.16). The algebra of interactions, or *bond algebra* in the language of Refs. 12–15, underlies their basic structure. In this case, this is the von Neumann algebra $\mathcal{A}_{\mathsf{XY}}$ generated by the bonds

$$L_{z,1}, \quad L_{z,i}, \quad e^{\pm i(\hat{\theta}_i - \hat{\theta}_{i-1})}, \quad i = 2, \ldots, N.$$

Notice that $T_1$, $T_2 \in \mathcal{A}_{\mathsf{XY}}$, since these operators are expressible as sums of products of the bonds listed in Eq. (3.32). $\mathcal{A}_{\mathsf{XY}}$ reflects the interactions present in the XY model and is at the same time easy to characterize in terms of relations. Then we can look for other dual realizations $\mathcal{A}_{\mathsf{XY}}^D$ that are isomorphic images of $\mathcal{A}_{\mathsf{XY}}$. By the general properties of von Neumann algebras, it must be that $\mathcal{A}_{\mathsf{XY}}^D = \mathcal{U}_{\mathsf{d}} \mathcal{A}_{\mathsf{XY}} \mathcal{U}_{\mathsf{d}}^\dagger$, with $\mathcal{U}_{\mathsf{d}}$ unitary, provided both algebras act on state Hilbert spaces of the same dimensionality. This is all we need to establish a duality for the XY model. The dual partition function is determined from the dual transfer operator $T_{\mathsf{XY}}^D = \mathcal{U}_{\mathsf{d}} T_{\mathsf{XY}} \mathcal{U}_{\mathsf{d}}^\dagger$.

The goal of this section is to look for a dual representation of the XY model in terms of integer-valued degrees of freedom, so we can expect the dual bond algebra $\mathcal{A}_{\mathsf{XY}}^D$ to act on the state space $\bigotimes_{i=1}^{N} \mathcal{L}^2(\mathbb{Z})$. Let us introduce the states $|n\rangle = \bigotimes_{i=1}^{N} |n_i\rangle$, and the operators $X_i$, $R_i$

$$X_i |n\rangle = n_i |n\rangle,$$
$$R_i |n\rangle = |\ldots, n_{i-1}, n_i - 1, n_{i+1}, \ldots\rangle, \quad (3.32)$$
$$R_i^\dagger |n\rangle = |\ldots, n_{i-1}, n_i + 1, n_{i+1}, \ldots\rangle,$$

that satisfy the algebra

$$[X_i, R_j] = -\delta_{i,j} R_j, \quad [X_i, R_j^\dagger] = \delta_{i,j} R_j^\dagger, \quad R_j R_j^\dagger = \mathbb{1}. \quad (3.33)$$

Then, the operators

$$X_1, \quad X_{i+1} - X_i, \quad R_i, \quad R_i^\dagger, \quad i = 1, \ldots, N-1, \quad (3.34)$$

generate an isomorphic dual representation $\mathcal{A}_{\mathsf{XY}}^D$ of the bond algebra of the XY model. The isomorphism $\Phi_{\mathsf{d}}$ connecting the two bond algebras

$$L_{z,1} \xrightarrow{\Phi_{\mathsf{d}}} X_1, \quad L_{z,i} \xrightarrow{\Phi_{\mathsf{d}}} X_i - X_{i-1}, \quad i = 2, \ldots, N,$$
$$e^{+i(\hat{\theta}_{i+1} - \hat{\theta}_i)} \xrightarrow{\Phi_{\mathsf{d}}} R_i, \quad e^{-i(\hat{\theta}_{i+1} - \hat{\theta}_i)} \xrightarrow{\Phi_{\mathsf{d}}} R_i^\dagger, \quad i = 1, \ldots, N-1, \quad (3.35)$$

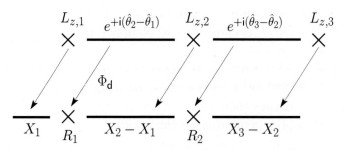

Fig. 3.2. The bond algebra isomorphism $\Phi_\mathsf{d}(\cdot) = \mathcal{U}_\mathsf{d} \cdot \mathcal{U}_\mathsf{d}^\dagger$ defined in Eq. (3.35), illustrated for $N = 3$ sites.

is illustrated in Fig. 3.2, for $N = 3$ sites. The resulting dual transfer operators are

$$T_1^D = \prod_{i=1}^{N-1} e^{\frac{K_1}{2}(R_i + R_i^\dagger)},$$
$$T_2^D = \int_0^{2\pi} d\theta\, e^{K_2 \cos\theta} e^{-i\theta X_1} \prod_{i=2}^{N} \int_0^{2\pi} d\theta\, e^{K_2 \cos\theta} e^{-i\theta(X_i - X_{i-1})}. \tag{3.36}$$

The next and last step is to compute the dual partition function

$$\mathcal{Z}_{\mathsf{XY}}^D = \mathrm{tr}[(T_2^D T_1^D)^N] = \mathrm{tr}[(T_{\mathsf{XY}}^D)^N], \tag{3.37}$$

in the product basis states of Eq. (3.32).

$T_2^D$ is already diagonal in that basis, leading simply to

$$T_2^D |n\rangle = \exp\left[-V_{K_2}(n_{1,j}) - \sum_{i=2}^{N} V_{K_2}(n_{i,j} - n_{i-1,j})\right] |n\rangle, \tag{3.38}$$

where

$$V_K(n) = -\ln \int_0^{2\pi} d\theta\, e^{K \cos\theta} e^{-i\theta n} = -\ln(2\pi I_n(K)). \tag{3.39}$$

The evaluation of $\langle n'|T_1^D|n\rangle$ factors into the computation of the one-body matrix elements,

$$\langle n_i'| e^{\frac{K_1}{2}(R_i + R_i^\dagger)} |n_i\rangle. \tag{3.40}$$

This is simplified by noticing that the Fourier transform operator of Eq. (A.2) maps

$$F_i^\dagger R_i F_i = e^{-i\hat\theta_i}, \quad F_i^\dagger R_i^\dagger F_i = e^{i\hat\theta_i} \tag{3.41}$$

thus putting the exponential in diagonal form. It follows that

$$\langle n'_i | e^{\frac{K_1}{2}(R_i+R_i^\dagger)} | n_i \rangle = 2\pi e^{-V_{K_1}(n_{i,j+1}-n_{i,j})}. \quad (3.42)$$

If we now put all the pieces together, we arrive at the conclusion that we have obtained the exact duality

$$\frac{\mathcal{Z}_{\mathsf{XY}}[K_\mu]}{(2\pi)^N}$$

$$= \sum_{\{n_{i,j}\}} e^{\left[-\sum_{j=1}^N \sum_{i=1}^{N-1} (V_{K_2}(n_{i+1,j}-n_{i,j})+V_{K_1}(n_{i,j+1}-n_{i,j}))+\sum_{j=1}^N V_{K_2}(n_{1,j})\right]}. \quad (3.43)$$

This illustrates a property typical of dualities: the coupling $K_1$ ($K_2$) in the $e_1$($e_2$)-direction of the XY model regulates the interaction in the orthogonal $e_2$($e_1$)-direction of the dual model.

It is standard to argue that the Villain model is an excellent approximation to the XY model, specially at low temperatures. As shown in Appendix C, the Villain model is dual to the SoS model. Thus, it must be that the SoS model is approximately related to $\mathcal{Z}_{\mathsf{XY}}^D$ defined by the right-hand side of Eq. (3.43), at least for low temperatures. Consider then the limit of large $K_1$, $K_2$ (i.e. low temperatures). We can then use the asymptotic expansion of Eq. (3.27) to obtain an asymptotic form of the dual potential of Eq. (3.39),

$$V_K(n) \approx \frac{n^2}{2K} + c(K), \quad K \to \infty \quad (3.44)$$

where $c(K)$ is independent of $n$ and can be computed from Eqs. (3.39) and (3.27). It follows that

$$\mathcal{Z}_{\mathsf{XY}}^D \propto \sum_{\{n_{i,j}\}} e^{\left[-\sum_{j=1}^N \sum_{i=1}^{N-1} (n_{i+1,j}-n_{i,j})^2/2K_2+(n_{i,j+1}-n_{i,j})^2/2K_1+\sum_{j=1}^N (n_{1,j})^2/2K_2\right]}$$

(3.45)

to the same level of approximation. Thus we have recovered the well known result that the XY model at very low temperatures (strong coupling) is well represented by the (approximately dual) SoS model at very high temperatures (weak coupling).

The action of the duality of Eq. (3.35) can be extended to act on the operator $e^{i\hat{\theta}_N}$ as $\Phi_{\mathsf{d}}(e^{-i\hat{\theta}_N}) = R_N$. It follows that $\Phi_{\mathsf{d}}$ generates the following set of $N$ dual variables (below, we distinguish a dual variable by an overtilde),

$$\widetilde{e^{-i\hat{\theta}_i}} \equiv \Phi_{\mathsf{d}}(e^{-i\hat{\theta}_i}) = \prod_{m=i}^N R_m, \quad i = 1, \ldots, N \quad (3.46)$$

$$\widetilde{L_{z,1}} \equiv \Phi_{\mathsf{d}}(L_{z,1}) = X_1, \quad \widetilde{L_{z,i}} \equiv \Phi_{\mathsf{d}}(L_{z,i}) = X_i - X_{i-1}, \quad i = 2, \ldots, N.$$

The dual variables satisfy the algebra of Eq. (3.7), confirming that $\Phi_\mathsf{d}$ defines an algebra isomorphism. Since this is also the algebra of the variables $R_i$, $R_i^\dagger$, $X_i$, we see that the dual variables $e^{\pm i\hat\theta_i}$, $\widetilde{L_{z,i}}$ afford an alternative representation of the elementary degrees of freedom. But what is their thermal behavior? This crucial question can be answered easily because $\Phi_\mathsf{d}$ amounts to a unitary transformation. It follows that

$$\langle e^{i\hat\theta_{m+r,n+s}} e^{-i\hat\theta_{m,n}}\rangle = \frac{\mathrm{tr}(T_{\mathsf{XY}}^{(N-n-s)} e^{i\hat\theta_{m+r}} T_{\mathsf{XY}}^s e^{i\hat\theta_m} T_{\mathsf{XY}}^n)}{\mathcal{Z}_{\mathsf{XY}}}$$

$$= \frac{\mathrm{tr}(T_{\mathsf{XY}}^{D(N-n-s)} \widetilde{e^{i\hat\theta_{m+r}}} T_{\mathsf{XY}}^{Ds} \widetilde{e^{i\hat\theta_m}} T_{\mathsf{XY}}^n)}{\mathcal{Z}_{\mathsf{XY}}^D}$$

$$= \langle e^{i\tilde\theta_{m+r,n+s}} e^{-i\tilde\theta_{m,n}}\rangle, \qquad (3.47)$$

that should be compared to

$$\langle n_{m+r,n+s} n_{m,n}\rangle = \frac{\mathrm{tr}(T_{\mathsf{XY}}^{(N-n-s)} X_{m+r} T_{\mathsf{XY}}^s X_m T_{\mathsf{XY}}^n)}{\mathcal{Z}_{\mathsf{XY}}^D}. \qquad (3.48)$$

The classical dual variables $e^{-i\tilde\theta_{m,n}}$ are difficult to define directly, but they are well defined in the sense that any correlator

$$\langle e^{(-1)^{\sigma_1} i\tilde\theta_{m_1,n_1}} e^{(-1)^{\sigma_2} i\tilde\theta_{m_2,n_2}} \cdots e^{(-1)^{\sigma_N} i\tilde\theta_{m_N,n_N}}\rangle, \quad \sigma_i = 0,1, \qquad (3.49)$$

in the ensemble $\mathcal{Z}_{\mathsf{XY}}^D$ can be computed by a straightforward generalization of (3.47).

The duality to a lattice Coulomb gas is of a very special nature (Poisson duality) and quite different from every other duality discussed in this chapter (or in the literature on dualities in general). Its general features are discussed in Ref. 13. Here, we briefly summarize the exact Coulomb gas-like dual model for the exact dual partition function $\mathcal{Z}_{\mathsf{XY}}^D$ computed above. According to Ref. 13, the lattice gas representation of $\mathcal{Z}_{\mathsf{XY}}^D$ is defined by

$$\mathcal{Z}_{\mathsf{XY}}^{DD} = \sum_{\{n_r\}} e^{-\mathcal{E}^D\{n_r\}}, \qquad (3.50)$$

with interaction energy $\mathcal{E}^D$ defined through the *global* Fourier transform

$$e^{-\mathcal{E}^D\{n_r\}} = \int \prod_r dx_r\, e^{i2\pi \sum_r n_r x_r} e^{\sum_r \sum_{\mu=1,2} V_{K_\mu}(x_{r+e_\mu}-x_r)}. \qquad (3.51)$$

The interaction energy $\mathcal{E}^D$ can be determined in closed-form in the limit in which $\mathcal{Z}_{\mathsf{XY}}^D$ reduces to the SoS model, and corresponds to the standard lattice

Coulomb gas result.[11] We see from Eq. (3.51) that Poisson dualities are in general only of practical use for models with Gaussian energy functionals.

## 3.3. The $p$-Clock Model: A Close Relative of XY

The $p$-clock model, also known as the vector Potts or $\mathbb{Z}_p$ model, allows for a systematic hierarchy of approximations to the XY model as a function of the positive integer $p$, and is a test ground for rich critical behavior. In $D = 2$ dimensions, the configurations of the classical $p$-clock model are described by a set of $p$ discretized angles $\theta_r$

$$\theta_r = \frac{2\pi s_r}{p}, \quad s_r = 0, 1, \ldots, p-1. \quad (3.52)$$

The partition function is given by

$$\mathcal{Z}_p[K_\mu] = \sum_{\{\theta_r\}} \exp\left[\sum_r \sum_{\mu=1,2} K_\mu \cos(\theta_{r+e_\mu} - \theta_r)\right]. \quad (3.53)$$

This is, in appearance, identical to the classical XY model ($h = 0$), except for the essential fact that the degrees of freedom $\{\theta_r\}$ are now discrete and countable. As in Eq. (3.3), $K_\mu = \beta J_\mu$. In the large $p$ limit ($p \to \infty$), however, one supposedly recovers the XY model. Since the $p$ points $e^{i2\pi s_r/p}$ on the unit circle close a $\mathbb{Z}_p$ subgroup of U(1) (the group of $p$'th roots of unity), the $p$-clock model manages to provide an approximation to the XY model that features a finite number of states per site, without sacrificing the XY's natural group structure. Also like the XY model, the $p$-clock has interesting but hard to uncover duality properties. We will address this problem by the same methods applied to the XY model. In fact, we will follow the discussion of the XY model as closely as possible, to highlight the connections between the two models.

### 3.3.1. *A transfer matrix for the p-clock model*

To introduce a transfer matrix for the $p$-clock model, we need to define a suitable Hilbert space and a set of basic kinematical operators. Let $\mathcal{Z}_p$ be defined on an $N \times N$ square lattice with open boundary conditions on the $e_1$-direction and periodic ones on the $e_2$-direction. The states on each site $i = 1, \ldots, N$ of a row can be described by orthonormal vectors

$$|s_i\rangle, \quad s_i = 0, \ldots, p-1 \quad (3.54)$$

such that $s_i$ represents the discrete angle $2\pi s_i/p$. They span the state space $\mathcal{H}_{p,i}$ at site $i$, so the total state space is just $\mathcal{H}_p = \bigotimes_{i=1}^{N} \mathcal{H}_{p,i}$. If we write

$|s\rangle \equiv \bigotimes_{i=1}^{N} |s_i\rangle$ for elements of the product basis of $\mathcal{H}_p$, then we can define a matrix

$$\langle s'|T_2|s\rangle = \exp\left[\sum_{i=1}^{N} K_2 \cos\left(\frac{2\pi s_{i,j+1}}{p} - \frac{2\pi s_{i,j}}{p}\right)\right], \quad (3.55)$$

related to any pair of adjacent rows $j$, $j+1$, and the diagonal matrix $T_1$ for which

$$T_1|s\rangle = \exp\left[\sum_{i=1}^{N-1} K_1 \cos\left(\frac{2\pi s_{i+1,j}}{p} - \frac{2\pi s_{i,j}}{p}\right)\right] |s\rangle. \quad (3.56)$$

These definitions guarantee that $\mathcal{Z}_p[K_\mu] = \text{tr}[(T_2 T_1)^N]$.

The degrees of freedom of the $p$-clock model (at any one site of the lattice) can take any value out of a discrete, equidistant subset of points of the unit circle. These points close a $\mathbb{Z}_p$ subgroup of the unit circle. To proceed in analogy to Sec. 3.2.1, we need to introduce position operators and their conjugate momenta in this discrete setting. The formalism that emerges was used extensively by Schwinger in his work on the foundations of quantum mechanics.[25] In what follows, we consider only one site (one degree of freedom), for the sake of clarity. We will consider all $N$ sites again near the end of the section.

It is easy to restrict the position operators $e^{\pm i\hat{\theta}}$ used for the XY model to the subset of configurations available to a *clock handle* in the $p$-clock model. The result is the operator $U$ satisfying

$$U|s\rangle = \omega^s |s\rangle, \quad s = 0, \ldots, p-1, \quad (3.57)$$

with $\omega \equiv e^{i2\pi/p}$ representing a $p$th root of unity. The position operator $U$ and its Hermitian-conjugate $U^\dagger$ satisfy $UU^\dagger = \mathbb{1} = U^p$. The momentum operator $V$ conjugate to $U$ rotates any state counter-clockwise to its *nearest-neighbor*

$$V|0\rangle = |p-1\rangle, \quad V|1\rangle = |0\rangle, \quad \ldots, \quad V|p-1\rangle = |p-2\rangle. \quad (3.58)$$

Momentum and position operators are represented, in $(p \times p)$ matrix form, as

$$V = \begin{pmatrix} 0 & 1 & 0 & \cdots & 0 \\ 0 & 0 & 1 & \cdots & 0 \\ \vdots & \vdots & \vdots & & \vdots \\ 0 & 0 & 0 & \cdots & 1 \\ 1 & 0 & 0 & \cdots & 0 \end{pmatrix}, \quad \text{and } U = \text{diag}(1, \omega, \omega^2, \ldots, \omega^{p-1}). \quad (3.59)$$

It follows that $V^\dagger$ implements a clockwise rotation, and that $VV^\dagger = \mathbb{1} = V^p$. The fundamental algebraic relation

$$VU = \omega UV \qquad (3.60)$$

follows directly from the definitions of $U$ and $V$.

As is well known from quantum mechanics, the ordinary position operator $\hat{x}$ and its conjugate momentum operator $\hat{p}$ are related by a Fourier transform $\mathcal{F}$, a unitary transformation in the space of wave functions. Essentially the same holds for the operators $U$, $U^\dagger$ and $V$, $V^\dagger$. The appropriate unitary transformation in this context is the discrete Fourier transform $F$, that in matrix form reads

$$F^\dagger = \frac{1}{\sqrt{p}} \begin{pmatrix} 1 & 1 & 1 & \cdots & 1 \\ 1 & \omega & \omega^2 & \cdots & \omega^{p-1} \\ 1 & \omega^2 & \omega^4 & \cdots & \omega^{2(p-1)} \\ \vdots & \vdots & \vdots & & \vdots \\ 1 & \omega^{p-1} & \omega^{(p-1)2} & \cdots & \omega^{(p-1)(p-1)} \end{pmatrix}. \qquad (3.61)$$

This is also known as Schur matrix.[29] It follows that[25]

$$FUF^\dagger = V^\dagger, \quad FVF^\dagger = U, \qquad (3.62)$$

and so the eigenvectors of $V$, $\tilde{s} = 0, 1, \ldots, p-1$,

$$V|\tilde{s}\rangle = \omega^{\tilde{s}}|\tilde{s}\rangle, \quad \text{with } |\tilde{s}\rangle = \frac{1}{\sqrt{p}} \sum_{s=0}^{p-1} \omega^{\tilde{s}s}|s\rangle, \qquad (3.63)$$

are easily determined via a Fourier transform of the eigenvectors of $U$.

In the mathematical literature $V$ is known as the fundamental circulant matrix. This is so as it generates the algebra of circulant matrices[26] (meaning that any circulant matrix $C$ is of the form $C = \sum_{m=0}^{p-1} a_m V^m$, $a_m \in \mathbb{C}$). Together, $U$ and $V$ generate the full algebra of $(p \times p)$ complex matrices,[25] that we continue to call the Weyl group algebra, to emphasize that we are working with a distinguished set of generators. This shows that they constitute a convenient basis set of kinematic operators, because we can write any other operator in terms of them.

We need to reintroduce the row spatial index $i$ to apply the technology just developed and rewrite the transfer matrices of Eqs. (3.56) and (3.55) in operator form. In what follows, $U_i$, $U_i^\dagger$, $V_i$, $V_i^\dagger$, for $i = 1, \ldots, N$, will be our basic set of operators. They *commute* at different sites, satisfy the relation (3.60) at any one site $i$, and act on the state space $\mathcal{H}_p = \bigotimes_{i=1}^N \mathcal{H}_{p,i}$. One

then obtains
$$T_1 = \prod_{i=1}^{N-1} e^{\frac{K_1}{2}(U_{i+1}^\dagger U_i + U_{i+1} U_i^\dagger)}, \quad T_2 = \prod_{i=1}^{N} \sum_{m=0}^{p-1} e^{K_2 \cos(2\pi m/p)} V_i^{\dagger m}. \quad (3.64)$$

This last expression for $T_2$ follows from the fact that $\langle s_i' | V_i^{\dagger m} | s_i \rangle = 0$ unless $s_i' - s_i \equiv m$ modulo $p$ (mod($p$)). It should be compared to the analogous expression for the continuum circle, Eq. (3.16).

### 3.3.2. Hamiltonian form of the p-clock model

In this section we compute the Hamiltonian form of the $p$-clock model following the strategy of Sec. 3.2.2. We start by computing $H_\mu = -\ln T_\mu$, $\mu = 1, 2$, with $T_1$, $T_2$ as defined in Eq. (3.64).

Since $T_1$ is diagonal, we can write
$$H_1 = -\sum_{i=1}^{N-1} \frac{K_1}{2}(U_{i+1}^\dagger U_i + U_i^\dagger U_{i+1}). \quad (3.65)$$

On the other hand, $H_2 = \sum_{i=1}^{N} H_{2,i}$ is not as easy to write down. $H_{2,i}$ is defined as
$$e^{-H_{2,i}} = \sum_{m=0}^{p-1} e^{K_2 \cos(2\pi m/p)} V_i^{\dagger m}. \quad (3.66)$$

As explained in Appendix A, we can solve this equation to obtain
$$H_{2,i} = -\sum_{m=0}^{p-1} a_m(K_2) V_i^{\dagger m}, \quad (3.67)$$

with
$$a_m(K_2) = \frac{1}{p} \sum_{s=0}^{p-1} \cos\left(\frac{2\pi m s}{p}\right) \ln\left(\sum_{l=0}^{p-1} e^{K_2 \cos(2\pi l/p)} \cos\left(\frac{2\pi l s}{p}\right)\right). \quad (3.68)$$

Then, the Hamiltonian $H_p$ for the $p$-clock model follows
$$H_p = -\sum_{i=1}^{N-1} \frac{K_1}{2}(U_{i+1}^\dagger U_i + U_i^\dagger U_{i+1}) - \sum_{i=1}^{N} \sum_{m=0}^{p-1} a_m(K_2) V_i^{\dagger m}, \quad (3.69)$$

provided we truncate the BCH expansion of $\ln T_p$ to linear order (see the discussion in Sec. 3.2.2). We notice for future reference that the discrete Fourier transform $\hat{F} = \prod_{i=1}^{N} F_i$ maps $H_p \to \hat{F}^\dagger H \hat{F} = \tilde{H}_p$, with

$$\tilde{H}_p = -\sum_{i=1}^{N-1} \frac{K_1}{2}(V_{i+1}^\dagger V_i + V_i^\dagger V_{i+1}) - \sum_{i=1}^{N} \sum_{m=0}^{p-1} a_m(K_2) U_i^{\dagger m}. \quad (3.70)$$

As discussed in Appendix A, the coefficients $a_m(K_2)$ have simple asymptotic forms in the limit $K_2 \to \infty$. The corresponding approximation to $H_p$ reads

$$H_p \approx -Na_0(K_2) - K_1 H_U - 2a_1(K_2) H_V, \qquad (3.71)$$

with

$$\begin{aligned} H_U &= \frac{1}{2}(U_N + U_N^\dagger + \sum_{i=1}^{N-1}(U_{i+1}^\dagger U_i + U_i^\dagger U_{i+1})), \\ H_V &= \frac{1}{2}\sum_{i=1}^{N}(V_i + V_i^\dagger), \end{aligned} \qquad (3.72)$$

and (see Eq. (A.12))

$$a_1(K_2) \approx e^{K_2(\cos(2\pi/p)-1)}, \quad a_0(K_2) \approx K_2. \qquad (3.73)$$

Equation (3.71) shows a boundary term $(-K_1(U_N + U_N^\dagger)/2)$ not present in Eq. (3.69), and that we include to make this approximation to $H_p$ exactly self-dual.[13]

The approximation made in going from Eq. (3.69) to Eq. (3.71), that keeps only $V_i^\dagger$ and $V_i^{\dagger(p-1)} = V_i$, is reminiscent of the one introduced in Sec. 3.2.2 based on Eq. (3.26), whereby we replaced the operator $\cos(\theta L_{z,i})$ for the simpler $L_{z,i}^2$ in Eq. (3.31). Indeed, the two approximations coincide in the $p \to \infty$ limit. A simple way to see this is to notice that we can realize the operator $V_i^\dagger$ directly in the Hilbert space of the XY model as $V_i^\dagger \to e^{-i2\pi L_{z,i}/p}$. Then, in the limit $p \to \infty$,

$$V_i + V_i^\dagger \to e^{i2\pi L_{z,i}/p} + e^{-i2\pi L_{z,i}/p} \approx 2 - (2\pi/p)^2 L_{z,i}^2. \qquad (3.74)$$

### 3.3.3. Dualities of the p-clock model

The dualities of the $p$-clock model appear as isomorphic representations of the bond algebras associated to the transfer matrices defined in Eq. (3.64). The bond algebra $\mathcal{A}_p$ generated by

$$V_1, \quad V_1^\dagger, \quad V_i, \quad V_i^\dagger, \quad U_i U_{i-1}^\dagger, \quad U_i^\dagger U_{i-1}, \quad i = 2, \ldots, N, \qquad (3.75)$$

is simple to work with and adequate for our purposes. It has a dual (isomorphic) representation $\mathcal{A}_p^D$ generated by the same bonds listed in Eq. (3.75), except for $V_N$, $V_N^\dagger$ that have to be removed from the set of generators, and replaced by $U_1$, $U_1^\dagger$. The duality isomorphism $\Phi_d : \mathcal{A}_p \to \mathcal{A}_p^D$ reads

$$\begin{aligned} U_{i+1}^\dagger U_i &\xrightarrow{\Phi_d} V_i^\dagger, & i &= 1, \ldots, N-1, \\ V_1 &\xrightarrow{\Phi_d} U_1, \quad V_i \xrightarrow{\Phi_d} U_{i-1}^\dagger U_i, & i &= 2, \ldots, N, \end{aligned} \qquad (3.76)$$

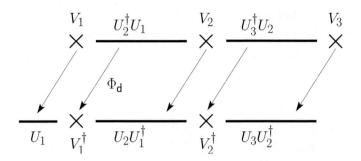

Fig. 3.3.  The duality isomorphic mapping $\Phi_{\sf d}$ of Eq. (3.76), for $N = 3$ sites.

together with the corresponding Hermitian-conjugate entries (since it must happen that $\Phi_{\sf d}(\mathcal{O}^\dagger) = \Phi_{\sf d}(\mathcal{O})^\dagger$). $\Phi_{\sf d}$ is illustrated in Fig. 3.3, and should be compared with the duality of Sec. 3.2.3 for the XY model, illustrated in Fig. 3.2.

The dual form $\mathcal{Z}_p^D = \mathrm{tr}[(T_2^D T_1^D)^N]$ of the $p$-clock model that follows from Eq. (3.76) is defined by the dual transfer matrices

$$T_1^D = \prod_{i=1}^{N-1} e^{\frac{K_1}{2}(V_i + V_i^\dagger)},$$

$$T_2^D = \sum_{m=0}^{p-1} e^{K_2 \cos(2\pi m/p)} U_1^{\dagger m} \prod_{i=2}^{N} \sum_{m=0}^{p-1} e^{K_2 \cos(2\pi m/p)} (U_i^\dagger U_{i-1})^m . \qquad (3.77)$$

Clearly, the $p$-clock model *is not self-dual* for arbitrary $p$ and arbitrary couplings. However, the model is approximately self-dual in the extreme anisotropic limit with $K_2 \gg K_1$, and it is exactly self-dual for $p = 2, 3, 4$, and any coupling. We study these aspects of the $p$-clock model in the next sections and in Appendix D.

We mention, in closing, that there is a $p$-state model that approximates the $p$-clock in the same sense in which the Villain model approximates the XY model. This $\mathbb{Z}_p$ Villain model[10] is exactly self-dual for any $p$, but is otherwise quite different from the self-dual $p$-clock model to be introduced next.

### 3.3.4. *Self-dual classical p-clock model*

In this section, we introduce a classical model $\mathcal{Z}_{\sf sdp}$ that we call the self-dual $p$-clock. It is closely related to the $p$-clock model (identical for $2 \leq p \leq 4$), yet it is exactly self-dual for any value of $p$, and has the distinct advantage over the $\mathbb{Z}_p$ Villain model[10] that its transfer matrix is remarkably simple.

To define the self-dual $p$-clock, we introduce the transfer matrices

$$T_1[K_1] = e^{\frac{K_1}{2}(U_N + U_N^\dagger)} \prod_{i=1}^{N-1} e^{\frac{K_1}{2}(U_{i+1}^\dagger U_i + U_{i+1} U_i^\dagger)},$$

$$T_2[a_0, a_1] = \prod_{i=1}^{N} e^{a_0 + a_1(V_i + V_i^\dagger)}.$$

(3.78)

Then, the partition function, $\mathcal{Z}_{\mathsf{sdp}}[a_0, a_1, K_1] = \text{tr}[T_{\mathsf{sdp}}^N]$, with $T_{\mathsf{sdp}} = T_2[a_0, a_1]T_1[K_2]$, and where $a_0$ and $a_1$ are free parameters of the model, to be determined for instance by the requirement that the approximation

$$\sum_{m=0}^{p-1} e^{K_2 \cos(2\pi m/p)} V_i^{\dagger m} \approx e^{a_0 + a_1(V_i + V_i^\dagger)},$$

(3.79)

be as good as possible for arbitrary $K_2$.

$\mathcal{Z}_{\mathsf{sdp}}$ is self-dual due to the existence of a unitary transformation that maps

$$T_1[K_1] \to T_2[0, K_1], \quad T_2[a_0, a_1] \to e^{Na_0} T_1[a_1].$$

(3.80)

This fact results from a bond-algebraic analysis, but we omit the details which can be found in Ref. 13. It follows from Eq. (3.80) that the self-dual line is specified by $K_1 = 2a_1$. The next issue then is to understand the structure of $\mathcal{Z}_{\mathsf{sdp}}$ in terms of classical variables. On one hand, it is clear that the interaction energy in the $e_1$-direction is still of the form $K_1 \cos(2\pi(s_{i+1,j} - s_{i,j})/p)$. On the other hand, the interaction energy in the $e_2$-direction, $u(s_{i,j} - s_{i,j+1})$, is determined by the relation

$$\sum_{m=0}^{p-1} e^{u(m)} V_i^{\dagger m} = e^{a_0 + a_1(V_i + V_i^\dagger)}.$$

(3.81)

Then, from Eq. (A.8),

$$e^{u(m)} = \frac{1}{[p/2]} \sum_{s=0}^{[p/2]} \cos\left(\frac{2\pi m s}{p}\right) e^{a_0 + 2a_1 \cos(2\pi s/p)}$$

(3.82)

($[p/2]$ denotes the largest integer smaller than or equal to $p/2$). With the interaction potential $u$ defined in this way,

$$\mathcal{Z}_{\mathsf{sdp}}[a_0, a_1, K_1] = \sum_{\{s_{i,j}\}} e^{\sum_{i,j}[u(s_{i,j} - s_{i,j+1}) + K_1 \cos(2\pi(s_{i+1,j} - s_{i,j})/p)]}$$

(3.83)

is exactly self-dual under the exchange $2a_1 \leftrightarrow K_1$.

One can see directly from Eq. (3.82) that $u(p-m) = u(m)$. It follows that

$$u(m) = \sum_{r=0}^{[p/2]} K_{2,r} \cos\left(\frac{2\pi r m}{p}\right), \qquad (3.84)$$

with couplings

$$K_{2,r} = a_0 \delta_{r,0} + \frac{1}{[p/2]} \sum_{m=0}^{[p/2]} \cos\left(\frac{2\pi r m}{p}\right) \ln\left[\frac{1}{[p/2]} \sum_{s=0}^{[p/2]} \cos\left(\frac{2\pi m s}{p}\right) e^{2a_1 \cos(2\pi s/p)}\right] \qquad (3.85)$$

determined by Eq. (3.82) and the orthogonality relation of Eq. (A.13). The point to notice is that to make the $p$-clock self-dual, we need to add higher-order harmonics (i.e., the terms $\cos(2\pi m(s_{i,j} - s_{i,j+1})/p)$, with $m = 2, \ldots, [p/2]$) to the basic cosine interaction. On the other hand, we show next that $\mathcal{Z}_{\text{sdp}}$ becomes a very good approximation to the standard $p$-clock model in a suitable limit.

A comparison of the self-dual $p$-clock to the standard $p$-clock model shows that the latter has an approximate self-duality for $p \geq 5$. As explained in Appendix A, in the limit in which $K_2$ grows large, Eq. (3.79) becomes almost exact, with

$$a_0 \approx K_2, \qquad a_1 \approx e^{K_2\left(\cos\frac{2\pi}{p} - 1\right)} \qquad (3.86)$$

(see Eq. (A.12)), so that

$$\begin{aligned}
\mathcal{Z}_p[K_\mu] &\approx e^{N^2 K_2} \operatorname{tr}\left[\left(e^{2a_1 H_V} e^{K_1 H_U}\right)^N\right] \\
&= e^{N^2 K_2} \operatorname{tr}\left[\left(e^{2a_1 H_U} e^{K_1 H_V}\right)^N\right] \\
&\approx e^{N^2(K_2 - K_2^*)} \mathcal{Z}_p[K_\mu^*]
\end{aligned} \qquad (3.87)$$

(see Eq. (3.72)), with dual couplings

$$K_1^* = 2e^{K_2\left(\cos\frac{2\pi}{p} - 1\right)}, \qquad K_2^* = \frac{\ln(K_1/2)}{\cos\frac{2\pi}{p} - 1}. \qquad (3.88)$$

We emphasize that this approximate self-duality, in the extreme anisotropic limit, is valid for *any* value of $p$. We consider *exact* self-dualities for the particular cases $p = 2, 3, 4$ in Appendix D.

### 3.3.5. *Exact and emergent symmetries of the p-clock model*

3.3.5.1. *Non-Abelian, discrete symmetries*

The representation of the transfer matrix $T_p = T_2 T_1$, Eq. (3.64), is very convenient for understanding the internal, global symmetries of the $p$-clock model. It is apparent that the model has an Abelian $\mathbb{Z}_p$ symmetry, but, as it turns out, when $p \geq 3$, its full group of symmetries is considerably larger and non-Abelian. To prove this we will show that there are two Hermitian operators

$$\mathcal{C}_0 = \prod_{i=1}^{N} \mathcal{C}_{0i}, \quad \mathcal{C}_1 = \prod_{i=1}^{N} \mathcal{C}_{1i}, \qquad (3.89)$$

that commute with $T_p$ and satisfy

$$\mathcal{C}_0^2 = \mathcal{C}_1^2 = (\mathcal{C}_0 \mathcal{C}_1)^p = \mathbb{1}. \qquad (3.90)$$

These relations show that, if $p \geq 3$, $\mathcal{C}_0$ and $\mathcal{C}_1$ generate a unitary representation of the so-called polyhedral group $P(2,2,p)$,[30] of order $2p$, and so the group of internal symmetries of the $p$-clock model is at least as big as this non-Abelian group. Notice that $\mathcal{C}_0 \mathcal{C}_1 \equiv \hat{Q}$, known as $\mathbb{Z}_p$ charge, generates a $\mathbb{Z}_p$ subgroup of $P(2,2,p)$. This is the standard Abelian symmetry of the $p$-clock model that gets broken in the low-temperature ordered phase (see Sec. 3.4). It becomes a U(1) symmetry in the limit $p \to \infty$, and corresponds to the usual continuous symmetry of the classical XY model $\theta_r \to \theta_r + \alpha$, with $\alpha$ an arbitrary real number.

As $\mathcal{C}_0$ and $\mathcal{C}_1$ are products of one-site operators, let us focus on a single site $i$ for now. We define the operators $\mathcal{C}_{0i}$ and $\mathcal{C}_{1i}$ by specifying their action on the basis of Eq. (3.54),

$$\mathcal{C}_{0i}|s_i\rangle = |-s_i\rangle, \quad \mathcal{C}_{1i}|s_i\rangle = |1 - s_i\rangle, \quad s_i = 0, \ldots, p-1. \qquad (3.91)$$

The arithmetic in these definitions is modular, mod($p$). For example, if $p = 5$, then $\mathcal{C}_{0i}|0\rangle = |-0\rangle = |0\rangle$, $\mathcal{C}_{0i}|1\rangle = |-1\rangle = |4\rangle$, and so on. Keeping this in mind, one can check that $\mathcal{C}_0$ and $\mathcal{C}_1$ are Hermitian, $\mathcal{C}_{0i}^\dagger = \mathcal{C}_{0i}$, $\mathcal{C}_{1i}^\dagger = \mathcal{C}_{1i}$, and satisfy the relations listed in Eq. (3.90) for $\mathcal{C}_0$ and $\mathcal{C}_1$. In particular, as $\mathcal{C}_{0i}\mathcal{C}_{1i} = V_i$, $(\mathcal{C}_{0i}\mathcal{C}_{1i})^p = \mathbb{1}$. If $p = 2$, then $\mathcal{C}_0 = \mathbb{1}$, and $\mathcal{C}_1$ generates the Abelian $\mathbb{Z}_2$ symmetry of the Ising model which is different from the non-Abelian $P(2,2,2)$.

A routine calculation shows the action of $\mathcal{C}_{0i}$, $\mathcal{C}_{1i}$ on the discrete position $U_i$ and momentum $V_i$ operators, is given by

$$\mathcal{C}_0 V_i \mathcal{C}_0 = V_i^\dagger, \quad \mathcal{C}_1 V_i \mathcal{C}_1 = V_i^\dagger, \qquad (3.92)$$

$$\mathcal{C}_0 U_i \mathcal{C}_0 = U_i^\dagger, \quad \mathcal{C}_1 U_i \mathcal{C}_1 = \omega U_i^\dagger. \qquad (3.93)$$

It is easy to check that $\mathcal{C}_0$ and $\mathcal{C}_1$ commute with $T_p$. The operator $\mathcal{C}_0$ is known in the literature as the "charge-conjugation" operator.[27] However, as we alluded to earlier, this is something of a misnomer. Geometrically speaking, $\mathcal{C}_0$ is the exact analogue of the parity operator $\mathcal{P}|x\rangle = |-x\rangle$ on the real line. In fact, $\mathcal{C}_0$ is related to the discrete Fourier transform as $\hat{F}^2 = (\hat{F}^\dagger)^2 = \mathcal{C}_0$, just as its counterpart on the real line $\mathcal{F}$ is connected to the parity operator as $\mathcal{F}^2 = \mathcal{P}$.

We wish to emphasize that these non-Abelian symmetries are shared by a large number of classical and quantum $p$-state models besides the $p$-clock, including the self-dual $p$-clock introduced in Sec. 3.3.4 and the $\mathbb{Z}_p$ Villain model.

### 3.3.5.2. *Emergent* U(1) *symmetry*

For $p \geq 5$ the *discrete* charge symmetry $\hat{Q}$ gets enhanced into a *continuous* U(1) symmetry. In reality, this is not an exact symmetry it is an *emergent* one,[31] but it is essential to establish the intermediate BKT critical phase (see Sec. 3.4). Let us derive this emergent symmetry.

Given the generators of the SU(2) algebra in the spin $S = (p-1)/2$ representation

$$S^z = \begin{pmatrix} \frac{p-1}{2} & 0 & 0 & \cdots & 0 & 0 \\ 0 & \frac{p-3}{2} & 0 & \cdots & 0 & 0 \\ \vdots & \vdots & \vdots & & \vdots & \\ 0 & 0 & 0 & \cdots & \frac{3-p}{2} & 0 \\ 0 & 0 & 0 & \cdots & 0 & \frac{1-p}{2} \end{pmatrix},$$

$$S^+ = \begin{pmatrix} 0 & \sqrt{p-1} & 0 & \cdots & 0 & 0 \\ 0 & 0 & \sqrt{2(p-2)} & \cdots & 0 & 0 \\ \vdots & \vdots & \vdots & & \vdots & \\ 0 & 0 & 0 & \cdots & \sqrt{2(p-2)} & 0 \\ 0 & 0 & 0 & \cdots & 0 & \sqrt{p-1} \end{pmatrix},$$

and $S^- = (S^+)^\dagger$, one may study the transformation properties of the Weyl's group generators $U$ and $V$ under the U(1) mapping

$$\mathcal{U}_\phi = e^{-i\phi S^z}. \tag{3.94}$$

Since $U = \omega^{\frac{p-1}{2}} \mathcal{U}_{2\pi/p}$, it commutes with $\mathcal{U}_\phi$. The transformation of $V$ requires some thinking: Let us rewrite $V$, Eq. (3.59), as the sum of two operators

$$V = \hat{V} + \hat{\Delta}, \quad \text{with } \hat{\Delta} = \begin{pmatrix} 0 & 0 & 0 & \cdots & 0 \\ 0 & 0 & 0 & \cdots & 0 \\ \vdots & \vdots & \vdots & & \vdots \\ 0 & 0 & 0 & \cdots & 0 \\ 1 & 0 & 0 & \cdots & 0 \end{pmatrix}, \quad (3.95)$$

i.e. the matrix that has only a 1 in the lower-left corner. Then,

$$\mathcal{U}_\phi \hat{V} \mathcal{U}_\phi^\dagger = e^{-i\phi} \hat{V}, \quad \text{and } \mathcal{U}_\phi \hat{\Delta} \mathcal{U}_\phi^\dagger = e^{i(p-1)\phi} \hat{\Delta}. \quad (3.96)$$

We are interested in analyzing how the transfer matrices $T_1$ and $T_2$ of Eq. (3.64) transform under $\hat{\mathcal{U}}_\phi = \prod_{i=1}^N \mathcal{U}_{\phi,i}$. It is indeed easier to analyze the Fourier transform transfer matrices, $\tilde{T}_\mu = \hat{F}^\dagger T_\mu \hat{F}$,

$$\tilde{T}_1 = \prod_{i=1}^{N-1} e^{\frac{K_1}{2}(V_{i+1}^\dagger V_i + V_{i+1} V_i^\dagger)}, \quad \tilde{T}_2 = \prod_{i=1}^{N} \sum_{m=0}^{p-1} e^{K_2 \cos(2\pi m/p)} U_i^{\dagger m}. \quad (3.97)$$

Clearly, $\hat{\mathcal{U}}_\phi$ commutes with $\tilde{T}_2$ but does not with $\tilde{T}_1$ unless $\phi = 2\pi/p$, which not surprisingly corresponds to the (Fourier transform) discrete $\hat{Q}$ symmetry. However, $\hat{\mathcal{U}}_\phi$ is an *exact continuous* symmetry of the modified transfer matrix

$$\widehat{T}_1 = \prod_{i=1}^{N-1} e^{\frac{K_1}{2}(\hat{V}_{i+1}^\dagger \hat{V}_i + \hat{V}_{i+1} \hat{V}_i^\dagger + \hat{\Delta}_{i+1}^\dagger \hat{\Delta}_i + \hat{\Delta}_{i+1} \hat{\Delta}_i^\dagger)}, \quad (3.98)$$

and becomes the usual U(1) symmetry of the XY model when $p \to \infty$. This emergent symmetry may allow for the construction of spin-wave excitations in the critical region. Note that in the original transfer matrix representation $T_1, T_2$, the continuous emergent symmetry is represented by $\hat{F} \hat{\mathcal{U}}_\phi \hat{F}^\dagger$. Moreover, it is an emergent symmetry of *both* p-clock and self-dual p-clock models.

## 3.4. Phase Diagram: From the p-Clock to the XY Model

We are thus left with the task of establishing the phase diagram of the p-clock model, the nature of its phase transitions and excitations, and its behavior as $p \to \infty$. One may argue that the phase structure of the model is well understood[10] (see Fig. 3.4). At very low temperatures, there is a ferromagnetic phase characterized by long-range order of the spin-spin correlation function $G(|\boldsymbol{r} - \boldsymbol{r}'|) = \langle e^{i\theta_r} e^{-i\theta_{r'}} \rangle$, and the breakdown of the $\mathbb{Z}_p$ symmetry $\hat{Q}$. At

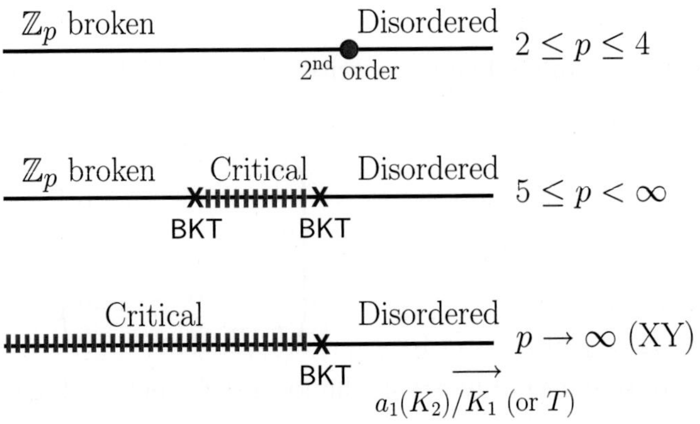

```
ℤ_p broken                    Disordered
━━━━━━━━━━━━━━━━●━━━━━━━━━━━━━━━━   2 ≤ p ≤ 4
                  2nd order

ℤ_p broken   Critical   Disordered
━━━━━━━━━╳┼┼┼┼┼┼┼╳━━━━━━━━━━━━━━   5 ≤ p < ∞
          BKT       BKT

           Critical             Disordered
┼┼┼┼┼┼┼┼┼┼┼┼┼┼┼┼┼┼┼┼┼┼┼╳━━━━━━━━━━━   p → ∞ (XY)
                BKT              ⟶
                              a_1(K_2)/K_1 (or T)
```

Fig. 3.4. Phase diagram of the $p$-clock model. For $p \geq 5$ there are three phases, the broken $\mathbb{Z}_p$ (low-temperature) phase disappearing in the limit $p \to \infty$ (XY limit). A transition is of BKT-type whenever it is associated to an essential singularity of the free energy. The critical phase is characterized by power-law correlations, i.e. quasi-long-range order, with non-universal exponents.

very high temperatures, the system is in a disordered phase with $G(|\bm{r}-\bm{r}'|)$ decaying as an exponential function of the distance. For $2 \leq p \leq 4$, these two phases are separated by a continuous second-order phase transition of the Ising ($p=2,4$) or Potts ($p=3$) type. (It is very easy to prove that the $p=4$ case is *identical* to two uncoupled $p=2$ Ising models,[32] see Appendix D.) For $p \gtrsim 5$ there is an additional intermediate *critical* phase separating the ferromagnetic from the disordered phase. It is characterized by a power-law behavior of $G(|\bm{r}-\bm{r}'|) \sim |\bm{r}-\bm{r}'|^{-\eta}$ with a non-universal exponent $\eta$, and by the absence of symmetry breakdown and quasi-long-range order. In the $p \to \infty$ limit the broken-symmetry phase disappears as one recovers the $D=2$ XY model with a continuous U(1) symmetry. This qualitative picture leaves several issues unresolved that numerical simulations have not been able to resolve either:

- What is the nature of the two phase transitions for $p \gtrsim 5$?
- What is the nature of the relevant topological excitations in each phase?
- What is the physical origin of the critical (massless) phase?
- What is the minimum $p$ after which the transitions are of the BKT type?

To understand qualitatively the nature of the phases of the model, consider the ground state of the self-dual quantum Hamiltonian $H_p$ defined in Eq. (3.71), in the large (low-temperature) and small (high-temperature) $K_1$

limits ($a_0(K_2) = 0$). Let us start with the broken $\mathbb{Z}_p$ symmetry, low temperature sector that corresponds to the line $(K_1, a_1(K_2) = 0)$. Then, the $p$-fold degenerate subspace of ground states is trivial to describe in terms of the simultaneous eigenvectors of the $U_i$ of Eq. (3.57),

$$|\Psi_0^s\rangle = \prod_{i=1}^{N} |s_i\rangle, \quad \text{with same } s \text{ for all } i. \tag{3.99}$$

The ground state energy is $E_0 = -K_1 N$ for periodic boundary conditions ($E_0 = -K_1(N-1)$ for open boundary conditions), and $\langle \Psi_0^r | \Psi_0^s \rangle = \delta_{rs}$.[33] The fully disordered, high-temperature phase is defined by the sector ($K_1 = 0, a_1(K_2)$). The ground state ($E_0 = -2a_1(K_2)N$) is unique and given by

$$|\Phi_0\rangle = \prod_{i=1}^{N} |\tilde{0}_i\rangle, \tag{3.100}$$

(in terms of the eigenstates of $V_i$, Eq. (3.63)), and satisfies $\mathcal{C}_0|\Phi_0\rangle = +|\Phi_0\rangle$. It is difficult to obtain exact results for arbitrary couplings. There is, however, an interesting exact relation that holds at the self-dual line $K_1 = 2a_1(K_2) \equiv K^*$ and follows from the fact that the self-duality unitary $\mathcal{U}_\mathsf{d}$ becomes a new symmetry of the problem on that line. It is clear from Eq. (3.71) that $H_p[K^*] = -K^*(H_U + H_V)$, and $\mathcal{U}_\mathsf{d} H_U \mathcal{U}_\mathsf{d}^\dagger = H_V$, $\mathcal{U}_\mathsf{d} H_V \mathcal{U}_\mathsf{d}^\dagger = H_U$.[13] Since $[H_p[K^*], \mathcal{U}_\mathsf{d}] = 0$, we can choose the energy eigenstates $|\Psi_n\rangle$, $n = 0, 1, \ldots$, to be also eigenstates of $\mathcal{U}_\mathsf{d}$, $\mathcal{U}_\mathsf{d}|\Psi_n\rangle = e^{i\phi_n}|\Psi_n\rangle$. Then

$$E_n = \langle \Psi_n|H_p[K^*]|\Psi_n\rangle = -2K^*\langle \Psi_n|H_U|\Psi_n\rangle = -2K^*\langle \Psi_n|H_V|\Psi_n\rangle. \tag{3.101}$$

For $2 \le p \le 4$, the $p$-clock model is *exactly* self-dual and the transition from the ferromagnetic to the disordered phase happens at the self-dual point $K_1 = 2a_1(K_2)$. For $p \ge 5$, Eq. (3.69) shows that the $p$-clock model is no longer exactly self-dual, but the self-dual approximation of Eq. (3.71) or (3.78) allow us to establish the following *self-dual equation* for arbitrary $p$

$$\frac{b_1}{b_0} = e^{K_2\left(\cos\frac{2\pi}{p} - 1\right)} = \frac{1}{2}\frac{\partial \ln B_p(a_1)}{\partial a_1}, \quad \text{where } B_p(a_1) = \sum_{m=0}^{p-1} e^{2a_1 \cos\left(\frac{2\pi}{p}m\right)}. \tag{3.102}$$

From the self-dual condition $K_1 = 2a_1(K_2)$ one can determine the *self-dual temperature* $T^*$. The self-dual point is a point of non-analyticity of the free energy for $2 \le p \le 4$, but for $p \ge 5$ it is analytic. Some values are indicated in Table 3.1, assuming isotropic couplings $K_1 = K_2 = J/(k_B T)$. It follows from very general considerations (see Sec. 8 of Ref. 13) that the two

Table 3.1. Critical, $T_c$, and self-dual, $T^*$, temperatures. For $p \geq 5$, there are two critical temperatures. The lowest one, $T_{\text{BKT}}^{(1)}$, goes to zero when $p \to \infty$, as $T_{\text{BKT}}^{(1)} \sim 1/p^2$, and the highest critical temperature $T_{\text{BKT}}^{(2)} \sim \mathcal{O}(1)$.

| $p$ | $T_c$ $[J/k_B]$ | $T^*$ $[J/k_B]$ |
| --- | --- | --- |
| 2 | $2/\ln(1+\sqrt{2})$ | $2/\ln(1+\sqrt{2})$ |
| 3 | $3/(2\ln(1+\sqrt{3}))$ | $3/(2\ln(1+\sqrt{3}))$ |
| 4 | $1/\ln(1+\sqrt{2})$ | $1/\ln(1+\sqrt{2})$ |
| 6 | $\cdots$ | $1/(2\ln(2\cos(\frac{\pi}{9})))$ |
| large $p$ | $\cdots$ | $2\pi/p$ |

critical points $c_1$ and $c_2$ bounding the self-dual point when $p \geq 5$ are exactly related by

$$\left.\frac{K_1}{2a_1(K_2)}\right|_{c_1} \cdot \left.\frac{K_1}{2a_1(K_2)}\right|_{c_2} = 1. \qquad (3.103)$$

It is interesting to analyze the large-$p$ limit of the self-dual equation (3.102). In that limit

$$\lim_{p\to\infty} \frac{1}{p} B_p(a_1) = I_0(2a_1), \quad \lim_{p\to\infty} \frac{1}{2p}\frac{\partial B_p(a_1)}{\partial a_1} = I_1(2a_1), \qquad (3.104)$$

and from the asymptotic expansion of the modified Bessel functions $I_{0,1}$, Eq. (3.27), one gets the relation between the transfer matrix and direct couplings

$$\frac{1}{2a_1(K_2)} = \frac{4\pi^2}{p^2} K_2. \qquad (3.105)$$

One can then use Eq. (3.103) to obtain a relation between the two critical temperatures $T_{\text{BKT}}^{(1)}$ and $T_{\text{BKT}}^{(2)}$ ($J_1 = J_2 = J$)

$$k_B T_{\text{BKT}}^{(1)} = \frac{4\pi^2}{p^2} \frac{J^2}{k_B T_{\text{BKT}}^{(2)}}. \qquad (3.106)$$

The Peierls argument developed in Appendix E, on the other hand, provides a rigorous scaling for the lowest transition temperature, $T_{\text{BKT}}^{(1)} = \mathcal{O}(1/p^2)$, as $p \to \infty$. Thus, this lowest critical temperature ($T_{\text{BKT}}^{(1)} < T_{\text{BKT}}^{(2)}$) vanishes for the classical XY model (where a broken-symmetry phase is not allowed) and, according to Eq. (3.106), $T_{\text{BKT}}^{(2)} \sim \mathcal{O}(1)$ for large $p$. The self-dual temperature $T^*$ has its own "intermediate" scaling with $p$, $k_B T^* = (2\pi/p)J$, so that it

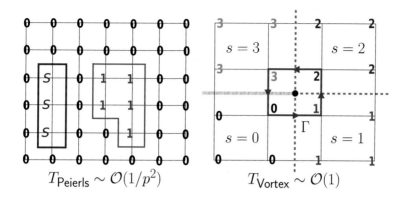

Fig. 3.5. Two types of topological excitations: domain wall (left panel) and *discrete* vortex-like (right panel) excitations. Integer numbers $s = 0, 1, 2, 3$ indicate the angle variables $\theta = 2\pi s/p$. For $p \geq 5$ the energy cost of domain walls depends on $|\Delta s|$ as opposed to the $2 \leq p \leq 4$ case where the cost is independent of $|\Delta s|$. Vortex-like configurations start appearing at $p = 5$, and the example above shows a vortex in $\mathbb{Z}_5$ of strength $k = 1$.

also vanishes in the $p \to \infty$ limit. This is to be expected since the XY model is not self-dual nor has a natural self-dual approximation.

To understand what makes $p \geq 5$ different from $p < 5$ and explain the appearance of the intermediate critical phase, one needs to analyze the nature of the topological excitations. For $p \geq 5$ there are two main types of topological excitations (see Fig. 3.5): (i) domain wall excitations that dominate the low-temperature physics, and (ii) discrete vortex-like excitations of relevance in the critical and high-temperature phases. The key distinction between the two is that domain walls exist for any $p$, while vortex-like excitations do not exist for $p < 4$, becoming manifest only for $p \geq 5$. [We remark here that the $p > 4$ system harbors vortex-like excitations. However, due to a special coincidence (see [32]), the $p = 4$ clock model is really an Ising model (or a $p = 2$ clock model) in disguise.] Also, if $2 \leq p \leq 4$ the energy cost to create a domain wall is independent of $|\Delta s| = |s_{in} - s_{out}|$, with $s_{in(out)}$ indicating the angular configurations at the two sides of the wall. This changes for $p \geq 5$, allowing for twists of the spin of size $|\Delta s| = 2, \ldots, (p-2)$.

Employing a Peierls argument provided in Appendix E, we will derive an upper bound for the probability of having a domain wall corresponds to a change of orientation between two domains of $(\pm 2\pi/p)$ (or equivalently $|\Delta s| \sim \mathcal{O}(1)$). Such a change *always appears* in all domain walls in systems with $p = 2, 3$. This, however, is not the case for $p \geq 5$ where there exist general domain wall topologies that do not allow for a uniform twist of the angle between neighboring domains. In such instances, $|\Delta s|$ can be $\mathcal{O}(p)$ and, as shown in Fig. 3.5, two types of excitations are generally possible. More precisely, in $p \geq 5$ systems, a *vorticity* arises. The topological invariant characterizing these configurations, that we call *discrete winding number* $k$ (quantifying this vorticity), is given by the circulation sum (defined for $p \geq 5$)

$$k = \frac{1}{2\pi} \sum_{\Gamma} \Delta\theta_{rr'}, \qquad (3.107)$$

taken around an oriented loop $\Gamma$, with the argument $\Delta\theta_{rr'} \in [-\pi, \pi)$ given by

$$\Delta\theta_{rr'} \equiv (\theta_{r'} - \theta_r) \mod(2\pi). \qquad (3.108)$$

In contrast, the sum in Eq. (3.107) *should not be taken* $\mod(2\pi)$ but just as an ordinary sum of real numbers (otherwise it would vanish identically). To make this definition lucid, consider a ($p = 5$) configuration such as the one shown in Fig. 3.5, with $\theta = 2\pi s/p$, and a loop $\Gamma$. Herein, the circulation sum explicitly reads $(\Delta\theta_{01} + \Delta\theta_{12} + \Delta\theta_{23} + \Delta\theta_{30}) = 2\pi/5 + 2\pi/5 + 2\pi/5 + 4\pi/5 = 2\pi$. Thus, the configuration in Fig. 3.5 has a vortex of strength $k = 1$ at its origin. As $\theta_0 - \theta_3 = -6\pi/5$, we set $\Delta\theta_{30} = 4\pi/5$. This shift in the value of $\Delta\theta_{rr'}$ (that must lie in the interval $[-\pi, \pi)$) leads to the non-zero value of $k$ in this case.

We may now use energy-versus-entropy balance considerations to argue for the relevance of these topological excitations in establishing the two phase transitions. The Peierls argument presented in Appendix E rigorously establishes that domain walls oriented relative to one another by the minimal energy cost (i.e. twists of $(\pm 2\pi/p)$) are responsible for the existence of a low-temperature ferromagnetic broken $\mathbb{Z}_p$ symmetry phase. Both the energy penalties and entropic costs associated with such *minimal cost* domain walls scale with $\ell$ (the domain wall length), and the analysis leads to a transition temperature that behaves as $k_B T_{\text{BKT}}^{(1)} \sim (1 - \cos\frac{2\pi}{p})J$. The second transition temperature, $T_{\text{BKT}}^{(2)}$ ($p \geq 5$), is associated with the proliferation of vortex-like excitations, and as indicated in Appendix E should scale as $T_{\text{BKT}}^{(2)} \sim \mathcal{O}(1)$. Note that this energy-versus-entropy balance argument does not rely on the existence or non-existence of the self-dual property of the model.

It is important to mention that while the physics of the low-temperature phase is associated with the exact discrete $\mathbb{Z}_p$ symmetry, the existence of vortex-like excitations is directly related to the *emergence* of the continuous symmetry $\hat{\mathcal{U}}_\phi$ unveiled in Sec. 3.3.5. This U(1) symmetry becomes *more* *exact* at high temperatures ($T \gtrsim T^*$ or $2a_1 \gtrsim K_1$), for a fixed $p \geq 5$, or it is exact at any temperature when $p \to \infty$. Thus, the physical origin of the critical phase and the *extended universality* concept introduced in Ref. 19 is simply our emergent $\hat{\mathcal{U}}_\phi$ continuous symmetry.

What is the nature of the phase transitions when $p \geq 5$? Given the current debates,[22] it is important to say what we mean by a "BKT-type phase transition". We simply mean a transition characterized by an essential singularity in the free energy (or the ground state energy in a quantum model). This includes those cases where there is an essential singularity but, for instance, the correlation function exponent $\eta \neq 1/4$ ($1/4$ is the exponent for the XY model[11]). It is very difficult to prove analytically the existence of an essential singularity but numerical simulations seem to indicate that for $p \geq 5$, the two phase transitions are continuous with continuous derivatives,[22] supporting the BKT scenario. Moreover, our new self-duality argument proves that the two transitions must be of the same nature.[13] In other words, *if there is an essential singularity (the function and all its derivatives remain continuous) in the free energy at $T_{\text{BKT}}^{(1)}$, then, there should be the same type of singularity at $T_{\text{BKT}}^{(2)}$.*[13] This, of course, does not mean that the self-duality fixes the value of, for instance, $\eta$, to be the same at the two transition points.

## Appendix A: Exponential of Shift Operators

In this appendix we collect some useful formulas to compute exponentials and logarithms of shift operators. Let us start with the operator $L_z$, the infinitesimal generator of translations on the circle. We have the general relation

$$e^{\int_0^{2\pi} d\theta\, a(\theta) e^{-i\theta L_z}} = \int_0^{2\pi} d\theta\, b(\theta) e^{-i\theta L_z}. \tag{A.1}$$

Our goal is to compute $a$ as a function of $b$, and *vice versa*. The first step is to notice that the Fourier transform operator

$$F = \sum_{n \in \mathbb{Z}} \int_0^{2\pi} d\theta \frac{e^{-i\theta n}}{\sqrt{2\pi}} |n\rangle\langle\theta|, \tag{A.2}$$

puts $L_z$ (and the expression of Eq. (A.1)) in diagonal form,

$$F e^{-i\theta L_z} F^\dagger = \sum_{n \in \mathbb{Z}} |n\rangle\langle n| e^{-i\theta n}. \tag{A.3}$$

This, together with the orthogonality relation $2\pi\delta(\theta' - \theta) = \sum_{n\in\mathbb{Z}} e^{i(\theta'-\theta)n}$, leads to

$$b(\theta) = \frac{1}{2\pi} \sum_{n\in\mathbb{Z}} e^{i\theta n} e^{\int_0^{2\pi} d\theta' a(\theta')e^{-i\theta' n}}, \qquad (A.4)$$

$$a(\theta) = \frac{1}{2\pi} \sum_{n\in\mathbb{Z}} e^{i\theta n} \ln\left(\int_0^{2\pi} d\theta' b(\theta')e^{-i\theta' n}\right). \qquad (A.5)$$

In Sec. 3.3.1 we introduced a diagonal matrix $U$, and a shift operator $V^\dagger$ that describes translations in a $p$-points discretization of the circle. This shift operator plays a role similar to that of $L_z$ (actually, $e^{-i\theta L_z}$). Now we have the general relation

$$e^{\sum_{m=0}^{p-1} a_m V^m} = \sum_{m=0}^{p-1} b_m V^{\dagger m}, \qquad (A.6)$$

that should be compared to Eq. (A.1), and our goal is to find closed-form expressions for the coefficients $a_m$ in terms of $b_m$ and *vice versa*. The unitary transformation that diagonalizes $V^\dagger$ is now given by the *discrete* Fourier transform of Eq. (3.62). Putting these pieces together, we get the solution to our problem,

$$b_m = \frac{1}{p}\text{tr}\left[U^m\, e^{\sum_{l=0}^{p-1} a_l U^{\dagger l}}\right], \quad a_m = \frac{1}{p}\text{tr}\left[U^m \ln\left(\sum_{l=0}^{p-1} b_l U^{\dagger l}\right)\right], \quad (A.7)$$

equations closely related to (A.4) and (A.5), as seen by expanding the trace.

In physical applications, the $a_m$ are Hermitian-symmetric, $a_{p-m} = a_m^*$ (to guarantee that $\sum_{m=0}^{p-1} a_m V^m$ is a Hermitian operator), and the $b_m$ are real and positive. Thus, it is convenient to assume that both sets of coefficients satisfy $a_{p-m} = a_m$, $b_{p-m} = b_m$, and the relations between them simplify to

$$b_m = \frac{1}{p}\sum_{s=0}^{p-1} \cos\left(\frac{2\pi m s}{p}\right) e^{\sum_{l=0}^{p-1} a_l \cos(\frac{2\pi l s}{p})}, \qquad (A.8)$$

$$a_m = \frac{1}{p}\sum_{s=0}^{p-1} \cos\left(\frac{2\pi m s}{p}\right) \ln\left(\sum_{l=0}^{p-1} b_l \cos\left(\frac{2\pi l s}{p}\right)\right). \qquad (A.9)$$

These are the expressions that are most useful in physical applications.

Suppose next that $b_m = e^{K u(m)}$, where $K$ is a positive constant, and $u(m)$ is a real function of $m = 0, \ldots, p-1$ (for example, $u(m) = \cos(\frac{2\pi m}{p})$ for the classical $p$-clock model). We would like to study the behavior of the

$a_m$ to next-to-leading order in $K$, in the limit that $K$ grows very large (this could happen at low temperature). Notice that in this limit

$$\sum_{l=0}^{p-1} b_l \cos\left(\frac{2\pi l s}{p}\right) \approx e^{Ku(0)}\left(1 + 2e^{K(u(1)-u(0))} \cos\left(\frac{2\pi s}{p}\right)\right) \quad (A.10)$$

to next-to-leading order, assuming that the inequalities

$$0 > (u(1) - u(0)) > (u(2) - u(0)) > \cdots \quad (A.11)$$

hold. The factor two in Eq. (A.10) is due to the symmetry $u(p-l) = u(l)$. Replacing expansion (A.10) into Eq. (A.9) leads to

$$a_m \approx Ku(0)\delta_{m,0} + e^{K(u(1)-u(0))}(\delta_{m,1} + \delta_{m,p-1}), \quad K \to \infty, \quad (A.12)$$

where we have used $\ln(1+x) \approx x$, and the orthogonality relation

$$\frac{1}{p}\sum_{s=0}^{p-1} \cos\left(\frac{2\pi m s}{p}\right) \cos\left(\frac{2\pi s l}{p}\right) = \frac{1}{2}(\delta_{m,l} + \delta_{m,p-l}). \quad (A.13)$$

## Appendix B: Duality of the XY Model to $q$-Deformed Bosons

In this section, we study a duality that illustrates the essential differences between compact, $\theta$, and non-compact, $x$, degrees of freedom. The algebraic tool of choice is the $q$-oscillator algebra,[34] specified by a positive real number $q$, a creation operator $a^\dagger$, its Hermitian conjugate $a$, and a Hermitian operator $\hat{n}$, satisfying

$$[\hat{n}, a] = -a, \quad [\hat{n}, a^\dagger] = a^\dagger, \quad aa^\dagger - qa^\dagger a = q^{-\hat{n}}. \quad (B.1)$$

If $q = 1$, this algebra reduces to the standard harmonic oscillator algebra, that is isomorphic to the Heisenberg algebra of translations on the line, $[\hat{x}, \hat{p}] = i$, provided $a = (\hat{x} + i\hat{p})/\sqrt{2}$, $a^\dagger = (\hat{x} - i\hat{p})/\sqrt{2}$. It was pointed out in Ref. 35 that the mapping

$$L_z \mapsto -\hat{n} + \ln\sqrt{2\sinh(1)},$$
$$e^{i\hat{\theta}} \mapsto \sqrt{2\sinh(2)}\, a\, e^{-\hat{n}}, \quad e^{-i\hat{\theta}} \mapsto \sqrt{2\sinh(2)}\, e^{-\hat{n}} a^\dagger, \quad (B.2)$$

affords a representation of the algebra of translations in the circle $[L_z, e^{\pm i\hat{\theta}}] = \pm e^{\pm i\hat{\theta}}$, provided we set $q = e^{-2}$ in Eq. (B.1).

Clearly, we can extend the mapping of Eq. (B.2) to a duality isomorphism $\Phi_d$ for the XY model. The dual transfer operators read

$$T_2^D = \prod_{i=1}^{N} \sqrt{2\sinh(1)} \int_0^{2\pi} d\theta\, e^{K_2 \cos\theta} e^{i\theta \hat{n}_i},$$

$$T_1^D = \prod_{i=1}^{N-1} \exp\left[K_1 \sinh(2)(a_{i+1} e^{-(\hat{n}_{i+1}+\hat{n}_i)} a_i^\dagger + \text{h.c})\right]. \tag{B.3}$$

To compute $\mathcal{Z}_{XY}^D$, one should take the trace in the eigenbasis of $\hat{n}_i$. This basis is described in Ref. 34.

This description of the XY model in terms of $q$-deformed bosons with $q = e^{-2}$ suggests that the algebra of Eq. (B.1) affords a continuous interpolation between the XY model and ordinary phonons (characterized by $q = 1$), but this is not the case: The XY model belongs to a representation of the algebra of Eq. (B.1) that is inequivalent to that describing phonons. The reason is that Eq. (3.7) is not enough to specify the algebra of translations in the circle. We must also have that

$$e^{\pm i\hat{\theta}} e^{\mp i\hat{\theta}} = \mathbb{1}. \tag{B.4}$$

The mapping of Eq. (B.2) will respect this constraint only if $a, a^\dagger$ satisfy

$$aa^\dagger - q^{-1} a^\dagger a = 0 \tag{B.5}$$

at least for $q = e^{-2}$, *including the relations listed in* Eq. (B.1). But the resulting set of four relations becomes inconsistent at $q = 1$. This shows that the $q$-oscillator algebra cannot interpolate continuously between canonical bosons and compact excitations.

## Appendix C: The Villain and Its Dual Solid-On-Solid Models

The Villain model[11]

$$\mathcal{Z}_V[K_\mu] = \sum_{\{n_{(r,\mu)}\}} \sum_{\{\theta_r\}} \exp\left[\sum_r \sum_{\mu=1,2} \frac{K_\mu}{2}(\theta_{r+e_\mu} - \theta_r - 2\pi n_{(r,\mu)})^2\right], \tag{C.1}$$

was introduced in Ref. 16 to provide a Gaussian approximation to the XY model that preserves the essential property of compactity, and is a good approximation at sufficiently low temperatures. We now show, by using our bond-algebraic approach, that it is dual to the solid-on-solid (SoS) model of the roughening transition,

$$\mathcal{Z}_{\mathsf{SoS}}[K_\mu] = \sum_{\{m_r\}} \exp\left[\sum_r \sum_{\mu=1,2} K_\mu^{-1}(m_{r+e_\mu} - m_r)^2\right], \qquad \text{(C.2)}$$

characterized by integer-valued degrees of freedom $m_r \in \mathbb{Z}$.[11] We work directly in the thermodynamic limit, $N \to \infty$, to avoid dealing with boundary terms.

The transfer operator $T_{\mathsf{SoS}} = T_2 T_1$ for the SoS model can we written as

$$T_1 = \prod_i e^{\frac{K_1}{2}(X_{i+1} - X_i)^2}, \quad T_2 = \prod_i \sum_m e^{\frac{K_2}{2} m^2} R_i^{\dagger m}, \qquad \text{(C.3)}$$

in terms of the operators $X_i$, $R_i$, $R_i^\dagger$ defined in Eq. (3.32). Now, however, $i \in \mathbb{Z}$ labels the sites of an infinite straight line. The duality of bond algebras

$$X_i - X_{i-1} \xrightarrow{\Phi_\mathsf{d}} L_{z,i}, \quad R_i \xrightarrow{\Phi_\mathsf{d}} e^{\mathrm{i}(\hat{\theta}_{i+1} - \hat{\theta}_i)}, \quad R_i^\dagger \xrightarrow{\Phi_\mathsf{d}} e^{-\mathrm{i}(\hat{\theta}_{i+1} - \hat{\theta}_i)} \qquad \text{(C.4)}$$

affords a dual representation of $T_{\mathsf{SoS}}$,

$$T_1^D = \prod_i e^{\frac{K_1}{2} L_{z,i}^2}, \quad T_2^D = \prod_i \sum_m e^{\frac{K_2}{2} m^2} e^{-\mathrm{i}(\hat{\theta}_{i+1} - \hat{\theta}_i) m}, \qquad \text{(C.5)}$$

in terms of compact degrees of freedom. The next step is to compute $\mathcal{Z}_{\mathsf{SoS}}^D = \mathrm{tr}[(T_2^D T_1^D)^N]$ in the basis introduced in Eq. (3.9). $T_2^D$ is already diagonal in that basis

$$T_2^D |\theta\rangle = \prod_i \sum_m e^{\frac{K_2}{2} m^2} e^{-\mathrm{i}(\theta_{i+1,j} - \theta_{i,j}) m} |\theta\rangle. \qquad \text{(C.6)}$$

At this point we could proceed by analogy to previous sections and rewrite this expression in terms of an interaction potential $V_K(\theta) \equiv -\ln \sum_m e^{\frac{K_2}{2} m^2} e^{-\mathrm{i}\theta m}$, but this will turn out to be not the most convenient approach. Instead, let us proceed to compute the matrix elements of $T_1^D$. This task reduces to computing the matrix elements of a one-body operator,

$$\langle \theta_i' | e^{\frac{K_1}{2} L_{z,i}^2} | \theta_i \rangle = \frac{1}{2\pi} \sum_{m_i} e^{\frac{K_1}{2} m_i^2} e^{-\mathrm{i}(\theta_i' - \theta_i) m_i}, \qquad \text{(C.7)}$$

which results from recalling that the orthonormal states of $L_{z,i}$ are the plane waves $\langle \theta_i | n_i \rangle = e^{\mathrm{i}\theta_i n_i}/\sqrt{2\pi}$. Notice that the function $e^{\frac{K}{2} x^2}$ is the Fourier transform of $e^{\frac{x^2}{2K}}/\sqrt{K}$. It then follows that we can use Poisson's summation formula to write

$$\sum_{m_i} e^{\frac{K}{2} m_i^2} e^{-\mathrm{i}\theta m_i} = \sqrt{\frac{2\pi}{K}} \sum_{m_i} e^{\frac{(\theta - 2\pi m_i)^2}{2K}}. \qquad \text{(C.8)}$$

Putting all the pieces together, we obtain

$$\mathcal{Z}_{\text{SoS}}^D = \left(\frac{2\pi}{K_2}\right)^N \sum_{\{\theta_r\}} \prod_r \prod_{\mu=1,2} \sum_m \exp\left[\frac{K_\mu}{2}(\theta_{r+e_\mu} - \theta_r - 2\pi m)^2\right]$$

$$= \left(\frac{2\pi}{K_2}\right)^N \sum_{\{n_{(r,\mu)}\}} \sum_{\{\theta_r\}} \exp\left[\sum_r \sum_{\mu=1,2} \frac{K_\mu}{2}(\theta_{r+e_\mu} - \theta_r - 2\pi n_{(r,\mu)})^2\right]. \quad (\text{C.9})$$

The last expression is exactly $(2\pi/K_2)^N \mathcal{Z}_V[K_\mu]$, and thus the Villain model is dual to the SoS model. Notice the reciprocal relation between the couplings: The Villain model is strongly coupled only if its dual SoS representation is weakly coupled.

## Appendix D: The $p$-Clock Model for $p = 2, 3,$ and $4$

Let us start with the simplest $p = 2$ case. Then, $U_i = U_i^\dagger = \sigma_i^z$ and $V_i = V_i^\dagger = \sigma_i^x$, and the transfer matrix $T_p = T_2 T_1$ of Eq. (3.64) reduces to

$$T_1 = \prod_{i=1}^{N-1} e^{K_1 \sigma_i^z \sigma_{i+1}^z}, \quad T_2 = \prod_{i=1}^{N} \left(e^{K_2} + e^{-K_2} \sigma_i^x\right). \quad (\text{D.1})$$

This finite Ising model is self-dual *up to boundary corrections*. The substitution $T_1 \to e^{K_1 \sigma_N^z} T_1$ renders the model *exactly* self-dual for any $N$.[13]

If $p = 3$, then $V_i^{\dagger 2} = V_i$, and $T_2$ becomes

$$T_2 = \prod_{i=1}^{N} \left(e^{K_2} + e^{-\frac{K_2}{2}}(V_i + V_i^\dagger)\right). \quad (\text{D.2})$$

$T_1$ is just as in Eq. (3.64), with $U$s appropriate for $p = 3$. It follows that if we introduce the boundary correction $T_1 \to e^{\frac{K_1}{2}(U_N + U_N^\dagger)} T_1$,[13] then $T_2 \xrightarrow{\Phi_d} T_1$ and $T_1 \xrightarrow{\Phi_d} T_2$, rendering $\mathcal{Z}_p$ exactly self-dual for any $N$.

The case $p = 4$ is special because it can be mapped onto two decoupled Ising models.[32] Since $V_i^2 = V_i^{\dagger 2}$ in this case, $T_2$ reads

$$T_2 = \prod_{i=1}^{N}(e^{K_2} + V_i + V_i^\dagger + e^{-K_2} V_i^2). \quad (\text{D.3})$$

Moreover, it is easy to check that the operator $2\mathbf{C}_i = (\mathbb{1} + \sigma_{1,i}^z) + (\mathbb{1} - \sigma_{1,i}^z)\sigma_{2,i}^x$

(known as a controlled-NOT gate in quantum computation) maps

$$U_i = e^{i\frac{\pi}{4}} \mathbf{C}_i \left( \frac{\sigma^z_{1,i} - i\sigma^z_{2,i}}{\sqrt{2}} \right) \mathbf{C}_i, \tag{D.4}$$

$$V_i + V_i^\dagger = \mathbf{C}_i(\sigma^x_{1,i} + \sigma^x_{2,i})\mathbf{C}_i, \quad V_i^2 = \mathbf{C}_i \sigma^x_{1,i} \sigma^x_{2,i} \mathbf{C}_i,$$

and thus it follows that $\mathbf{C} = \prod_{i=1}^N \mathbf{C}_i$ maps

$$\mathbf{C} T_1 \mathbf{C} = \prod_{i=1}^{N-1} e^{\frac{K_1}{2}(\sigma^z_{1,i}\sigma^z_{1,i+1} + \sigma^z_{2,i}\sigma^z_{2,i+1})}, \tag{D.5}$$

$$\mathbf{C} T_2 \mathbf{C} = \prod_{i=1}^{N} \left( e^{\frac{K_2}{2}} + e^{-\frac{K_2}{2}} \sigma^x_{1,i} \right)\left( e^{\frac{K_2}{2}} + e^{-\frac{K_2}{2}} \sigma^x_{2,i} \right), \tag{D.6}$$

that clearly defines two decoupled Ising models, with couplings that are half of those of the $p = 2$ model. In particular, the $p = 4$ clock model is exactly self-dual provided $T_1 \to e^{\frac{K_1}{2}(\sigma^z_{1,N} + \sigma^z_{2,N})} T_1$.

## Appendix E: Peierls Argument for the $p$-Clock Model

The Hamiltonian of the classical $p$-clock model (associated with the partition function of Eq. (3.53)) is given by $H = -\sum_r \sum_\mu J_\mu \cos(\theta_{r+e_\mu} - \theta_r)$. In what follows, we will consider the isotropic case and set the coupling constants $J_\mu = J = 1$ in this classical system.

We now use the Peierls argument to prove that there should be a broken symmetry phase (low-temperature ordered phase) in the $p$-clock model on the square lattice. The proof establishes the existence of a phase transition at a temperature $T^{(1)}$ below which global $\mathbb{Z}_p$ symmetry is broken. For large $p \gg 1$, $T^{(1)} = \mathcal{O}(1/p^2)$.

Specifically, our objective is to show that if uniform boundary conditions pertaining to one of the clock states $\theta = 2\pi s/p$, with $0 \le s \le p-1$ a fixed integer, are applied on the boundary of the square lattice, then there *provably* exists a temperature $T_{\text{Peierls}} > 0$ such that for temperatures $T < T_{\text{Peierls}}$, spontaneous symmetry breaking (SSB) of the global $\mathbb{Z}_p$ symmetry arises. (In the context of our discussions thus far, $T_{\text{Peierls}} < T^{(1)}$; asymptotically, for large $p$, both temperatures scale as $1/p^2$.) By SSB in this context, we refer to the lifting of the symmetry triggered by applying the uniform boundary conditions at spatial infinity. That is, when the aforementioned boundary conditions are introduced, then, for $T < T_{\text{Peierls}}$, the probability distribution $\mathcal{P}(\theta_0)$ for the angular orientation of the spin at the origin $\mathbf{S_0}$ is not symmetric between the $p$ possible values of $\theta_0$. In particular, we will demonstrate that

$\mathcal{P}(\theta_0)$ is maximal when $\theta_0$ has an orientation that matches that on the boundary, $\theta_\infty$. In other words, for temperatures $T < T_{\text{Peierls}}$,

$$\mathcal{P}(\theta_0 = \theta_\infty) \geq \frac{1}{p}. \tag{E.1}$$

To prove this inequality, we note that

$$\mathcal{P}(\theta_0 \neq \theta_\infty) \leq \text{Prob (outer domain wall } \Gamma), \tag{E.2}$$

where Prob(X) denotes the probability of the set of events X. The *domain wall* is defined as the boundary between differently oriented spins. The logic underlying Eq. (E.2) is clear: if $\theta_0 \neq \theta_\infty$ then, by its very definition, at least one domain wall must separate the spin at the origin from the spins on the boundaries of the lattice.

We will now bound the probability of having a particular domain wall. Specifically, let us denote by $\{C_\alpha\}$ the set of configurations that have $\Gamma$ as the outer-most domain wall surrounding the origin. That is, $\Gamma$ separates spins with an orientation $\theta = \theta_\infty$ from those having another (uniform) orientation $\theta_{\text{in}}$. [Note that, generally, more than one domain wall may be present and thus $\theta_{\text{in}}$ need not be the same as $\theta_0$.] The upper bound on the probabilities in Eq. (E.2) is a sum over the probabilities of having such different outer-most domain walls $\Gamma$. We furthermore define the partition function

$$\mathcal{Z}_\Gamma = \sum_{\{C_\alpha\}} \exp[-\beta E_\alpha], \tag{E.3}$$

where $E_\alpha \equiv E(C_\alpha)$ is the energy of the spin configuration $C_\alpha$. Thus, $\mathcal{Z}_\Gamma$ is smaller than the total partition function $\mathcal{Z}_p$ of the system. This is so as $\mathcal{Z}_\Gamma$ contains only a subset of the Boltzmann weights appearing in $\mathcal{Z}_p$. That is, in Eq. (E.3) we sum only over spin configurations with at least one domain wall surrounding the origin.

We next define a new configuration $\bar{C}_\alpha$ formed by rotating all of the spins inside $\Gamma$ by a uniform angle $\Delta\theta$ such that the outermost domain wall that surrounds the origin is removed. That is, for all spins $\mathbf{S}_r$ that (i) lie inside the region bounded by the domain wall $\Gamma$, we perform the transformation $\theta_r \to (\theta_r + \Delta\theta)$ with an angle of rotation

$$\Delta\theta = \theta_\infty - \theta_{\text{in}} \equiv \frac{2\pi}{p}\Delta s, \tag{E.4}$$

where $\Delta s$ is an integer. (ii) All spins lying outside the domain wall $\Gamma$ have an orientation $\theta = \theta_\infty$; these spins are not rotated. When present, any other more internal domain walls will remain unchanged by this uniform rotation of all the spins inside $\Gamma$. In order to bound, from above, the probability of having an outermost domain wall $\Gamma$, we now consider ($E_{\bar{\alpha}} = E(\bar{C}_\alpha)$)

$$\mathcal{Z}_{\bar{\Gamma}} = \sum_{\{\bar{C}_\alpha\}} \exp[-\beta E_{\bar{\alpha}}]. \tag{E.5}$$

The probability of having the domain wall $\Gamma$ is fixed by the ratio of the sum of Boltzmann weights associated with having the domain wall $\Gamma$ divided by the sum of Boltzmann weights associated with all spin configurations (i.e. the partition function $\mathcal{Z}_p$). As $\mathcal{Z}_{\bar{\Gamma}}$ contains a sum only over a subset of all Boltzmann weights that appear in $\mathcal{Z}_p$, we have

$$\text{Prob (outer domain wall } \Gamma) = \frac{\mathcal{Z}_\Gamma}{\mathcal{Z}_p} \leq \frac{\mathcal{Z}_\Gamma}{\mathcal{Z}_{\bar{\Gamma}}} = \frac{e^{-\beta E_{C_1}} + e^{-\beta E_{C_2}} + \cdots}{e^{-\beta E_{\bar{C}_1}} + e^{-\beta E_{\bar{C}_2}} + \cdots}. \tag{E.6}$$

The smallest energy difference between a configuration $C_\alpha$ and $\bar{C}_\alpha$ is bounded by

$$E_{C_\alpha} - E_{\bar{C}_\alpha} \geq \ell\left(1 - \cos\frac{2\pi}{p}\right), \tag{E.7}$$

where $\ell$ is the length of the domain wall $\Gamma$ (exchange constants are set to unity, $J_\mu = 1$). As Eq. (E.7) applies to all configuration pairs $C_\alpha$ and $\bar{C}_\alpha$ that appear in Eq. (E.6),

$$\frac{e^{-\beta E_{C_\alpha}}}{e^{-\beta E_{\bar{C}_\alpha}}} = e^{-\beta\ell(1-\cos\frac{2\pi\Delta s}{p})} \leq e^{-\beta\ell\left(1-\cos\frac{2\pi}{p}\right)}, \tag{E.8}$$

we have that

$$\frac{\mathcal{Z}_\Gamma}{\mathcal{Z}_{\bar{\Gamma}}} \leq e^{-\beta\ell\left(1-\cos\frac{2\pi}{p}\right)}. \tag{E.9}$$

It is important to emphasize that when the bound of Eq. (E.9) is saturated, $|\Delta s| = 1$. [Physically, for $p \gg 1$, only such domain walls (as opposed to far more energetically prohibitive domain walls with $|\Delta s| = \mathcal{O}(p)$) may appear at sufficiently low temperatures ($T \lesssim \mathcal{O}(1/p^2)$).]

Returning to the probability that (at least) one domain wall surrounds the origin in Eq. (E.2), we have that

$$\text{Prob (outer domain wall } \Gamma) \leq \sum_\ell N_\ell D_\ell, \tag{E.10}$$

with $N_\ell$ denoting an upper bound on the number of domain walls of perimeter $\ell$ that enclose the origin and $D_\ell$ an upper bound on the probability of having a domain wall of length $\ell$. Inserting Eq. (E.9) while taking note of an upper bound of $4 \times 3^{\ell-1}$ on the number of non-backtracking walks of length

$\ell$ on the square lattice, and an upper bound of $(\ell/4)^2$ on the maximum number of initial starting points for a walk of length $\ell$ that surrounds the origin, we have

$$\text{Prob (outer domain wall } \Gamma) \leq \sum_{\ell \geq 4} \left[ (\ell/4)^2 \times 4 \times 3^{\ell-1} e^{-\beta \ell \left(1-\cos \frac{2\pi}{p}\right)} \right]$$

$$\equiv w(\beta, p) \qquad (E.11)$$

(the minimal domain wall on the square lattice has length $\ell = 4$). The function $w$ is trivially bounded by performing the summation over all natural numbers $\ell$

$$w(\beta, p) \leq \sum_{\ell=1}^{\infty} \frac{3^\ell \ell^2}{12} e^{-\beta \ell \left(1-\cos \frac{2\pi}{p}\right)} = \frac{x(1+x)}{12(x-1)^3} \equiv \bar{w}(\beta, p), \qquad (E.12)$$

with $x = e^{\beta(1-\cos \frac{2\pi}{p})}/3$.

In performing the summation in Eq. (E.12), we assumed a sufficiently low temperature so that $x > 1$, and $\bar{w}$ is a monotonically decreasing function of $\beta$. Notably, $\bar{w}$ can be made arbitrarily close to zero for large enough $\beta$. Let us denote by $\beta_{\text{Peierls}}$ the solution to the equation $\bar{w}(\beta_{\text{Peierls}}, p) = \frac{p-1}{p}$. Then, for $\beta > \beta_{\text{Peierls}}$, the probability of the spin at the origin being the same as that on the boundary is $\mathcal{P}(\theta_0 = \theta_\infty) > 1/p$. In other words, for $T < T_{\text{Peierls}}$, we clearly have SSB. It is important to emphasize that $T_{\text{Peierls}}$ is only a lower bound to the transition temperature, and the actual SSB occurs for $T^{(1)} > T_{\text{Peierls}}$. An estimate for $T_{\text{Peierls}}$ resulting from this analysis is $T_{\text{Peierls}} \approx (1 - \cos \frac{2\pi}{p})/\ln 6$ ($\sim \mathcal{O}(1/p^2)$ for large $p$). As in the lower bound derived herein, the energy cost for a domain wall is anticipated to determine the actual ordering temperature. This bound is *rigorous*. Physically, in Eq. (E.10), the logarithms of the two terms $N_\ell$ and $D_\ell$ capture, respectively, bounds on the entropy and energy costs associated with domain walls of length $\ell$.

Discrete vortices such as the one shown in Fig. 3.5 with a typical change of angle $\Delta\theta = \mathcal{O}(1)$ (or $|\Delta s| = \mathcal{O}(p)$) across the intersecting domain walls that extend over a linear distance $\ell$ may entail, for all $p \geq 5$, an energy cost that scales as $\ell$. This is to be contrasted with the minimal energy penalty associated with a difference in angle of $|\Delta\theta| = 2\pi/p$ for which the corresponding energy penalty scales as $\ell/p^2$ (and that physically sets the bounds that we derived in the Peierls argument above). Thus, from energy-versus-entropy balance considerations, the temperature below which it is unfavorable to have vortices is $T_{\text{Vortex}} \sim \mathcal{O}(1)$ (or of order $J$ when units are restored): the energy for such domain walls scales as $\ell$ as does the entropy associated with a network of possible intersecting domain walls that have a total length $\ell$.

## References

1. R. Potts, *Proc. Camb. Philos. Soc.* **48**, 106 (1952).
2. F. Y. Wu, *Rev. Mod. Phys.* **54**, 235 (1982).
3. V. L. Berezinsky, *Sov. Phys. JETP* **32**, 493 (1971).
4. J. Kosterlitz and D. Thouless, *J. Phys. C* **6**, 1181 (1973).
5. J. Kosterlitz, *J. Phys. C* **7**, 1046 (1974).
6. J. V. José, L. P. Kadanoff, S. Kirkpatrick and D. R. Nelson, *Phys. Rev. B* **16**, 1217 (1977).
7. P. Minnhagen, *Rev. Mod. Phys.* **59**, 1001 (1987).
8. D. B. Abraham, *Surface Structures and Phase Transitions — Exact Results*, in *Phase Transitions and Critical Phenomena* Vol. 10, eds. C. Domb and J. L. Lebowitz (Academic Press, 1986), p. 1.
9. D. R. Nelson and Halperin, *Phys. Rev. B* **19**, 2457 (1979); A. P. Young, *Phys. Rev. B* **19**, 1855 (1979).
10. S. Elitzur, R. B. Pearson and J. Shigemitsu, *Phys. Rev. D* **19**, 3698 (1979).
11. H. Nishimori and G. Ortiz, *Elements of Phase Transitions and Critical Phenomena* (Oxford University Press, 2011).
12. E. Cobanera, G. Ortiz and Z. Nussinov, *Phys. Rev. Lett.* **104**, 020402 (2010).
13. E. Cobanera, G. Ortiz and Z. Nussinov, *Adv. in Phys.* **60**, 679 (2011).
14. Z. Nussinov and G. Ortiz, *Phys. Rev. B* **79**, 214440 (2009).
15. Z. Nussinov and G. Ortiz, *Europhysics Lett.* **84**, 36005 (2008).
16. J. Villain, *J. Phys.* (Paris) **36**, 581 (1975).
17. R. Savit, *Rev. Mod. Phys.* **52**, 453 (1980).
18. M. B. Einhorn, R. Savit and E. Rabinovici, *Nucl. Phys. B* **170**, 16 (1980).
19. C. M. Lapilli, P. Pfeifer and C. Wexler, *Phys. Rev. Lett.* **96**, 140603 (2006).
20. J. L. Cardy, *J. Phys. A: Math. Gen.* **13**, 1507 (1980).
21. J. Fröhlich and T. Spencer, *Comm. Math. Phys.* **81**, 527 (1981).
22. O. Borisenko, G. Cortese, R. Fiore, M. Gravina and A. Papa, *Phys. Rev. E* **83**, 041120 (2011). This work presents a summary of current debates in its introduction.
23. Y. Tomita and Y. Okabe, *Phys. Rev. B* **65**, 184405 (2002).
24. V. A. Malyshev and E. N. Petrova, *J. Math. Sciences* **21**, 877 (1983).
25. J. Schwinger, *Quantum Mechanics: Symbolism of Atomic Measurements* (Springer Verlag, Berlin, 2001).
26. P. J. Davis, *Circulant Matrices* (John Wiley, New York, 1979).
27. M. Henkel, *Conformal Invariance and Critical Phenomena* (Springer Verlag, Berlin, 1999).
28. M. Abramowitz and I. A. Stegun (eds.), *Handbook of Mathematical Functions* (Dover, New York, 1972).
29. V. B. Matveev, *Inverse Probl.* **17**, 633 (2001).

30. J. J. Rotman, *An Introduction to the Theory of Groups* (Springer Verlag, New York, 1999). See specially Chapter 11, Example 11.5 on page 347.
31. C. D. Batista and G. Ortiz, *Adv. in Phys.* **53**, 1 (2004).
32. For $p = 4$ we may represent the four equidistant spin orientations on the unit circle in terms of the normalized two component spin vectors $(\chi_r^1, \chi_r^2)/\sqrt{2}$ at sites $r$ with $\chi_r^{1,2} = \pm 1$. With this representation, the isotropic partition function reads

$$\mathcal{Z}_{p=4}[K] = \sum_{\{\chi_r^1, \chi_r^2\}} \exp\left[\sum_r \sum_{\mu=1,2} K(\chi_r^1 \chi_{r+e_\mu}^1 + \chi_r^2 \chi_{r+e_\mu}^2)/2\right], \quad (1.13)$$

which shows that the $p = 4$ clock model is equivalent to two decoupled Ising systems.

33. For $p \geq 3$, ground states that are also eigenstates of $\mathcal{C}_0$, $\mathcal{C}_0|\tilde{\Psi}_\pm^s\rangle = \pm|\tilde{\Psi}_\pm^s\rangle$, can be constructed:
    (i) $p \in$ odd:
    $$\begin{cases} |\tilde{\Psi}_+^0\rangle = |\Psi_0^0\rangle \\ |\tilde{\Psi}_\pm^s\rangle = (|\Psi_0^s\rangle \pm |\Psi_0^{p-s}\rangle)/\sqrt{2}, \text{ for } 1 \leq s \leq (p-1)/2 \end{cases},$$
    (ii) $p \in$ even:
    $$\begin{cases} |\tilde{\Psi}_+^0\rangle = |\Psi_0^0\rangle, \; |\tilde{\Psi}_+^{p/2}\rangle = |\Psi_0^{p/2}\rangle \\ |\tilde{\Psi}_\pm^s\rangle = (|\Psi_0^s\rangle \pm |\Psi_0^{p-s}\rangle)/\sqrt{2}, \text{ for } 1 \leq s \leq p/2 - 1 \end{cases}.$$

34. G. Rideau, *Lett. Math. Phys.* **24**, 147 (1992).
35. K. Kowalski, J. Rembielinski and L. C. Papaloucas, *J. Phys. A: Math. Gen.* **29**, 4149 (1996).

Chapter 4

# The Berezinskii–Kosterlitz–Thouless Transition in Superconductors

A. M. Goldman

*School of Physics and Astronomy, University of Minnesota, 116 Church St. SE, Minneapolis, MN 55455, USA*
*goldman@physics.umn.edu*

The Berezinskii–Kosterlitz–Thouless (BKT) transition occurs in thin superconducting films and Josephson junction arrays in a manner closely analogous to what is found for superfluid helium films. Initially it was believed that a BKT thermodynamic instability in which vortex–antivortex pairs, bound at low temperatures dissociate into free vortices at a characteristic temperature, would not occur in superconductors. The reason for this was that the vortex–antivortex interaction potential in bulk superconductors falls off as $1/r$, and the requirement for a BKT transition is that the interaction potential between vortex–antivortex pairs be a logarithmic function of their separation. Once it was realized that a logarithmic vortex–antivortex interaction occurs in thin films to a characteristic distance that can be macroscopic, systematic experiments were carried out on such systems establishing the existence of such a transition. In this chapter, we review the elementary theory of the BKT transition in superconductors along with the earliest experimental results leading to the establishment of its existence in thin films and Josephson junction arrays. At the present time, the BKT paradigm plays a central role in the understanding of thin film superconductors, Josephson junction arrays, and complex compounds which have weakly coupled layers, such as many of the high temperature superconductors. The limitations of space preclude a discussion of many issues that are of contemporary interest.

## 4.1. Introduction

Kosterlitz and Thouless[1,2] and Berezinskii[3] predicted that in two-dimensional neutral superfluids a thermodynamic instability in which vortex–antivortex pairs that are bound at low temperatures dissociate at a characteristic temperature. This phenomenon, known as the Berezinskii–Kosterlitz–Thouless (BKT) transition, depends explicitly on the fact that in such systems the interaction energy between vortex–antivortex pairs is a logarithmic function of their separation. In bulk superconductors, the interaction is of this form over a very small range, destroying the analogy with neutral superfluids. However, Pearl demonstrated some years ago that vortices in thin superconducting films exhibit a logarithmic interaction energy for a characteristic distance $\lambda_\perp = \lambda^2/d$ where $\lambda$ is the bulk penetration depth of the metal, and $d$ is the film thickness.[4] The quantity $\lambda_\perp$ is effectively the magnetic penetration depth for fields applied perpendicular to the film and when it becomes large, diamagnetism, which is responsible for the $1/r$ potential in bulk superconductors, becomes increasingly unimportant. This was first recognized by Beasley, Mooij and Orlando (BMO) in a work which opened the way for the study of the BKT transition in superconducting systems.[5]

In a short chapter, it is not possible to fully review the theory of the BKT transition as it is related to superconductors. Such reviews can be found in the articles by Mooij,[6] and Minnhagen.[7] We will begin with the earliest ideas, the recognition by BMO that a BKT transition in superconductors was realizable. This will be followed by a more or less heuristic discussion of the theory, mostly focused on films and largely following the treatment of Girvin.[8] Subsequent sections will be devoted to the early experimental evidence for the BKT transition in thin films, aspects of the phenomenology of Josephson junction arrays, and experiments involving arrays. The latter are important model systems, which are described by the two-dimensional XY model. The chapter will conclude with a brief section on renormalization effects and a brief summary section which enumerates topics not covered.

## 4.2. Phenomenological Theory — Mostly Films

The consideration of thin-film superconductors as candidate systems for the Berezinskii–Kosterlitz–Thouless transition began with the work of Beasley, Mooij, and Orlando (BMO).[5] Using the relationship between the BKT transition temperature, derived by Nelson and Kosterlitz,[9] which is universal, BMO derived an expression relating $T_c$ to the mean field transition temperature $T_{c0}$ that depended upon the sheet resistance of the film. This was done by parameterizing the expression relating $T_c$ to $n_s$, which is the areal superfluid particle density. This is of the form:

$$k_B T_c = \frac{1}{2}\hbar^2 n_s / m^*. \tag{4.1}$$

Here $m^*$ is the particle mass which is $2m$ for superconductors because of pairing. Then one can write the areal density in terms of the bulk superfluid density, and express the latter in terms of the penetration depth:

$$n_s = Nd = \frac{1}{2}\frac{mc^2}{4\pi e^2}\frac{d}{\lambda^2}. \tag{4.2}$$

Here $N$ is the volume particle density. Substituting Eq. (4.2) into Eq. (4.1), we find that $T_c$ is given by

$$k_B T_c = \frac{\Phi_0^2}{32\pi^2}\frac{d}{\lambda^2} = \frac{\Phi_0^2}{32\pi^2}\frac{1}{\lambda_\perp}. \tag{4.3}$$

Here $\Phi_0 = hc/2e$ is the flux quantum. We can then solve for $\lambda_\perp$ at $T_c$

$$\lambda_\perp(T_c) = \frac{\Phi_0^2}{32\pi^2}\frac{1}{k_B T_c} = \frac{0.98}{T_c} \text{ cm}. \tag{4.4}$$

Then employing the dirty-limit expression[10] for the penetration depth for very thin films

$$\lambda_\perp = \frac{\lambda^2}{d} = \frac{\lambda_L^2(0)}{d}\left(\frac{\xi_0}{l}\right)\left[\frac{\Delta(T)}{\Delta(0)}\tanh\left[\frac{\beta\Delta(T)}{2}\right]\right]^{-1}. \tag{4.5}$$

This can be recast in a more useful form as

$$\lambda_\perp = 1.78\frac{\Phi_0^2}{4\pi^3}\frac{e^2}{\hbar}\frac{R}{k_B T_{c0}}f^{-1}\left(\frac{T}{T_{c0}}\right). \tag{4.6}$$

Here $\lambda_L$ is the London penetration depth, $\Delta$ is the energy gap, $\xi_0$ is the BCS coherence length, and $\ell$ is the electron mean free path. Also $R = \rho/d$ is the sheet resistance, $T_{c0}$ is the BCS or mean field transition temperature, and $f(T/T_{c0})$ is the temperature dependent factor in brackets in Eq. (4.5). By substituting Eq. (4.6) into Eq. (4.3) one can arrive at an implicit relation between $T_c$ and $T_{c0}$ that depends on fundamental constants and the sheet resistance, which is of the form:

$$\frac{T_c}{T_{c0}}f^{-1}\left(\frac{T_c}{T_{c0}}\right) = 0.561\frac{\pi^2}{8}\frac{\hbar}{e^2}\frac{1}{R} = 2.18\frac{R_c}{R}. \tag{4.7}$$

Here $R_c = \hbar/e^2$ which corresponds to a resistance of 4.12 k$\Omega$/□.

Very near the transition temperature Eq. (4.7) takes the form

$$\frac{T_c}{T_{c0}} = \left[1 + 0.173\frac{R}{R_c}\right]^{-1}. \tag{4.8}$$

From Eq. (4.8), any substantial reduction of $T_c$ below $T_{c0}$ will require high sheet resistances. For thin and highly resistive films as temperature is reduced, the BCS or mean field transition temperature will be reached, and then somewhere below that temperature the BKT temperature will be found. This scenario involving two characteristic temperatures is a persistent feature of the application of these ideas to physical systems.

Having established that in principle there can be a BKT transition, we will now review further aspects of theory that are relevant to superconducting films. The main focus will be to present a development of the two major aspects of the phenomenology of the BKT transition in superconducting films, the temperature dependence of the resistance above the BKT transition temperature and the nonlinear current–voltage characteristics below. We will not treat in any detail the renormalization group analysis as it is presented in other sections of this book and in detailed review articles. We will also not discuss nonzero frequency experiments. The development will largely follow ideas presented by Girvin.[8]

First, we introduce some general preliminaries relating to the nature of the superconducting state. This state is usually described by a complex order parameter or macroscopic wave function of the form

$$\Psi(r) = |\Psi|e^{i\varphi(r)} \equiv \langle \psi_\uparrow^\dagger(r)\psi_\downarrow^\dagger(r) \rangle . \tag{4.9}$$

Here the quantities in brackets are field creation operators for electrons with spin up and spin down. In the usual theory, the amplitude of the order parameter becomes nonzero at the mean field transition temperature and is zero above. This ignores thermal fluctuations, which will result in a nonzero root mean square value for the order parameter in the normal state. In two dimensions, the fluctuations will suppress the transition temperature and actually the transition will be controlled by fluctuations of the phase of the order parameter. This enables us to set the amplitude of the order parameter to unity and develop a model for the superconducting transition in two dimensions, which considers only issues relating to the phase of the order parameter.

We note that the macroscopic Hamiltonian must commute with the electron number operator

$$[H, \hat{N}] = 0, \tag{4.10}$$

where the number operator in terms of field operators is

$$\hat{N} = \sum_\sigma \psi_\sigma^\dagger \psi_\sigma . \tag{4.11}$$

Then we can define the unitary transformation

$$U \equiv e^{i\theta \hat{N}}. \tag{4.12}$$

The Hamiltonian must be invariant under this transformation to insure particle number conservation. This is a special case of a more general gauge transformation with a position dependent phase angle $\theta$. This type of gauge invariance is equivalent to particle number conservation. On the other hand, the order parameter is not invariant under this transformation as can be seen from the expression

$$\langle U\psi_\uparrow^\dagger \psi_\downarrow^\dagger U^{-1} \rangle = e^{2i\theta} \langle \psi_\uparrow^\dagger \psi_\downarrow^\dagger \rangle = e^{i[\varphi(r)+2\theta]}. \tag{4.13}$$

Here the amplitude of the order parameter is set to unity. Now charge conservation requires that the energy be invariant under the transformation $\varphi(r) \to \varphi(r) + 2\vartheta$. The simplest form for the energy consistent with this requirement is

$$U = \frac{1}{2}\rho_s \int d^2r |\vec{\nabla}\varphi|^2. \tag{4.14}$$

This two-dimensional XY model with $\rho_s$ as the super-electron density is generally used to describe the behavior of superconducting films.

In order to adapt the BKT paradigm to superconducting films, as discussed at the beginning of this section, it is essential that the films be high resistance and thin so that magnetic screening effects that would spoil the logarithmic interaction of vortices, are irrelevant.

Proceeding further with the development of the behavior of two-dimensional systems, we note that the Mermin-Wagner[11] theorem as applied to this model with continuous symmetry, states that there cannot be true spontaneous symmetry breaking in 2D. True long-range order in the form of

$$\lim_{r\to\infty} \langle e^{-i\varphi(r)} e^{i\varphi(0)} \rangle \neq 0 \tag{4.15}$$

is not possible at any finite temperature. A second consequence is that it can be shown that the correlation function

$$G(r) = \langle e^{-i\varphi(r)} e^{i\varphi(0)} \rangle : \left(\frac{a}{r}\right)^{\eta(T)}, \tag{4.16}$$

where $\eta(T) = k_B T / 2\pi \rho_s$. Thus there is no long-range order and there is no characteristic length scale associated with exponential decay since the decay is always algebraic for $T > 0$. Here $a$ is a microscopic length in the problem such as the Cooper pair size. This form of order given by Eq. (4.16) is often referred to as quasi-long-range order. This is in contrast with long-range order where the correlation function is a constant.

The energy displayed in Eq. (4.14) is essentially what one might expect for a Gaussian model, which will never have correlations decaying exponentially. This model does not take into account the possibility of topological defects in the order, or vortex excitations. These can be included by adopting a lattice model in which vortex excitations are natural. Discrete models can be converted to continuum models by extrapolation. A possible Hamiltonian is then

$$H = -J \sum_{\langle i,j \rangle} \cos(\varphi_i - \varphi_j), \qquad (4.17)$$

where the subscripts label lattice sites and the sum is over nearest neighbors. As we will see later, the fact that the Hamiltonian for junction arrays in some instances is completely equivalent to Eq. (4.17) is the motivation for the study of such arrays in the context of the BKT transition. At low temperatures $\varphi_i - \varphi_j \ll 1$ and we can expand Eq. (4.17) to obtain the lattice equivalent of Eq. (4.14) with identification $J = \rho_s$.

$$H \approx \frac{1}{2} J \sum_{\langle i,j \rangle} (\varphi_i - \varphi_j)^2. \qquad (4.18)$$

The parameter $J$ is often referred to as the spin stiffness because Eq. (4.17), with the introduction of two-dimensional spin vectors, can be transformed to a model of a two-dimensional magnetic array.

The use of a lattice introduces an additional symmetry not found in the Gaussian approximation. In addition to being invariant under the unitary transformation given in Eq. (4.12), which is global, it is invariant under local transformations of the form

$$\varphi_i \to \varphi_i \pm 2\pi. \qquad (4.19)$$

This leads to the possibility of topological defects such as vortices, in which the phase changes by $\pm 2$ in going around the defect. The resultant phase change is

$$\int d\vec{r} \cdot \vec{\nabla} \varphi = 2\pi n_W. \qquad (4.20)$$

Here $n_W$ is the so-called winding number or topological charge.

We now consider a single vortex at the origin with $n_W = \pm 1$. Its phase should change uniformly as one traverses a circle enclosing the origin. Hence, we can write

$$e^{i\varphi} = \frac{x \pm iy}{|r|} \cdot e^{i\varphi_0}. \qquad (4.21)$$

Here $\varphi_0$ is an arbitrary constant and the alternating signs refer to the values of $n_W$. We can then note that $|\vec{\nabla}\varphi| = 1/r$. The energy of an isolated vortex in a system of size $L$ can then be written using Eq. (4.14) as

$$E = \int_a^L dr 2\pi r \frac{1}{2}\rho_s \frac{1}{r^2} \sim \pi \rho_s \ln\left(\frac{L}{a}\right) + E_c. \tag{4.22}$$

Here $E_c$ is the vortex core energy and $L$ is a characteristic length in the system. This integration is cut off by the core size, which is taken to be the coherence length. For superconductors, this length is the Ginzburg–Landau coherence length. The energy of an isolated vortex diverges in the thermodynamic limit. This is not true for a pair of vortices with opposite winding number or helicity. Again this result depends upon being able to ignore the out-of-plane magnetic fields associated with bulk vortices.

We now consider the entropy and the Helmholtz free energy of an isolated vortex. The number of independent locations for a vortex is proportional to $L^2/a^2$ so that the entropy is given by $S = \ln(L^2/a^2)$.

As a consequence the free energy of a vortex is

$$F = \pi \rho_s \ln\left(\frac{L}{a}\right) + E_c - 2k_B T \ln\left(\frac{L}{a}\right). \tag{4.23}$$

This expression changes sign as $T$ passes through the BKT temperature as long as the core energy is small. Thus, at the transition temperature, vortices are abundant.

$$T_c = \frac{\pi}{2}\rho_s. \tag{4.24}$$

Above $T_c$ the correlation function decays exponentially with a length scale, $\xi$ that is roughly proportional to the inverse of the square root of the vortex density.

$$\xi(T) \sim e^{-b|T-T_c|^{-1/2}}. \tag{4.25}$$

The latter result follows from a detailed renormalization group analysis. The actual behavior near the transition is far more intricate. Below $T_{\text{BKT}}$ vortices are found as bound pairs held together by a logarithmic confining potential. The condition for the transition is more complicated than Eq. (4.24) because the superfluid density is renormalized by bound pairs of vortices present at short distances. The expression is correct for the transition measured at long length scales. We will briefly return to this near the end of this chapter.

Above $T_{\text{BKT}}$ the finite density of unbound vortices causes the superelectron density to drop abruptly to zero. This universal jump in $\rho_s$ is given by Eq. (4.24). Also the algebraic decay parameter $\eta$ in the expression for the correlation function below the transition temperature [Eq. (4.16)] is equal to 1/4 at the transition.

The development of the phenomenology of the BKT transition in superconductors originates with the work of Halperin and Nelson.[12] This approach has been reviewed in detail by Mooij.[6] Minnhagen[7] has produced an extensive review of the Coulomb gas picture, which deals extensively with the problem.

Girvin developed a simple approach to the phenomenology in which the singularities associated with vortices play the role of point charges in analogy with two-dimensional electrostatics.[8] In his picture, the statistical mechanics of a collection of interacting vortices becomes that of a two-dimensional plasma of interacting charges. We start by defining an effective displacement field

$$\vec{D} = \vec{\nabla}\varphi \times \hat{z}. \tag{4.26}$$

Then the two-dimensional version of Laplace's equation for a point charge, which is actually the strength of a vortex, takes the form

$$\vec{\nabla} \cdot \vec{D} = \vec{\nabla} \times \vec{\nabla}\varphi = 2\pi n_W \delta(r). \tag{4.27}$$

Here $\delta(r)$ is a two-dimensional delta function and the equation is a Poisson equation in a representation in which the effective charge is the winding number of a vortex. In this picture, the displacement field from an effective charge of unit strength in two dimensions is $D = \hat{r}/r$ where $\hat{r}$ is a unit vector. We can then define an effective dielectric constant

$$\varepsilon = \frac{1}{2\pi\rho_s}. \tag{4.28}$$

Then the analogous electric field is related to the displacement in the usual manner

$$\vec{D} = \varepsilon \vec{E}. \tag{4.29}$$

These definitions then permit the energy [Eq. (4.14)] to be written in a familiar form

$$U = \frac{1}{4\pi} \int d^2 r \vec{E} \cdot \vec{D}. \tag{4.30}$$

The effective potential energy can be obtained by integrating $E = n_W \hat{r}/r$ from $a$, the vortex core radius out to a distance $r$ to yield

$$V(r) = -\frac{n_W}{\varepsilon} \ln\left(\frac{r}{a}\right). \tag{4.31}$$

This makes plausible the procedure of taking the interaction energy of a collection of interacting vortices as a two-dimensional plasma of charges with a potential energy of the form

$$U = -\frac{1}{\varepsilon} \sum_{i<j} q_i q_j \ln \left| \frac{\vec{r}_i - \vec{r}_j}{a} \right|. \qquad (4.32)$$

Here $q_i$ is the winding number of the $i$th vortex which would be $\pm 1$.

There are some interesting consequences of this picture. First the interaction potential becomes stronger with increasing separation of vortices. In effect, there is vortex confinement as a result of the nature of the interaction. In the electrostatics analogy, the system is an insulator at low temperatures because the vortices are bound. However vortex motion in a superconductor gives rise to voltage through the Gor'kov–Josephson relation so that the insulator in the electrostatics analog is a zero resistance superconductor. In the vicinity of the transition temperature, there are free vortices and the system is a conducting-plasma. However, when the vortices are free and can move, voltages will develop in response to the current and this regime corresponds to a resistive state. To summarize, when the vortices are bound the film is effectively superconducting, and when they are free, the film is resistive. The former is an insulator in the electrostatic analogy and the latter is a conductor.

At this point, we can develop the two aspects of the phenomenology of the BKT transition in superconducting films, the linear resistive regime above $T_c$ and the nonlinear regime below. The motion of unbound vortices above the transition temperature gives rise to a linear resistance. The escape of vortices from the confining potential in the presence of current gives rise to the nonlinear response below $T_c$.

Using the Ginzburg–Landau current expression, which is gauge invariant, a current density is the response to a phase gradient

$$\vec{J}_s = \frac{2e}{\hbar} \rho_s \vec{\nabla} \varphi. \qquad (4.33)$$

From Eqs. (4.26)–(4.29), the electric field in the electrostatic analogy is

$$\vec{E} = \frac{\hbar}{2e} \frac{1}{\varepsilon \rho_s} \vec{J}_s \times \hat{z} = \frac{h}{2e} \vec{J}_s \times \hat{z}. \qquad (4.34)$$

There is a force at right angles to the current on a stationary vortex with effective charge $n_W$, which is given by

$$\vec{F} = n_W \frac{h}{2e} \vec{J}_s \times \hat{z}. \qquad (4.35)$$

This force is equal in magnitude and opposite in sign for the members of a bound vortex–antivortex pair. The vortices can polarize and renormalize the dielectric constant, and at zero temperature they do not respond in any way. The renormalization increases the dielectric constant and decreases the superelectron density. At the transition, the superelectron density falls discontinuously to zero as the vortex–antivortex pairs become infinitely polarizable.

Above the transition temperature the vortices are free, and will move in response to the above force associated with the current. This is a dissipative process, which will lead to a nonzero voltage along the length of the sample. The motion of the vortices is at right angles to the current direction. In the limit of weak forces, the response is linear and the drift velocity of the vortices is given by $v = \mu F$ where $\mu$ is determined by details of the material including the normal resistance.

The motion of vortices at right angles to the current leads to a time dependent phase and a voltage. The phase changes by $2\pi$ whenever a vortex crosses the width of the sample. The time it takes for a vortex to cross the width of the sample is $W/v$. The number of vortices in the sample is $n_\nu L w$ where $L$ is the length of the sample. Hence, the time rate of phase change is

$$\varphi = 2\pi n_\nu L W \frac{1}{W/v} = 2\pi L n_\nu \mu \frac{h}{2e} J . \tag{4.36}$$

Here the total vortex density $n_\nu = n_+ + n_-$. The voltage along the film is then given by the Gor'kov–Josephson relation $\hbar \varphi Y = 2eV$ leading to

$$V = L n_\nu \mu \left(\frac{\hbar}{2e}\right)^2 J . \tag{4.37}$$

Then the resistivity of the film is

$$\rho = \left(\frac{h}{2e}\right)^2 n_\nu \mu = \sigma_\nu , \tag{4.38}$$

where the electrical resistivity is the vortex conductivity.

The vortex density is given by $n_\nu \sim \xi^{-2} \sim e^{2b|T-T_{\mathrm{BKT}}|^{-1/2}}$. The temperature dependence of the resistance is then of the form

$$R \sim e^{2b|T-T_{\mathrm{BKT}}|^{-1/2}} . \tag{4.39}$$

This is one of the two major phenomenological tests for the BKT transition, the temperature dependence of the resistance above the transition temperature.

The second phenomenological prediction involves consideration of the response of bound vortex–antivortex pairs to current. In addition to being polarized, bound pairs have a nonvanishing probability of being ionized by the force exerted on them by a current. As a consequence, in the presence of current below the transition temperature there are always free vortices and effectively the critical current is zero. This means that the resistance when measured by current is always nonzero and a two-dimensional superconductor is never really "superconducting" in the literal sense.

To understand this behavior it is necessary to consider the total energy of a bound vortex–antivortex pair that includes the interaction energy and the potential energy associated with the force on a vortex due to the applied current.

$$U = 2\pi\rho_s \ln\left(\frac{d}{a}\right) - \frac{h}{2e} Jd. \tag{4.40}$$

Here $d$ is the separation of the vortex and antivortex. The confining potential is eventually exceeded by the second term at some separation even for small $J$. In essence, the potential energy has a maximum (actually a saddle point), which can be found by taking the derivative with respect to $d$. The energy at the maximum is then

$$U(d_0) = 2\pi\rho_s \ln\left(\frac{J_0}{J}\right), \tag{4.41}$$

where $d_0 = a(J_0/J)$ and $J_0 = (2e\rho_s)/(\hbar a)$.

Next, it is assumed that thermal fluctuations can excite the system over this energy barrier and the excitation process is given by an Arrhenius activated form:

$$R_{\text{ionization}} \propto e^{-\beta U(d_0)}. \tag{4.42}$$

Here $R_{\text{ionization}}$ is the rate of ionization. In steady state, the rates of recombination and ionization are equal. The rate of recombination is given by the expression

$$R_{\text{recombination}} \propto n_+ n_- = n_\nu^2. \tag{4.43}$$

Here, it is assumed that there is zero external magnetic field so that all vortices are produced by dissociation of vortex–antivortex pairs resulting in equal numbers of cortices with opposite helicities or winding numbers. Then, we can equate the two rates and solve for $n_\nu$

$$n_\nu \propto R_{\text{ionization}}^{1/2} \propto e^{-\beta U(d_0)/2} \propto \left(\frac{J}{J_0}\right)^{\frac{1}{2\eta(T)}}. \tag{4.44}$$

Here $\eta(T) = \frac{k_B T}{2\pi\rho_s}$. Now since the resistivity is proportional to the number density of free vortices, this leads to a result for the nonlinear current–voltage characteristic of the form

$$V \sim J^a, \qquad (4.45)$$

where $a = 1 + (2T_c/T)$ with $T_c = \pi \rho_s/2$.

This then leads to the following phenomenological signature of the BKT transition. As previously discussed, above the transition temperature, there are thermally dissociated vortices and the current–voltage characteristic is linear for small currents. However, approaching the BKT transition temperature from below, the exponent in Eq. (4.45) will decrease to 3, and then upon crossing the transition falls to 1.

In the next section, we will discuss the experimental support for both phenomenological signatures of the BKT transition in superconducting films, the temperature dependence of the resistivity above the transition and the nonlinear current–voltage characteristics below the transition temperature.

### 4.3. Experimental Evidence — Films

As we have seen, there are two aspects of phenomenology, which, can be used to establish the existence of a Berezinskii–Kosterlitz–Thouless transition in thin film superconductors. First, there is a mean-field transition temperature $T_{c0}$ that is greater than the BKT temperature $T_c$. The latter separates two regimes of parameter space as a function of temperature. Above $T_c$, the order parameter correlation function decays exponentially, and the response is dominated by a plasma of thermally excited free vortices with equal number with opposite helicities or charges, in zero magnetic field. Below $T_c$, the correlation function decays algebraically according to Eq. (4.16) with the transition occurring when $\eta = \frac{k_B T}{2\pi \rho_s} = 1/4$. The superconducting state is stable at lower temperatures. In this regime, the electrical response is controlled by the behavior of bound vortex–antivortex pairs.

The specific predictions for the electrical response in these two regimes have been explored in a variety of thin film systems including granular Al,[13] granular NbN,[14] quench-condensed Hg-Xe films,[15] indium/indium oxide composite films.[16] The work on proximity-coupled superconductors[17–19] and Josephson junction arrays[20,21] will be discussed in a later section. Also finite frequency measurements of the super-electron density have been effectively used to infer $T_c$ and the value of $\eta$.[22] These will not be treated for lack of space.

Most experiments, with the exception of the work on granular Al, support the existence of a BKT transition.[13] An example of data, which demonstrates both aspects of the phenomenology is found in the work of Hebard

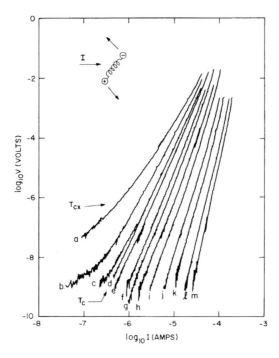

Fig. 4.1. Plot on a log–log scale of the $V$–$I$ characteristic at thirteen successively lower temperatures ranging from 1.939 K (curve a) to 1.460 (curve m). The curves are low-frequency (24.2 Hz) digitized data representing approximately 200 points per voltage decade. The nonlinear vortex–antivortex pair breaking process is illustrated in the inset. The symbol $T_c$ is used for $T_{\text{BKT}}$. (From Ref. 16.)

and Fiory on indium/indium oxide composite thin films.[16] These films, which have a significant amorphous component appear to be homogeneous over the longest length-scales probed, which are of order 0.05 cm. In this work, the samples were 100 Å thick and were 0.01 cm in width and 0.05 cm in length. The ambient magnetic field was less than $10^{-6}$ T. Measurements of the $I$–$V$ characteristics were carried out using currents ranging from $10^{-7}$ to $10^{-4}$ A and voltages from $10^{-9}$ to $10^{-4}$ V. The main feature of this experiment was that the transition could be identified by the onset of linearity at low current levels, which was independent of the condition involving the power-law exponent becoming equal to 3.

In Fig. 4.1, the power law dependencies are displayed. They are a direct consequence of the current-induced unbinding of vortex–antivortex pairs. The more detailed theory of Halperin and Nelson predicts that for $T < T_{\text{BKT}}$

$$V = 2R_N I_0 [a(T) - 3] \left(\frac{I}{I_0}\right)^{a(T)}. \tag{4.46}$$

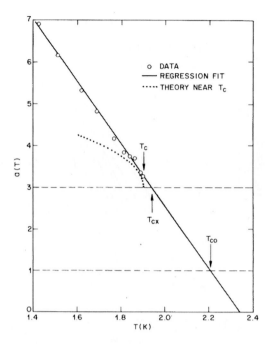

Fig. 4.2. Plot of the temperature dependence of the power-law slopes extracted from curves d through m of Fig. 4.1. From Ref. 16.

Here $a(T) = 1+0.5/\eta(T)$ is the exponent, $I_0 = (wek_BT_c)/(\hbar\xi_c)$ is an effective critical current, $w$ is the width of the film and $\xi_c$ is the vortex core size. If one takes the superfluid density to have a mean field temperature dependence, then the temperature dependence of $a(T)$ is of the form $1+2(T_{c0}-T)/(T_{c0}-T_{\text{BKT}})$. This is true except very near $T_{\text{BKT}}$, where the theory predicts a square root cusp of the form

$$a(T) = 3 + \pi[(T_{\text{BKT}} - T)/b(T_{c0} - T_{\text{BKT}})]^{1/2}. \quad (4.47)$$

The constant $b$ is determined from experiment and is non-universal. Values of $\eta(T)$ can be determined from measurements of $a(T)$. One expects a rapid crossover from nonlinear to linear characteristics at $T_{\text{BKT}}$ where $a(T) = 3$. This crossover is easily seen in Fig. 4.1. The curve at $T = 1.903$ K is the highest temperature at which no deviation from power-law behavior is seen at low current. This corresponds to curve d, which has a slope $a(T) = 3.281(3)$. For the lower temperature data, the straight-line portions below $10^{-5}$ V were used to obtain $a(T)$.

Figure 4.2 shows a plot of the temperature dependence of $a(T)$ with an extrapolated value of 3 occurring at $T_{cx} = 1.939$ K. At this temperature the voltage at small current is a linear function of voltage.

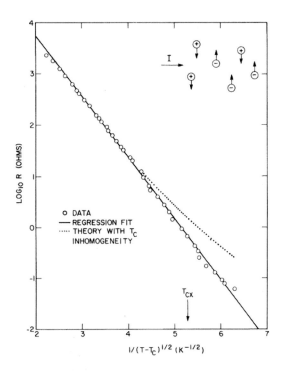

Fig. 4.3. Plot for $T > T_c$ of the logarithm of the resistance as a function of $(T - T_c)^{-1/2}$ where $T = 1.903$ K. This is the regime of flux flow resistance of thermally excited vortices, a process indicated schematically in the inset. The various fits are discussed in Ref. 16 from which this figure is derived.

The sharp crossover from linear to nonlinear serves to ascertain the transition temperature and confirm that it is associated with $a(T) = 3$. An extrapolation of $a(T)$ to unity in Fig. 4.2 gives the result that $T_{c0} = 2.206$ K is in reasonable agreement with the value $T_{c0} \cong 2.3$ K determined from a fit of the Aslamazov–Larkin theory of fluctuation conductivity to the resistive transition in the regime above $T_{c0}$.

In the presence of free vortices above the Berezinskii–Kosterlitz–Thouless transition the current–voltage characteristics are linear and are predicted to have the form

$$R = 10.8 b R_n \exp\{-2[b(T_{c0} - T_c)/(T - T_c)]^{1/2}\}. \qquad (4.48)$$

Here the parameter $b$ is the same as that in Eq. (4.47). The data plotted in Fig. 4.3 display the agreement with this functional form.

From this data set a parameterization can be completed. This is discussed in detail by Hebard and Fiory.[16] This set of data provides a fairly complete, *albeit* qualitative test of the theory developed by Halperin and

Nelson.[12] Fiory, Hebard and Glaberson have given a more complete account of further work, including finite frequency studies involving indium/indium oxide composites.[17]

## 4.4. Phenomenological Theory — Arrays

In parallel with work on thin films various investigators constructed large two-dimensional arrays of superconducting weak links to study the problem in its lattice version. The approach depends upon being able to prepare identical numbers of weak links with nearly identical characteristics, which is possibly easier than preparing an ultrathin film that is uniformly disordered. This involved investigations of Josephson junction arrays as well as proximity-effect coupled arrays, which are simpler to fabricate. The latter because of their relatively large dimensions can be reproduced. This problem has been treated in a comprehensive manner by Lobb, Abraham and Tinkham.[18] Some of the results are different from those developed earlier for the continuum case, but they may be similar at temperatures close to the BKT transition. Arrays turn out to be nearly ideal realizations of the X–Y model in two dimensions. Tunneling junction arrays have advantages over proximity coupled arrays because the current phase relationship of each element is clearly sinusoidal, which means that the array maps directly onto the X–Y model.

For a square array of superconducting islands connected by weak links, the Hamiltonian is

$$H = \sum_{\langle i,j \rangle} E_J(T)[1 - \cos(\phi_i - \phi_j)]. \qquad (4.49)$$

Here the argument of the cosine is the difference in the phases of the order parameters on nearest neighbor islands. The coupling energy $E_j(T)$ is associated with a super-current between nearest neighbor islands given by

$$i = i_c(T)\sin(\phi_i - \phi_j), \qquad (4.50)$$

where $i_c(T) = e^* E_J(T)/\hbar$ and $e^* = 2e$. In the absence of fluctuations $i_c(T)$ would be a critical current. It can be shown that the energy of a vortex in a region bounded by radii $r_1$ and $r_2$ is given by

$$U = \pi E_J \ln(r_2/r_1). \qquad (4.51)$$

This can be compared with a continuum calculation, which yields

$$U = n_s^*(T)\frac{\hbar^2}{m^*}\ln(r_2/r_1). \qquad (4.52)$$

Here $m^* = 2m$ and $n_s^*$ is the renormalized super-electron density. If Eqs. (4.51) and (4.52) are compared with the renormalized density replaced by the unrenormalized one, then

$$n_0^*(T) = m^* E_J(T)/\hbar^2 = m^* i_c(T)/\hbar e^* \,. \tag{4.53}$$

This illustrates one of the differences between a lattice model and a Ginzburg–Landau continuum model. In the continuum picture, the super-electron density is assumed to vary as $T_{c0} - T$ whereas in the lattice it will depend upon the temperature dependence of the critical currents of the junctions of the array, $i_c(T)$.

The expression for the penetration depth in perpendicular magnetic field can be written in the form

$$\lambda_\perp = m^* c^2 / 4\pi n_s^* e^{*2} \approx \Phi_0 c / 8\pi^2 i_c(T) \,. \tag{4.54}$$

For most weak link systems, this result will differ from the continuum thin film form. As an example for S–N–S junctions characterized by a length $d$ and a normal metal coherence length $\xi_N$,

$$i_c(T) \propto (1 - T/T_{cs})^2 e^{-d/\xi_N} \,. \tag{4.55}$$

Thus near the transition temperature of the islands, $T_{cs}$ the penetration depth will take a different form from that of a thin film. However, if the temperature dependence is that of an ideal tunnel junction, the standard dirty-limit result presented in the first section is obtained.

In the continuum thin film picture the transition temperature can be determined from

$$k_B T_c = \pi \hbar^2 n_s^*(T_c)/2m^* = \frac{\Phi_0^2}{32\pi^2} \frac{1}{\lambda_\perp(T_c)} \,. \tag{4.56}$$

This is the same as the thin film result, although it has not been assumed that the links are ideal tunnel junctions. This result demonstrates the universality of relations based on measured (renormalized) properties of the superconducting state.

It is useful to note the temperature dependence of the sheet resistance of an array above the BKT transition. The correlation length of an array of junctions is similar to the correlation length of the lattice XY model and is given by

$$\xi_+ = c_1 \exp\{[c_2/(T - T_c')]^{1/2}\} \,, \tag{4.57}$$

where $c_1$ and $c_2$ are the order of unity, and the primes over the temperature denote the dimension the temperature-dependent coupling parameter on which the system partition function depends. This is given by

$T' \equiv e^* k_B T / \hbar i_c(T)$. The correlation length is equal to the average distance between vortices. The prefactor is approximately equal to the lattice spacing.

The argument for the resistance in the regime above the transition is the same as that used for the continuum thin film case. The result is of the form

$$R = b_1 [I_0(E_J(T)/10 k_B T)]^{-2} \times \exp\{-[b_2/(T' - T'_c)]^{1/2}\} r_n. \quad (4.58)$$

Here $b_1$ is the order of unity, $b_2 \equiv 4 c_2$ and $r_n$ is the normal resistance of one link in the array. Also $I_0(x)$ is the hyperbolic Bessel function of order zero. Sufficiently close to the transition temperature where $E_J(T) \cong E_J(T_c) = $ const., this result is of the same form as that developed by Halperin and Nelson.[12]

$$R = b_3 \exp\{-[b_4 (T_{c0} - T_c)/(T - T_c)]^{-1/2}\} r_n. \quad (4.59)$$

Here $b_3$ and $b_4$ are the order of unity and $T_{c0}$ is the mean field transition temperature. Lobb, Abraham, and Tinkham go on to develop detailed expressions for the nonlinear regime.[18] The details can be found in their work. At small current $V \propto I^{a(T)}$ and $a$ jumps from 1 to 3 at the transition temperature.

These authors also develop translation relations between the continuum and lattice behaviors.

## 4.5. Experiments — Arrays

Several groups have carried out measurements on junction arrays, which consist of superconductor–insulator–superconductor (SIS),[19,20] superconductor–normal-metal–superconductor (SNS),[21] and superconductor–weaker-superconductor–superconductor (SS'S) junctions.[22,23] All of these samples are possible because optical lithography permits the fabrication of arrays of nearly identical junctions. The SNS and SS'S configurations exploit the proximity effect, which have the complication of a strong temperature-dependent coupling, but the advantage is that because of the large length scale of the lithography, there is a higher probability of smaller fluctuation in link-to-link coupling energy. The study of arrays obviates the criticism of transport measurements on films to the effect that some of the results may be influenced by inhomogeneity.[13]

In the work of Resnick et al.,[22] a 1000 Å thick layer of Sn was grown on an oxidized Si substrate. This film was then decorated with $10^6$ Pb disks in the form of a triangular lattice. These arrays were considerably larger than the similar structures of In dots on a Au/In plane studied by Sanchez

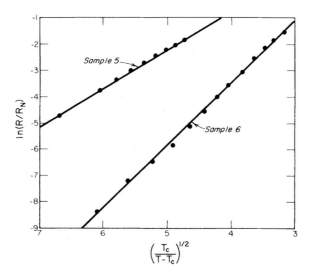

Fig. 4.4. Logarithmic plot of the resistance of two samples in the temperature range $T_c < T < T_{c0}$. For Sample 5, $T_c = 4.07$ K and for Sample 6, $T_c = 4.25$ K. (From Ref. 22.)

and Berchier.[23] The narrow layer of Sn between adjacent Pb disks forms a Pb–Sn–Pb junction (weak link). The Pb disks were 1300 Å thick and 13 microns in diameter. The distance of closest approach of adjacent Pb disks was 0.13 microns. The samples were $5 \times 20$ mm$^2$ and had normal state sheet resistances between 0.05 and 0.2 $\Omega/\square$. The resistive transitions of these samples exhibited the temperature dependence discussed for the regime where $T_c < T < T_{c0}$ as shown in Fig. 4.4.

The low-current $I$–$V$ characteristics, as shown in Fig. 4.5, are also consistent with the BKT interpretation. Above the mean field transition temperature they are strictly linear. Below, they increasingly acquire a nonlinear component. At $T_c$, the linear term is decreased to zero and the curves satisfy $V \propto I^3$ as is observed in other systems. Below $T_c$, the current exponent increases with decreasing temperature, but at a rate greater than predicted theoretically. This is clearly a consequence of the temperature dependence of the coupling as discussed by Abraham et al.[19]

Large proximity junction arrays of PbBi$_5$ at%/Cu/PbBi$_5$ at% with $\sim 10^6$ junctions were fabricated by Abraham et al.[21] The larger number of junctions relative to other experiments mitigates size effects. Also, in contrast with previously discussed arrays, Cu is a truly normal metal. This ensures that the transition to zero resistance is a consequence of inter-island coupling and is not a signature of the onset of superconductivity in the underlayer itself.

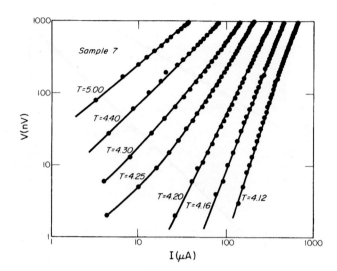

Fig. 4.5. Log–log plot of the voltage–current characteristics of one of the samples at temperatures just above $T_c$. At $T_{c0}$ the $I$–$V$ characteristic is linear and at $T_c$ the voltage $V \propto I^3$. (From Ref. 22.)

In this study, the conductance voltage characteristics far below the BKT temperature were dominated by noninteracting junction behavior. This enabled parameters such as the critical current and the normal coherence length to be determined for temperatures well below $T_c$, and using the theory, extrapolated into the regime near $T_c$ where interaction and renormalization effects are important. The resistive transition of a typical sample is shown in Fig. 4.6. The designator HN refers to the theory of Ref. 12. The proximity effect model also labeled in the figure is discussed in Abraham et al.[21] Below the transition temperature, the $I$–$V$ characteristics follow the form $V \propto I^{\alpha(T)}$. Here $\alpha$ is the same as the exponent $a$ used earlier for the exponent. Actually a plot of $\log V$ versus $\log I$ is nonlinear, even above the BKT temperature. The parameter $\alpha(T)$ is taken to be the exponent best fitting the low-current part of the curve. Its dependence on temperature is shown in Fig. 4.7.

The complication in the analysis, which was discussed in the previous section, is dealt with by Abraham et al.[21] It is the temperature dependence of the Josephson coupling energy. This is especially important for the regime above the BKT temperature. The work-around is to define an effective temperature $\tilde{T}$ in the following manner: $\tilde{T}/E_J(T_c) = T/E_J(T)$. When this analysis is carried out, the Halperin–Nelson form describes the data over two decades of resistance. The upshot of this work is that the resistive tran-

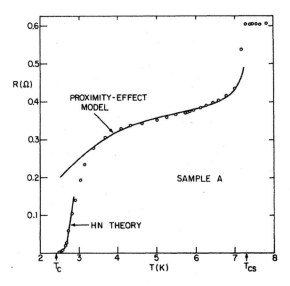

Fig. 4.6. Resistive transition of a proximity-effect junction array. The high-temperature part is described by a proximity-effect model, while the BKT theory for superconductors applies to the regime near but above $T_c$. (From Ref. 21.)

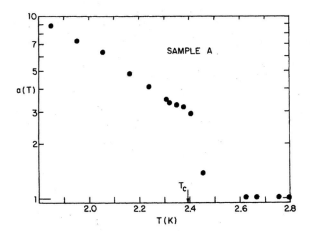

Fig. 4.7. Plot of $\alpha(T)$ for the sample of Fig. 4.6, where $V \propto I^{\alpha(T)}$. The transition temperature is defined implicitly by $\alpha(T_c) = 3$. (From Ref. 21.)

sition of proximity-junction arrays, suitably adjusted for the temperature dependence of the coupling energy, is in excellent agreement with the BKT theory for superconductors.

A third example of a BKT transition in an array structure is in the nonlinear voltage–current characteristic of true Josephson junction arrays.

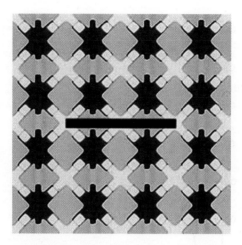

Fig. 4.8. Optical micro-photograph of a small section of the array. The junctions are found at the overlap of the Nb electrodes, which are the light-colored crosses. The scale bar is 80 microns long. (From Ref. 20.)

These are a nearly ideal realization the X–Y model. Gordon et al. fabricated square arrays of one million Josephson junction using the Selective Niobium Anodization Process (SNAP).[20] In zero magnetic field, there was clear evidence of the BKT transition in the power law dependence of the voltage on current in the vicinity of and immediately below the transition temperature. However, in the region above the transition, where the temperature-dependent resistance is determined by the free-vortex density, surprisingly, only activated behavior was observed. This data, which suggests unrenormalized vortex interactions will be discussed below. These 1000×1000 arrays were 2.54 cm × 2.54 cm in area with junction areas of $(2.5~\mu\text{m})^2$. Figure 4.8 is an optical micrograph of a small section of an array.

An exponential dependence of resistance on $1/T$ was observed near but above, the BKT transition temperature over nearly three decades of resistance. This implies that the vortices are thermally activated and independent. The free vortex density is then of the form:

$$n_f \propto e^{-U/k_B T} = e^{-A/\tau}. \tag{4.60}$$

Here $U = \pi E_J \ln N$ and $\tau = T/E_J$. From the I–V characteristics of the array at very low temperatures, the critical current can be determined and from (4.60) calculate $A = 22$. The fit of Eq. (4.60) to the data yielded $A = 24$. The normalized Halperin–Nelson form was expected near the transition temperature. Within the sensitivity of the measurements, this was not observed.

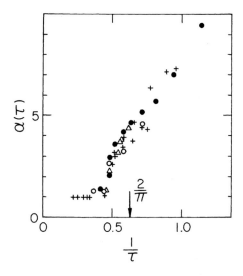

Fig. 4.9. Plot of the temperature dependence of the exponent from $I \propto V^{\alpha(T)}$. A jump is found at $\tau = \pi/2$. (From Ref. 20.)

On the other hand, the nonlinear features of the $I$–$V$ characteristic predicted by theory were found below the transition temperature. In particular, the universal jump from Ohmic behavior above the transition to $\alpha = 3$ just below $T_c$ was clearly observed as shown in Fig. 4.9.

The slopes in Fig. 4.9 were all measured at voltages less than 10 nV. The rather abrupt jump in $\alpha$ is of a magnitude consistent with the theoretical interpretation and is found at a temperature close to the theoretically expected $\tau = \pi/2$. The increase of the slope below the transition temperature is somewhat faster than what is expected from the predicted temperature dependence not including renormalization effects. The anomalous feature of this data, which was discussed above, is that the resistance versus temperature above the transition temperature is governed by an Arrhenius form suggesting *independent* free vortex excitations. More recent measurements carried out at lower measuring currents on similar arrays have resulted in data above the transition temperature consistent with the Halperin–Nelson formula.[24]

## 4.6. Comments on Renormalization Effects

The theory of topological phase transitions predicts a universal relation between the vortex unbinding transition temperature and the areal superelectron density at the transition, in two-dimensional films. The evaluation of the latter using the theory of dirty superconductors as done first by Beasley, Mooij, and Orlando[5] must be modified to account for the distinction between the normalized and unrenormalized densities in the critical region near the transition temperature.[25,26] This also applies to all of the other relationships discussed earlier. The renormalization can be expressed using a non-universal parameter, an effective dielectric constant, which depends upon the density of statistically independent vortices within an area given by the square of the coherence length.

The idea is that at low temperatures well below $T_c$ thermally excited bound vortex–antivortex pairs are rather rare. If the members of a pair are separated by a distance $r$, then the ineraction is logarithmic and is of the form

$$U(r) = 2\pi K_0 \ln(r/\xi), \qquad (4.61)$$

where $K_0 = n_s^0(T)\hbar^2/4m$, the Ginzburg–Landau coherence length $\xi(T)$ is the effective radius of the vortex core, $m$ is the electronic mass and $n_s^0(T)$ is the unrenormalized superelectron density per unit area.

However, close to the transition temperature, a given vortex–antivortex pair with separation $r \gg \xi$ is likely to contain between its members many similar, but smaller bound pairs, which act as a polarizable medium which partially screens the primary vortex–antivortex interaction. This can be expressed in terms of renormalized quantities

$$K(r) = n_s(r,T)\hbar^2/4m = K_0/\varepsilon(r,T), \qquad (4.62)$$

where

$$U(r) = \int_\xi^r 2\pi K(r')d(\ln r'). \qquad (4.63)$$

The quantities $K(r)$, $n_s(r,T)$, and the effective dielectric constant $\varepsilon(r,T)$ all incorporate the effects of vortex pairs with separation $\leq r$. They have renormalized these vortex pairs out of the problem. If $K^R \equiv K(r \to \infty)$ and $n_s^R \equiv n_s(r \to \infty)$ are the fully renormalized quantities, then the vortex unbinding temperature is determined by the universal relation

$$\pi K^R(T_c) = \pi \hbar^2 n_s^R/4m = 2k_B T_c \qquad (4.64)$$

since unbinding will occur first for vortices with separation $r \to \infty$. The issue is then the determination of $\varepsilon(r)$. The appropriate approach and its influence on the details of the parameters of the theory are discussed in many places. The actual value of the vortex dielectric constant can be inferred by an analysis of the full set of current–voltage characteristics in the critical region of the transition.[25,26]

## 4.7. Summary

This chapter has focused on discussions of the earliest and simplest experimental evidence for the BKT transition in superconducting films and junction arrays. As a consequence, there are many topics that have not been addressed. These include a rigorous presentation of the theory rather than a largely heuristic discussion. For the interested reader, there is a comprehensive review article on this subject focusing on a two-dimensional Coulomb gas model which can be consulted.[7] We have also not discussed dynamical effects which have been measured[22,27] and which are the subject of current research. Matters relating to finite size effects have also not been discussed.[28] There are experiments related to the BKT transition in high temperature superconductors which have also been omitted.[29] Another major omission is the consideration of effects associated with superconductors in magnetic fields, such as vortex lattice melting, which is also a BKT transition. Phenomena associated with magnetic fields have played an important role in the understanding of the magnetic properties of cuprate superconductors. A key to the literature in this field is the extensive review by Blatter *et al.*[30]

## Acknowledgments

This work has been supported by the Condensed Matter Physics Program of the National Science Foundation under various grants. The author would like to thank Alan Kadin and Kenneth Epstein, who were his collaborators during the period in which research in this area was being pursued actively. This chapter is dedicated to the later James Gordon, who was also a collaborator.

## References

1. J. M. Kosterlitz and D. J. Thouless, *J. Phys. C: Solid State Physics* **5**, L124 (1972).
2. J. M. Kosterlitz and D. J. Thouless, *J. Phys. C: Solid State Physics* **6**, 1181 (1973).
3. V. L. Berezinskii, *Zh. Eksp. Teor. Fiz.* **61**, 1144 (1971) [*Sov. Phys. JETP* **34**, 610 (1972)].
4. J. Pearl, *Appl. Phys. Lett.* **5**, 65 (1964).
5. M. R. Beasley, J. E. Mooij and T. P. Orlando, *Phys. Rev. Lett.* **42**, 1165 (1979).
6. J. E. Mooij, in *Percolation, Localization and Superconductivity*, eds. A. M. Goldman and S. A. Wolf (Plenum Press, New York and London, 1984), p. 325.
7. P. Minnhagen, *Rev. Mod. Phys.* **59**, 1001 (1987).

8. S. Girvin, Lecture notes on the Berezinskii–Kosterlitz–Thouless Transition, unpublished.
9. D. R. Nelson and J. M. Kosterlitz, *Phys. Rev. Lett.* **39**, 1201 (1977).
10. M. Tinkham, *Introduction to Superconductivity* (McGraw-Hill, New York, 1975), p. 80.
11. N. D. Mermin and H. Wagner, *Phys. Rev. Lett.* **17**, 1133 (1966).
12. B. I. Halperin and D. R. Nelson, *J. Low Temp. Phys.* **36**, 599 (1979); Also see: S. Doniach and B. Huberman, *Phys. Rev. Lett.* **42**, 1169 (1979).
13. P. A. Bancel and K. E. Gray, *Phys. Rev. Lett.* **46**, 148 (1981).
14. S. A. Wolf, D. U. Gubser, W. W. Fuller, J. C. Garland and R. S. Newrock, *Phys. Rev. Lett.* **47**, 1071 (1981).
15. K. Epstein, A. M. Goldman and A. M. Kadin, *Phys. Rev. Lett.* **47**, 534 (1981).
16. A. F. Hebard and A. T. Fiory, *Phys. Rev. Lett.* **50**, 1603 (1983).
17. A. T. Fiory, A. F. Hebard and W. I. Glaberson, *Phys. Rev. B* **28**, 5075 (1983).
18. C. J. Lobb, D. W. Abraham and M. Tinkham, *Phys. Rev. B* **27**, 150 (1983).
19. R. F. Voss and R. A. Webb, *Phys. Rev. B* **25**, 3446 (1982).
20. J. M. Gordon, A. M. Goldman, M. Bhushan and R. H. Cantor, *Jpn. J. Appl. Phys.* **26**, Supplement 26-2, 1425 (1987).
21. D. W. Abraham, C. J. Lobb, M. Tinkham and T. M. Klapwijk, *Phys. Rev. B* **26**, 5268 (1982).
22. D. J. Resnick, J. C. Garland, J. T. Boyd, S. Shoemaker and R. S. Newrock, *Phys. Rev. Lett.* **47**, 1542 (1981).
23. D. H. Sanchez and Jena-Luc Berchier, *J. Low Temp. Phys.* **43**, 65 (1981).
24. I. Elsayed, K. H. Sarwa, B. Tan, K. A. Parendo and A. M. Goldman, *Phys. Rev. B* **80**, 052502 (2009).
25. A. M. Kadin, K. Epstein and A. M. Goldman, *Phys. Rev. B* **27**, 6691 (1983).
26. K. Epstein, A. M. Goldman and A. M. Kadin, *Phys. Rev. B* **26**, 3950 (1982).
27. A. F. Hebard and A. T. Fiory, *Phys. Rev. Lett.* **44**, 291 (1980).
28. S. Pierson and O. T. Valls, *Phys. Rev. B* **61**, 663 (2004).
29. D. H. Kim, A. M. Goldman, J. H. Kang and R. T. Kampwirth, *Phys. Rev. B* **40**, 8834 (1989).
30. G. Blatter, M. V. Feigel'man, V. B. Geshkenbein, A. I. Larkin and V. M. Vinokur, *Rev. Mod. Phys.* **66**, 1125 (1994) and references cited therein and citing articles.

Chapter 5

# Berezinskii–Kosterlitz–Thouless Transition within the Sine-Gordon Approach: The Role of the Vortex-Core Energy

Lara Benfatto[*,†] and Claudio Castellani[*,†]

[*]*Institute for Complex Systems (ISC), CNR, U.O.S. Sapienza, P.le A. Moro 2, 00185 Rome, Italy*
[†]*Department of Physics, Sapienza University of Rome, P.le A. Moro 2, 00185 Rome, Italy*

Thierry Giamarchi

*DPMC-MaNEP University of Geneva, 24 Quai Ernest-Ansermet CH-1211 Genève 4, Switzerland*

One of the most relevant manifestations of the Berezinskii–Kosterlitz–Thouless transition occurs in quasi-two-dimensional superconducting systems. The experimental advances made in the last decade in the investigation of superconducting phenomena in low-dimensional correlated electronic systems have raised new questions on the nature of the BKT transitions in real materials. A general issue concerns the possible limitations of theoretical predictions based on the XY model, that was studied as a paradigmatic example in the original formulation. Here, we review the work we have done in revisiting the nature of the BKT transition within the general framework provided by the mapping into the sine-Gordon model. While this mapping was already known since long, we recently emphasized the advantages of such an approach to account for new variables in BKT physics. One such variable is the energy needed to create the core of the vortex, that is fixed within the XY model, while it attains substantially different values in real materials. This has interesting observable consequences, especially in the case when additional relevant perturbations are present, as a coupling between stacked two-dimensional superconducting layers or a finite magnetic field.

## 5.1. Introduction

Almost 40 years after the pioneering work by Berezinskii[1] and Kosterlitz and Thouless[2,3] the Berezinskii–Kosterlitz–Thouless (BKT) transition is still the subject of an intense experimental and theoretical research. On very general grounds, what makes it so fascinating is the possibility of a phase transition that is not driven by the explicit breaking of a given symmetry, but is based on the emergence of a finite (and measurable) rigidity of the system. The BKT transition was originally formulated within the context of the two-dimensional (2D) XY model, which describes the exchange interaction between classical two-component spins with fixed length $S = 1$:

$$H_{XY} = -J \sum_{\langle ij \rangle} \cos(\theta_i - \theta_j), \tag{5.1}$$

where $J$ is the spin–spin coupling constant and $\theta_i$ is the angle that the $i$th spin forms with a given direction, and $i$ are the sites of a square lattice. This model admits a continuous $U(1)$ symmetry (encoded in the transformation $\theta_i \to \theta_i + \chi$), that cannot be broken at finite temperature by an average magnetization $\langle \mathbf{S} \rangle = \langle e^{i\theta} \rangle$ different from zero because of the Mermin–Wagner theorem. Nonetheless, the system can become "stiff" at low temperature with respect to fluctuations of the $\theta$ variable, leading to a power-law decay of the spin correlation functions, i.e. $\langle e^{i(\theta(r)-\theta(0))} \rangle \simeq r^{-T/2\pi J}$, in contrast to the exponential one expected in the truly disordered state. Such a change of behavior cannot be "smooth", i.e. a phase transition occurs in between, which appears to be controlled by the emergence of vortex-like excitations. The original argument used in Ref. 2 to capture the temperature scale of such a transition is rather intuitive: in two dimensions both the energy $E$ and the entropy $S$ of a single vortex excitation depend logarithmically on the size $L$ of the system, so that the free energy reads:

$$F = E - TS = (\pi J - 2T) \ln \frac{L}{a}. \tag{5.2}$$

As a consequence, at temperatures larger than

$$T_{\text{BKT}} \simeq \frac{\pi J}{2}, \tag{5.3}$$

free vortices start to proliferate and destroy the quasi-long-ranged order of the correlation functions.

As it was observed already in the original papers,[2,3] several physical phenomena are expected to belong to the same universality class than the XY model (5.1), as for example, the superfluid transition in two dimensions.

Later on, it was realized that the same idea can be applied also to superconducting (SC) thin films,[4,5] even though in the charged superfluid the logarithmic interaction between vortices is screened by the supercurrents at a finite distance $\Lambda = \lambda^2/d$, where $\lambda$ is the penetration depth of the magnetic field and $d$ is the film thickness.[6] In practice, for sufficiently thin films with large disorder (so that $\lambda$ is also large) the electromagnetic screening effects are weak enough to expect the occurrence of the BKT transition.

As a matter of fact, the case of SC thin films represented one of the most studied applications of the BKT physics. In principle, in this case, one has also several possibilities to access experimentally the specific signatures of the BKT physics. For example, by approaching the transition from below, the superfluid density $n_s$ is expected to go to zero discontinuously at the BKT temperature $T_{BKT}$, with a "universal" relation between $n_s(T_{BKT})$ and $T_{BKT}$ itself,[7–9] that is, the equivalent of the relation (5.3), since $J \propto n_s$ (see Eq. (5.5) below). Approaching instead the transition from above, one has in principle the possibility to identify the BKT transition from the temperature dependence of the SC fluctuations. Indeed, in 2D the temperature dependence of both the paraconductivity $\Delta\sigma \equiv \sigma - \sigma_n$ and the diamagnetism $\chi_d$ is encoded in the SC correlation length $\xi(T)$,[5] which increases approaching the transitions due to the increase of SC fluctuations:

$$\Delta\sigma \propto \xi^2(T), \quad \chi_d \propto -\xi^2(T). \tag{5.4}$$

Within BKT theory $\xi(T)$ diverges exponentially at $T_{BKT}$ as $\xi_{BKT}(T) \simeq ae^{b/\sqrt{t}}$,[3,8] where $t = T/T_{BKT} - 1$, in contrast to the power-law $\xi_{GL}^2(T) \simeq 1/(T - T_c)$ expected within Ginzburg–Landau (GL) theory.[10] As a consequence, by direct inspection of the paraconductivity or diamagnetism near the transition, one could identify the occurrence of vortex fluctuations, which lead to an exponential temperature dependence of the SC correlation length.

Quite interestingly, very direct measurements of the BKT universal jump in the superfluid density of SC films became available only recently,[11–16] due to the improvement of the experimental techniques, triggered mostly by the investigation of high-temperature superconductors in the late nineties. In particular, the use of the two-coil mutual inductance technique[17] turned out to be crucial to obtain the absolute value of the superfluid density at zero temperature, which is needed to compare the experimental data with the BKT predictions. Recently a great deal of information has come also from Tera-hertz spectroscopy,[18–20] which probes the finite-frequency analogue to the superfluid-density jump. At the same time, in the last decade new 2D or quasi-2D SC systems emerged where the BKT transition is expected to occur. To this category belong, for example, the nanometer-thick layers of SC electron systems formed at the interface between artificial heterostructures made of insulating oxides as $LaAlO_3/SrTiO_3$[21,22] or $LaTiO_3/SrTiO_3$,[23] or at the liquid/solid interface of field-effect transistors made with organic electrolytes.[24] A second remarkable example is provided by layered 3D systems as cuprate high-temperature superconductors, where the weak interlayer coupling makes it plausible that at least in the same regions of the phase diagram a BKT transition could be at play.[25,26] Even though the existence of a BKT transition in bulk (3D) samples is still controversial, as we shall discuss below, nonetheless, in cuprates the proximity of the SC phase to the Mott insulator leads to a large penetration depth without the need of introducing strong disorder, making in principle thin-films of cuprate superconductors the best candidate to study BKT physics.[12,13,16,20,27,28] Recently, much attention has been devoted also to artificial heterostructures made of cuprates at different doping level,[29,30] where the observation of BKT physics in the SC films is complicated even more by the proximity to the non-SC correlated insulator.

A common characteristic of the cases mentioned above is that the BKT transition is expected to occur in systems where electronic correlations are not necessarily in the weak-coupling limit. This can be due to the presence of strong disorder, as it is the case for thin disordered films of conventional superconductors, to the artificial spatial confinement, as in the SC interfaces, or to the intrinsic nature of the system, as it occurs in cuprate superconductors. As a consequence, several experimental results seem to point towards a kind of "unconventional" BKT physics, which needs to be addressed using a wider perspective than the one proposed in the original formulation. A paradigmatic example of an apparent failure of the standard BKT approach is posed by recent measurements of the universal superfluid-density jump in $InO$[18,19] and $NbN$[14,15] films. In a quasi-2D superconductor of thickness $d$ the energy scale corresponding to the coupling $J$ of the XY model (5.1) is the so-called superfluid stiffness, which is connected to the (areal) density of superfluid electrons $\rho_s^{2d} \equiv n_s d$, which in turn is measured via the inverse penetration depth $\lambda$ of the magnetic field:

$$J = \frac{\hbar^2 \rho_s^{2d}}{4m} = \frac{\hbar^2 c^2 d}{16\pi e^2 \lambda^2}. \qquad (5.5)$$

This coupling has itself a bare temperature dependence $J(T)$ due to the presence of quasiparticle excitations: however, one would expect that at a temperature scale corresponding to the relation (5.3) free vortices start to proliferate, so that $n_s(T)$ jumps discontinuously to zero. In the experiments of Refs. 14 and 15 one can clearly see that as the film thickness decreases $n_s(T)$ starts to deviate abruptly from its BCS temperature dependence. However, such a deviation seems to occur at a temperature *lower* than the one predicted by Eq. (5.3). The same observation holds for finite-frequency measurements of $n_s(T)$ in InO films,[18,19] casting some doubt on that "universal" relation between the superfluid density and the critical temperature that is one of the hallmarks of the BKT transition.

It is worth noting that while the measurement of the superfluid-density behavior gives access to the most straightforward manifestation of BKT physics, its identification via SC fluctuations is much more subtle. Indeed, according to the general result (5.4), one needs in this case a controlled procedure to first extract the SC fluctuations contribution, and then to fit it with the BKT expression for the correlation length. Such a procedure is, in general, applied to the paraconductivity, even though much care should be used to disentangle GL from BKT fluctuations, as it has been discussed in a seminal paper long ago by Halperin and Nelson[5] (HN). Indeed, the BKT fit should be applied only in the region between the BCS mean-field temperature $T_c$ and the true $T_{BKT}$, that do not differ considerably in thin films of conventional superconductors. In contrast, recent applications to paraconductivity measurements in thin films of cuprate superconductors[28] or in SC interfaces[21,31] seem to suggest that in these systems the whole fluctuation regime above $T_{BKT}$ is dominated by BKT vortex fluctuations, and deviations only occur near the transition because of finite-size effect. Also in this case, one would like to distinguish unconventional effects due possibly to the nature of the underlying system, from spurious results due to an incorrect application of BKT theory. As we discussed recently in Ref. 32 a BKT fit of the paraconductivity must be done taking into account from one side the existence of unavoidable constraints on the values of the fitting parameters, and from the other side the existence of inhomogeneity on mesoscopic scales, that can partly mask the occurrence of a sharp BKT transition. A very interesting example of application of such a procedure has been recently provided by NbN thin films,[15] where the direct comparison between superfluid-density data below $T_{BKT}$ and resistivity data above $T_{BKT}$ provided a paradigmatic example of BKT transition in a real system.

The issue of the inhomogeneity emerges also within the context of high-temperature cuprate superconductors. It is worth noting that in the literature, the discussion concerning the occurrence of BKT physics in these layered anisotropic superconductors has been often associated with a somehow related issue, i.e. the nature of the pseudogap state above $T_c$. Indeed, despite the intense experimental and theoretical research devoted to it, no consensus has been reached yet concerning its origin, with two main lines of interpretation based either on a preformed-pair scenario, or on the existence of a competing order, associated with fluctuations in the particle-hole channel.[26] The preformed-pair scenario has in turn triggered the attention on the role of SC phase fluctuations, which are expected to be very soft in these materials having a small superfluid density. Finally, the experimental observation of a large Nerst effect[33–35] and diamagnetism[35–37] well above $T_c$ has been used to support the notion that phase fluctuations have a vortex-like character, as it is indeed the case within the BKT picture. However, the overall interpretation of the experimental data in terms of BKT physics is not so straightforward, despite several theoretical attempts based both on the mapping into the Coulomb-gas problem[38,39] or on numerical simulations for the XY model.[40,41] On the other hand, a large Nerst effect arises also from the Fermi-surface reconstruction associated with stripe order,[42] or from ordinary GL fluctuations,[43] as observed, for example, in thin films of conventional superconductors.[44] Moreover, while the SC fluctuations contribution to the diamagnetism seems to fit the BKT behavior of the SC correlation length,[36] the paraconductivity shows usually more direct evidence of GL fluctuations.[45,46] Also the issue of the superfluid-density jump is controversial: while it has been clearly identified in very thin films,[13,16,20] it seems to be absent in bulk materials even for highly underdoped samples,[47] where the low superfluid density and the weak interplane coupling would make more plausible the presence of BKT physics. The aim of this paper is not to give a detailed overview of all the arguments in favor or against the occurrence of the BKT transition in cuprates, but to focus on the correct identification of the BKT signatures in a non-conventional quasi-2D superconductor. In general, a layered system with very low in-plane superfluid density and very weak interlayer coupling is one of the best candidates to observe those signatures of BKT physics that we mentioned above, i.e. a rapid downturn of the superfluid density coming from below $T_c$ and a regime of vortex-like excitations above $T_c$. However, since the underlying superconductor is an unconventional one, the occurrence of BKT physics could be masked by other effects, making its identification more subtle than in films of conventional superconductors. We notice that addressing this issue does not solve the more general problem of the nature of the pseudogap phase: for example, to estimate the extension in temperature of the regime of BKT fluctuations above $T_c$, one needs to consider physical ingredients that are beyond the BKT problem addressed here. Nonetheless, a deeper understanding of what could be the signatures of BKT physics can help discriminating its occurrence or not in unconventional superconductors.

On the light of the above discussion, we will review in the present manuscript, the work we have done in the last years to investigate the outcomes of the BKT transition in the presence of some additional ingredients that influence its occurrence in real systems without invalidating the basic physical picture behind it. The first ingredient that we will consider is the role played by the vortex-core energy $\mu$, both in quasi-2D systems and in layered ones. As we shall see, while in the original formulation of BKT theory based on the XY model (5.1) $\mu$ is just a fixed constant times the coupling $J$, in real materials $\mu/J$ can depend crucially on the microscopic nature of the underlying system. Since it represents the energy scale needed to create the core of the vortex, having "cheap" or "expensive" vortices can influence in a non-trivial way the tendency of vortex formation below and above the transition. Thus, while in 2D the critical behavior will not change, in the layered 3D case the existence of expensive vortices can move the vortex-unbinding transition away from the temperature where it would occur in each (uncoupled) layer. The second aspect that we will discuss is the presence of inhomogeneity on a mesoscopic scale. This issue is somehow related to the effect of disorder on the BKT transition: however, instead of considering a model of microscopic disorder, we will implement a simpler approach where the spatial inhomogeneity of the superfluid density can be mapped in a probability distribution of the possible realizations of the superfluid-density values. This issue is in part motivated by several experimental[48–54] and theoretical[55–58] suggestions that inhomogeneity on a mesoscopic scale can occur both in highly-disordered films of conventional superconductors[49–54] and in layered cuprate superconductors,[48] making then timely to investigate its effect on the BKT transition. Finally, we will discuss the role of a finite external magnetic field, an issue that is strictly related to the peculiarity of the BKT transition in a charged superfluid. In this case, the motivation comes in part from recent experiments in cuprate superconductors, where anomalous nonlinear magnetization effects have been reported[34–36] even for those samples[36] where apparently clear signatures of BKT physics appear. However, the main focus here is to establish a clear theoretical framework to deal with this complicated problem, based on the mapping into the sine-Gordon model. This last methodology aspect will serve as a general guideline for the present review paper. Indeed, we shall argue in the present manuscript that the mapping between the BKT transition in two dimensions and the quantum phase transition within the sine-Gordon model in one spatial dimension is the most powerful approach to explore the outcomes of BKT physics beyond the standard results based on the original XY model. In particular, such a mapping provides us with a straightforward framework to explore the role of arbitrary values of $\mu/J$, to include the effects of relevant perturbations (in the RG sense) as the interlayer coupling or the magnetic field, and to account at a basic level for the presence of inhomogeneities. Due to the several subtleties of such a mapping, we shall first review the basic steps of its derivation in Sec. 5.2, taking the point of view of the longitudinal *versus* transverse current decoupling in the XY model. Once the general formalism is established and the role of the vortex-core energy is clarified we shall address in Sec. 5.3 the consequences for the universal versus non-universal behavior of the superfluid density. In relation to the superfluid-density and paraconductivity behavior, we shall give in Sec. 5.4 a short account about the role of inhomogeneity and its observation in recent experiments in several systems. Finally, in Sec. 5.5, we give a detailed derivation of the sine-Gordon mapping in the presence of an external magnetic field, that completes our overview on the approach to BKT physics in a real superconductor, and we discuss only briefly some physical outcomes related to the previous sections. The concluding remarks are reported in Sec. 5.6.

## 5.2. Mapping on the Sine-Gordon Model and the Vortex-Core Energy

As it is well known, the BKT transition occurs in three different physical phenomena, that belong to the same universality class: the vortex-unbinding transition within the XY model (5.1), the charge-unbinding transition in the 2D Coulomb gas, and the quantum metal-insulator transition in the 1D Luttinger liquid, as described by the sine-Gordon model. In the first two cases, we are dealing with a *classical* model of point-like objects (vortices or charges) interacting via a logarithmic potential, that becomes short-ranged when the objects are free to move and to screen the interaction. In the latter case, we deal with a quantum 1D model, that becomes effectively a 2D one at $T = 0$ where dynamic degrees of freedom provide the extra dimension. Even though all these analogies have been reviewed several times in the literature (see for example, Refs. 8, 9 and 59, just to mention a few references), nonetheless, we will recall in this section the main steps of the mapping between the classical XY model and the quantum 1D sine-Gordon model. We shall take as a starting point of view the separation between longitudinal and transverse excitations of the phase, and we shall derive as an intermediate step the mapping on the Coulomb-gas problem, to make the physical aspects of the problem more evident. Instead in Sec. 5.5, where we discuss the case of a finite magnetic field, we shall use a more formal approach, that has however the great advantage to provide us with an elegant and powerful formalism to discuss the case of a superconductor embedded in an external electromagnetic potential.

As a starting point, we shall consider a low-temperature limit for the XY model (5.1), where one could expect that the difference in angle between neighboring spins varies very slowly on the scale $a$ of the lattice, so that one can approximate $\theta_i - \theta_{i+\hat{\delta}} \approx a\partial\theta(\mathbf{r})/\partial\hat{\delta}$ where $\theta(\mathbf{r})$ is a smooth function and $\hat{\delta} = x, y$. Moreover, by retaining the leading powers in the phase differences from the cosine in Eq. (5.1), we find that in the low-temperature phase the model reduces to:

$$H_{XY} = \frac{J}{2}\int d\mathbf{r}(\nabla\theta)^2 = \frac{J}{2}\int d\mathbf{r}\,\mathbf{j}^2(\mathbf{r}), \qquad (5.6)$$

where we introduced, in analogy with the case of the superfluid, a current proportional to the phase gradient, $\mathbf{j} = \nabla\theta$. Because of the smoothness assumption, the approximation (5.6) accounts only for the longitudinal component $\mathbf{j}_\parallel$ of the current, while vortices, i.e. singular configurations of the phase, can be associated to a transverse current component $\mathbf{j}_\perp$. Indeed, a vortical configuration for the phase $\theta_i$ of the XY model (5.1) corresponds to a non-vanishing circuitation of the phase gradient along a closed line, that is nonzero only for a transverse current $\mathbf{j}_\perp$:

$$\oint \mathbf{j}\cdot d\boldsymbol{\ell} = \int_S (\nabla\times\mathbf{j})\cdot d\mathbf{s} = \int_S (\nabla\times\mathbf{j}_\perp)\cdot d\mathbf{s} = 2\pi\sum_i q_i, \qquad (5.7)$$

where $q_i = \pm m$ is the vorticity of the $i$th vortex, with $m$ integer. We can then decompose, in general, the current of Eq. (5.6) as $\mathbf{j} = \mathbf{j}_\parallel + \mathbf{j}_\perp$, where $\nabla\times\mathbf{j}_\parallel = 0$ and $\nabla\cdot\mathbf{j}_\perp = 0$. One can easily see that the mixed terms $\int d\mathbf{r}\,\mathbf{j}_\parallel\cdot\mathbf{j}_\perp = 0$ in Eq. (5.6) vanish, so that longitudinal and transverse degrees of freedom decouple $H = H_\parallel + H_\perp$, and we can focus on the term $H_\perp = (J/2)\int d\mathbf{r}\,\mathbf{j}_\perp^2$ to describe the interaction between vortices. By introducing a scalar function $W$ the transverse current can be written as $\mathbf{j}_\perp = \nabla\times(\hat{z}W) = (\partial_y W, -\partial_x W, 0)$, so that $\nabla\times\mathbf{j}_\perp = (0, 0, -\nabla^2 W)$ and inserting it into Eq. (5.7) we conclude that $W$ must satisfy the equation:

$$\nabla^2 W(\mathbf{r}) = -2\pi\rho(\mathbf{r}), \quad \rho(\mathbf{r}) = \sum_i q_i \delta(\mathbf{r}-\mathbf{r}_i). \qquad (5.8)$$

Equation (5.8) is exactly the Poisson equation in 2D for the potential $W$ generated by a distribution of point-like charges $q_i$ at the positions $\mathbf{r}_i$. Its solution, is in general:

$$W(\mathbf{r}) = 2\pi \int d\mathbf{r}' V(\mathbf{r} - \mathbf{r}')\rho(\mathbf{r}'), \qquad (5.9)$$

where $V(\mathbf{r})$ is the solution of the homogeneous equation for the Coulomb potential in 2D

$$\nabla^2 V(\mathbf{r}) = -\delta(\mathbf{r}) \Rightarrow V(\mathbf{r}) = \int \frac{d\mathbf{k}}{(2\pi)^2} \frac{e^{i\mathbf{k}\cdot\mathbf{r}}}{\mathbf{k}^2}, \qquad (5.10)$$

so that $V(r) \simeq -\ln r$ at large distances. Thanks to the results (5.8) and (5.9) $H_\perp$ can be written as:

$$H_\perp = \frac{J}{2}\int d\mathbf{r}\, \mathbf{j}_\perp^2 = \frac{J}{2}\int d\mathbf{r}(\nabla \times \hat{z}W)^2 = \frac{J}{2}\int d\mathbf{r}(\nabla W)^2 = -\frac{J}{2}\int d\mathbf{r}\, W\nabla^2 W$$

$$= \pi J \int d\mathbf{r}\, W(\mathbf{r})\rho(\mathbf{r}) = 2\pi^2 J \int d\mathbf{r}d\mathbf{r}'\rho(\mathbf{r})V(\mathbf{r}-\mathbf{r}')\rho(\mathbf{r}')$$

$$= 2\pi^2 J \sum_{ij} q_i q_j V(\mathbf{r}_i - \mathbf{r}_j). \qquad (5.11)$$

Equation (5.11) expresses the electrostatic energy for a Coulomb gas with charge density $\rho(\mathbf{r})$, completing thus the analogy between the system of vortices and the system of charges. The 2D Coulomb potential (5.10) shows the characteristic infrared divergence which reflects on the divergence of $V(\mathbf{r} = 0)$ in the thermodynamic limit, leading to the neutrality constraint for the gas. Indeed, by close inspection of Eq. (5.10) one sees that $V(\mathbf{r} = 0) = \int_{1/L} dk(1/k) \sim \ln L \to \infty$ as $L \to \infty$. If we separate this divergent term by defining

$$V(\mathbf{r}) = V(0) + G(\mathbf{r}), \qquad (5.12)$$

where now $G(r=0) = 0$, in Eq. (5.11), we obtain:

$$2\pi^2 J \sum_{ij} q_i q_j [V(0) + G(\mathbf{r}_i - \mathbf{r}_j)]$$

$$= 2\pi^2 JV(0)\left(\sum_i q_i\right)^2 + 2\pi^2 J \sum_{ij} q_i q_j G(\mathbf{r}_i - \mathbf{r}_j). \qquad (5.13)$$

Since the Boltzmann weight of each configuration is $e^{-\beta H_\perp}$, the divergence of $V(0)$ in the thermodynamic limit leads to a vanishing contribution to the partition function, unless

$$\sum_i q_i = 0, \qquad (5.14)$$

which means that only neutral configurations are allowed. A second consequence of the above discussion is that one should include a cut-off for the smallest possible distance between two vortices. Starting from the lattice XY model (5.1) a natural cut-off is provided by the lattice spacing $a$ in the original model, which translates to the correlation length $\xi_0$ when applied to SC systems. The exact form of the function $G(\mathbf{r})$ at short distances defines then the energetic cost of a vortex on the smallest scale of the system, i.e. the so-called vortex-core energy. By computing $G(\mathbf{r})$ on the lattice (see also Eq. (5.60) below), one can see that at distances $r \geq a$ it can be well approximated as

$$G(r) \simeq -\frac{1}{4} - \frac{1}{2\pi} \ln\left(\frac{r}{a}\right). \tag{5.15}$$

Using the neutrality condition (5.14), the fact that $G(0) = 0$ (so that in the last term of Eq. (5.13) one can use $i \neq j$) and the form (5.15), Eq. (5.11) can be written as:

$$H_\perp = 2\pi^2 J \sum_{i \neq j} q_i q_j G(\mathbf{r}_i - \mathbf{r}_j) = -2\pi^2 J \sum_{i \neq j} \left[\frac{1}{4} + \frac{1}{2\pi} \ln\left(\frac{r_{ij}}{a}\right)\right] q_i q_j$$

$$= -\frac{\pi^2 J}{2} \sum_{i \neq j} q_i q_j - \pi J \sum_{i \neq j} \ln\left(\frac{r_{ij}}{a}\right) q_i q_j$$

$$= \mu \sum_i q_i^2 - \pi J \sum_{i \neq j} \ln\left(\frac{r_{ij}}{a}\right) q_i q_j, \tag{5.16}$$

where we used $\sum_{i \neq j} q_i q_j = -\sum_i q_i^2$ from Eq. (5.14) and we identified the vortex-core energy $\mu$ with

$$\mu = \mu_{XY} \equiv \frac{\pi^2 J}{2}. \tag{5.17}$$

Finally, we can use the neutrality condition (5.14) by imposing that there are $n$ pairs of vortices of opposite vorticity. Moreover, we shall consider in what follows, only vortices of the lower vorticity $q_i = \pm 1$, so that $H_\perp$ reads:

$$H_\perp = 2n\mu - \pi J \sum_{i \neq j}^{2n} \ln\left(\frac{r_{ij}}{a}\right) \varepsilon_i \varepsilon_j, \quad \varepsilon_i = \pm 1. \tag{5.18}$$

Equation (5.18) describes the interaction between vortices in a given configuration with $n$ vortex pairs. In the partition function of the system, we must consider all the possible values of $n$, taking into account that interchanging the $n$ vortices with same vorticity gives the same configuration (so one should divide by a factor $1/(n!)^2$. In conclusion, $Z$ reads:

$$Z = \sum_{n=1}^{\infty} \frac{1}{(n!)^2} \int d\mathbf{r}_1 \cdots d\mathbf{r}_{2n} e^{-\beta 2n\mu} e^{\pi \beta J \sum_{i \neq j}^{2n} \ln\left(\frac{r_{ij}}{a}\right) \varepsilon_i \varepsilon_j}$$

$$= \sum_{n=1}^{\infty} \frac{1}{(n!)^2} y^{2n} \int d\mathbf{r}_1 \cdots d\mathbf{r}_{2n} e^{\sum_{i<j}^{2n} 2\pi \beta J \ln\left(\frac{r_{ij}}{a}\right) \varepsilon_i \varepsilon_j}, \quad (5.19)$$

where we introduced the vortex fugacity

$$y = e^{-\beta \mu}. \quad (5.20)$$

The explicit derivation of the partition function $Z$ has the great advantage to allow us to recognize immediately the analogy with the quantum sine-Gordon model, defined by the Hamiltonian:

$$H_{\text{sg}} = \frac{v_s}{2\pi} \int_0^L dx \left[ K(\partial_x \theta)^2 + \frac{1}{K}(\partial_x \phi)^2 - \frac{2g_u}{a^2} \cos(2\phi) \right], \quad (5.21)$$

where[9] $\theta$ and $\partial_x \phi$ represent two canonically conjugated variables for a 1D chain of length $L$, with $[\theta(x'), \partial_x \phi(x)] = i\pi \delta(x' - x)$, $K$ is the Luttinger-liquid (LL) parameter, $v_s$ the velocity of 1D fermions, and $g_u$ is the strength of the sine-Gordon potential. In this formulation, the role of the SC phase is played by the field $\theta$. Indeed, when the coupling $g_u = 0$, one can integrate out the dual field $\phi$ to get the action

$$S_0 = \frac{K}{2\pi} \int dx d\tau [(\partial_x \theta)^2 + (\partial_\tau \theta)^2], \quad (5.22)$$

equivalent to the gradient expansion (5.6) of the model (5.1), once considered that the rescaled time $\tau \to v_s \tau$ plays the role of the second (classical) dimension. The dual field $\phi$ describes instead the transverse vortex-like excitations. This can be easily understood by considering the quantum nature of the operators within the usual language of the sine-Gordon model. Indeed, a vortex configuration requires that $\oint \nabla \theta = \pm 2\pi$ over a closed loop, see Eq. (5.7) above. Since $\phi$ is the dual field of the phase $\theta$, a $2\pi$ kink in the field $\theta$ is generated by the operator $e^{i2\phi}$,[9] i.e. by the sine-Gordon potential in the Hamiltonian (5.21). More formally, one can show that the partition function of the $\phi$ field in the sine-Gordon model corresponds to the (5.19) derived above. To see this, let us first of all integrate out the $\theta$ field in Eq. (5.21), to obtain

$$S_{\text{SG}} = \frac{1}{2\pi K} \int d\mathbf{r} (\nabla \phi)^2 - \frac{g}{\pi} \int d\mathbf{r} \cos(2\phi). \quad (5.23)$$

The overall factor $Z_\parallel = \Pi_{q>0}(1/\beta J\mathbf{q}^2)$ due to the integration of the $\theta$ field (corresponding to the longitudinal excitations $Z_\parallel = \int \mathcal{D}\theta_\parallel e^{-\beta H_\parallel}$ in Eq. (5.6) above) will be omitted in what follows. We can treat the first term of the above action as the free part $S_0$, and we can expand the exponential of the interacting part in series of powers, so that

$$Z = \int \mathcal{D}\phi e^{-S_0} \sum_{p=0}^{\infty} \frac{1}{p!} d\mathbf{r}_1 \cdots d\mathbf{r}_p \left(\frac{g}{\pi}\right)^p \cos(2\phi(\mathbf{r}_1)) \cdots \cos(2\phi(\mathbf{r}_p)). \quad (5.24)$$

Here $\int \mathcal{D}\phi$ is the functional integral over the $\phi$ field. When we decompose each cosine term as

$$\cos(2\phi(\mathbf{r}_i)) = \frac{e^{i\phi(\mathbf{r}_i)} + e^{-i\phi(\mathbf{r}_i)}}{2} = \sum_{\epsilon=\pm 1} \frac{e^{i\epsilon\phi(\mathbf{r}_i)}}{2}, \quad (5.25)$$

we recognize that in Eq. (5.24) we are left with the calculation of average value of exponential functions over the Gaussian weight $S_0$, i.e. of factors

$$\langle e^{2i \sum_i \epsilon_i \phi(\mathbf{r}_i)} \rangle = e^{2K \sum_{i<j} \ln\left(\frac{r_{ij}}{a}\right) \epsilon_i \epsilon_j}. \quad (5.26)$$

Here we used the well-known properties of Gaussian integrals[9,59] that impose that the above expectation value is nonzero only for neutral configurations $\sum_{i=1}^{p} \epsilon_i = 0$, in full analogy with the result found above for the vortices. We then take again $p = 2n$. For instance, if $\epsilon_1, \ldots \epsilon_n = +1$ while $\epsilon_{n+1}, \ldots \epsilon_{2n} = -1$ the combinatorial prefactor $1 = p! \equiv 1/(2n)!$ in Eq. (5.24) should be multiplied times the number $\binom{2n}{n} = (2n)!/(n!)^2$ of possibilities to choose the $n$ positive $\epsilon_i$ values over the $2n$ ones. Thus, Eq. (5.24) reduces to:

$$Z = \sum_{n=1}^{\infty} \frac{1}{(n!)^2} \left(\frac{g}{2\pi}\right)^{2n} \int d\mathbf{r}_1 \cdots d\mathbf{r}_{2n} e^{2K \sum_{i<j}^{2n} \ln\left(\frac{r_{ij}}{a}\right) \epsilon_i \epsilon_j}. \quad (5.27)$$

By comparing Eqs. (5.19) and (5.27), we see that the vortex problem (as well as the Coulomb-gas problem) is fully mapped into the sine-Gordon model, provided that we identify:

$$K = \frac{\pi J}{T}, \quad (5.28)$$

$$g = 2\pi e^{-\beta\mu}. \quad (5.29)$$

As it is clear from the above derivation, within the XY model there exists a precise relation (5.17) between the value of the vortex-core energy $\mu$ and the value of the superfluid coupling $J$. This is somehow a natural consequence of the fact that the XY model (5.1) has only *one* coupling constant, $J$. Thus, when deriving the mapping on the continuum Coulomb-gas problem (5.18) $\mu$ is fixed by the short length-scale interaction, that fixes the behavior of $G(r)$ in Eq. (5.15) and consequently the vortex-core energy value (5.17). In contrast, within the sine-Gordon language $\mu$ is determined by the value of the interaction $g$ for the model (5.23), that can attain, in principle, arbitrary values. Thus, such a mapping is the more suitable one to investigate situations where $\mu$ actually deviates from the XY-model value, as it is suggested by the physics of various systems.

A typical example is provided by the case of ordinary films of superconductors, where usually the BCS approximation — and its dirty-limit version — reproduces quite well the values of the SC quantities, like the gap and the transition temperature.[14,15] In this case, one would also expect that $\mu$ has a precise physical analogous with the loss in condensation energy within a vortex core of size of the order of the coherence length $\xi_0$,

$$\mu = \pi \xi_0^2 \varepsilon_{\text{cond}}, \tag{5.30}$$

where $\varepsilon_{\text{cond}}$ is the condensation-energy density. In the clean case, Eq. (5.30) can be expressed in terms of $J_s$ by means of the BCS relations for $\varepsilon_{\text{cond}}$ and $\xi_0$. Indeed, since $\varepsilon_{\text{cond}} = dN(0)\Delta^2/2$, where $N(0)$ is the density of states at the Fermi level and $\Delta$ is the BCS gap, and $\xi_0 = \xi_{\text{BCS}} = \hbar v_F/\pi\Delta$, where $v_F$ is the Fermi velocity, and assuming that $n_s = n$ at $T = 0$, where $n = 2N(0)v_F^2 m/3$, one has

$$\mu_{\text{BCS}} = \frac{\pi \hbar^2 n_s d}{4m} \frac{3}{\pi^2} = \pi J_s \frac{3}{\pi^2} \simeq 0.95 J_s, \tag{5.31}$$

so that it is much smaller than in the XY model case (5.17). This can have profound physical consequences on the manifestation of the BKT signatures in real materials, as we shall discuss in more details in the next section.

## 5.3. The Universal Jump of the Superfluid Density

The above derivation of the mapping between the XY model and the sine-Gordon model can be used directly to identify the two relevant running couplings $K$ and $g$ that must be considered under renormalization group. The coupling $K$ (5.28) is connected to the superfluid behavior of the system, thanks to the identification (5.5) of $J$ with the superfluid stiffness of the system. Such an identification relies on the analogy between Eq. (5.6) and the kinetic energy of a superfluid. As a consequence, when vortex excitations are also present, the physical $J_s$ must account also for vortex–antivortex pairs at short distances.[7] This physical picture has a precise correspondence on the values of the coupling constants under RG flow, whose well-known equations are[3,8,9]:

$$\frac{dK}{d\ell} = -K^2 g^2, \qquad (5.32)$$

$$\frac{dg}{d\ell} = (2-K)g, \qquad (5.33)$$

where $\ell = \ln a/a_0$ is the rescaled length scale. The superfluid stiffness is then identified by the limiting value of $K$ as one goes to large distances, i.e.

$$J_s \equiv \frac{TK(\ell \to \infty)}{\pi}. \qquad (5.34)$$

Even though the behavior of the RG equations (5.32) and (5.33) has been described at length in several papers, we want to recall here the basic ingredients needed to describe the BKT transition. There are two main regimes: for $K \gtrsim 2$ the r.h.s. of Eq. (5.33) is negative, so that $g \to 0$ and $K$ tends to a finite value $K \to K^*$ that determines the physical stiffness $J_s$, according to Eq. (5.34). Instead for $K \lesssim 2$ the vortex fugacity grows under RG flow, $K$ in Eq. (5.32) scales to zero, and $J_s = 0$. The BKT transition temperature is defined as the highest value of $T$ such that $K$ flows to a finite value. This occurs at the fixed point $K = 2, g = 0$, so that at the transition one always has

$$K(\ell \to \infty, T_{\text{BKT}}) = 2, \Rightarrow \frac{\pi J_s(T_{\text{BKT}})}{T_{\text{BKT}}} = 2. \qquad (5.35)$$

As soon as the temperature grows above $T_{\text{BKT}}$ $K \to 0$, so also $J_s \to 0$. As a result, one finds $J(T_{\text{BKT}}^+) = 0$, i.e. the superfluid density jumps discontinuously to zero right above the transition. Equation (5.35) describes the so-called universal relation between the transition temperature $T_{\text{BKT}}$ and the value of the superfluid stiffness $J_s$ at the transition, and represents a more refined version of the relation (5.3) based on the balance between the energy and the entropy of a single-vortex configuration.

It should be noticed that the BKT RG equations account only for the effect of vortex excitations, so that any other excitation that contributes to the depletion of the superfluid stiffness must be introduced by hand in the initial values of the running couplings. For example, in real superconductors, there are also quasiparticle excitations, while in the XY model there are also longitudinal phase fluctuations, that give rise to a linear depletion to the superfluid stiffness $J_0(T) = J(1 - T/4J)$. As a consequence, the hand-waving argument usually adopted in the literature to estimate $T_{\text{BKT}}$ in a system that is expected to have a BKT transition is to look for the intersection between the universal line $2T/\pi$ and the $J(T)$ expected from

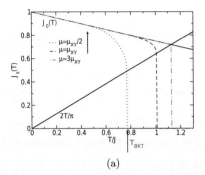 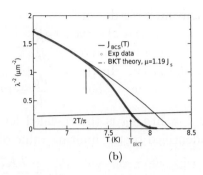

Fig. 5.1. (a) Temperature dependence of the superfluid density for different values of the ratio $\mu/J$, measured in units of the value (5.17) it has within the XY model. Notice that for small $\mu$ values the deviation from $J_0(T)$ starts much before (blue arrow) than the temperature where the universal jump (red arrow) occurs. This behavior is indeed observed in thin NbN films, panel (b). Here we report data from Ref. 15 along with the theoretical BKT fit. Notice that the jump here is further smeared out by the inhomogeneity, see Sec. 5.4 below.

the remaining excitations other than vortices. However, such a procedure can only be approximate, since in the relation (5.35) the temperature dependence of $J_s(T)$ is determined also by the presence of bound vortex–antivortex pairs, which can renormalize $J_s$ already *below* $T_{BKT}$. This effect is usually negligible when $\mu$ is large, as it is the case for superfluid films[60] or within the standard XY model. However, as $\mu$ decreases, the renormalization of $J_s$ due to bound vortex pairs increases, and consequently $T_{BKT}$ is further reduced with respect to the mean-field critical temperature $T_c$. As an example, we show in Fig. 5.1 the behavior of $J_s(T)$ using a bare temperature dependence as in the XY model and switching the vortex-core energy from the value (5.17) to values smaller or larger. As one can see, for decreasing $\mu$ the effect of bound vortex–antivortex pairs below $T_{BKT}$ is significantly larger, moving back the transition temperature to smaller values. The very same effect has been recently observed in thin films of NbN,[14,15] where it has been shown experimentally that the deviation of $J_s(T)$ from the BCS curve starts significantly below the transition temperature. Interestingly, the direct comparison between the experimental $J_s(T)$ and the results of the RG equations allowed the authors to show that the vortex-core energy in this system attains indeed a value of the order of the BCS estimate (5.31).

It must be emphasized that the case of thin NbN films must be seen as a paradigmatic example of manifestation of the *universal* relation (5.35), despite the fact that $T_{BKT}$ is lower than expected for a standard view based on the XY model results. Indeed, in this system once that the *renormalized* stiffness is of the order of $2T/\pi$, the transition actually occurs. A different behavior is instead observed in the case we discussed recently[61] within the context of cuprate superconductors, that can be well modeled as weakly coupled 2D layers. In this case, an additional energy scale exists, i.e. the Josephson coupling $J_c$ between layers, which is also a relevant coupling under RG flow, so that a vortex-core energy different from the XY model value can lead to a qualitatively different behavior. We notice that the case of weakly-coupled layered superconductors has been widely investigated in the past within the framework of a layered version of the XY model (5.1). In this case the presence of a finite interlayer coupling $J_c \ll J$ cuts off the logarithmic divergence of the in-plane vortex potential at scales $\sim a/\sqrt{\eta}$,[62] where $\eta = J_c/J$, so that the superfluid phase persists above $T_{\rm BKT}$, with $T_c$ at most few percent larger than $T_{\rm BKT}$.[63–65] As far as the superfluid density is concerned, there is some theoretical[63] and numerical[66] evidence that even for moderate anisotropy the universal jump at $T_{\rm BKT}$ is replaced by a rapid downturn of $\rho_s(T)$ at a temperature $T_d$ larger than the $T_{\rm BKT}$ of each layer, but still very near to it, $T_d \simeq T_{\rm BKT}$. Once more this result must be seen as an indication that the (only) scale $J$ dominates the problem: the result can be different when $\mu$ is allowed to vary, making the competition with the interlayer coupling more subtle.

The analysis of the more general case has been done in a very convenient way in Ref. 61 within the framework of the sine-Gordon model (5.21), that must be suitably extended to include the interlayer coupling. As we said, in the sine-Gordon model (5.21) the variable $\theta$ represents the SC phase. Since the phases in neighboring layers are coupled via a Josephson-coupling like interaction, the most natural assumption is an additional cosine term of strength $J_c$ for the interlayer phase difference, which translates to an interchain hopping term in the language of the 1D quantum model. A similar model has been also derived recently in Ref. 67 by using as the starting point the Lawrence–Doniach model for the layered superconductor. The full Hamiltonian that we consider is[61]:

$$H = \sum_m H_{sg}[\phi_m, \theta_m] - \frac{v_s g_{J_c}}{2\pi a^2} \sum_{\langle m,m' \rangle} \int_0^L dx \cos[\theta_m - \theta_{m'}], \qquad (5.36)$$

where $m$ is the layer (chain) index and $g_{J_c} \equiv \pi J_c/T$. In Ref. 61 we derived the perturbative RG equations for the couplings of the model (5.36) by means of the operator product expansion, in close analogy with the analysis of Refs. 68 and 69 for the multichain problem. Under RG flow, an additional coupling $g_\perp$ between the phase in neighboring layers is generated:

$$\frac{g_\perp}{2\pi} \sum_{\langle m,m' \rangle} \int dx \left[ -K(\partial_x\theta_m)(\partial_x\theta_{m'}) + \frac{1}{K}(\partial_x\phi_m)(\partial_x\phi_{m'}) \right], \quad (5.37)$$

which contributes to the superfluid stiffness $K_s$, defined as usual as the second-order derivative of the free energy with respect to an infinitesimal twist $\delta$ of the phase, $\partial_x\theta_m \to \partial_x\theta_m - \delta$. Thus, one immediately sees that Eq. (5.37) represents an interlayer current–current term, which contributes to the in-plane stiffness $J_s$ as:

$$K_s = K - nKg_\perp, \quad J_s = \frac{K_s(\ell \to \infty)T}{\pi}, \quad (5.38)$$

where $n = 2$ corresponds to the number of nearest-neighbors layers. The full set of RG equations for the couplings $K, K_s, g_u, g_{J_c}$ reads:

$$\frac{dK}{d\ell} = 2g_J^2 - K^2 g_u^2, \quad (5.39)$$

$$\frac{dg_u}{d\ell} = (2 - K)g_u, \quad (5.40)$$

$$\frac{dK_s}{d\ell} = -g_u^2 K_s^2, \quad (5.41)$$

$$\frac{dg_{J_c}}{d\ell} = \left(2 - \frac{1}{4K} - \frac{K_s}{4K^2}\right) g_{J_c}. \quad (5.42)$$

Observe that for $g_{J_c} = 0$ the first two equations reduce to the standard ones (5.28) and (5.29) of the BKT transition, and $K_s$ coincides with $K$. Instead, as an initial value $g_{J_c} \neq 0$ is considered, the interlayer coupling increases under RG,[63–65] contributing to stabilize the $K$ parameter in Eq. (5.39), with a consequent slowing down of the increase of $g_u$ coupling in Eq. (5.40). As in the pure 2D case, in the regime where $K$ goes to zero, $g_u$ increases, see Eq. (5.40), and vortices proliferate. However, in contrast to the single-layer case where $g_u$ becomes always relevant near $K \simeq 2$, here thanks to the $g_{J_c}$ term in Eq. (5.39), $K$ can become smaller than 2 before $g_u$ starts to increase. Since $K_s$ is controlled by $g_u$ coupling alone, this means that the system remains superfluid in a range of temperature above $T_{\rm BKT}$ that depends on the competition between vortices ($g_u$) and interlayer coupling ($g_{J_c}$).

The resulting temperature dependence of the superfluid stiffness $J_s$ for different values of the vortex-core energy is reported in Fig. 5.2(a), where we added for the sake of completeness a temperature dependence of the bare coupling $J_0(T) = J(1 - T/4J)$, which mimics the effect of long-wavelength

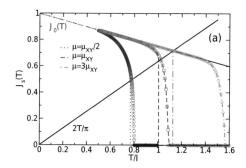
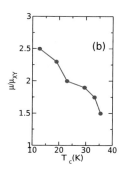

Fig. 5.2. (a) Temperature dependence of the superfluid stiffness $J_s(T)$ in the layered 3D case (symbols), taken from Ref. 61. Here $J_0(T) = J(1-T/4J)$ (solid line) mimics the effect of longitudinal excitations within the XY model. Also shown for comparison is the 2D case (lines), see Fig. 5.1. For $\mu \lesssim \mu_{XY}$ $J_s(T) = 0$ at $T_d \simeq T_{BKT}$. As $\mu$ increases, $T_d$ increases as well, so that at $T_{BKT}$ no signature is observed in $J_s(T)$ of the jump present in the 2D case. (b) Vortex-core energy in units of $\mu_{XY}$ as a function of the critical temperature of several $YBa_2Cu_3O_{6+x}$ samples, as derived in Ref. 70.

phase fluctuations in the XY model (5.1). Here the lines represent the pure 2D case (already shown in Fig. 5.1(a) above), while the symbols are the result of the RG equations (5.32)–(5.42) for a fixed value $J_c/J = 10^{-4}$ of the interlayer coupling. As one can see, as soon as a finite interlayer coupling is switched on, the jump of $J_s(T)$ at $T_{BKT}$ disappears and it is replaced by a rapid bending of $J_s(T)$ at some temperature $T_d$. However, while for $\mu \lesssim \mu_{XY}$, $T_d$ coincides essentially with $T_{KT}$, for a larger vortex-core energy $T_d$ rapidly increases and deviates significantly from the temperature scale where $J_s(T)$ intersects the universal line $2T/\pi$.

These results offer a possible interpretation of the measurements of superfluid density reported in strongly-underdoped samples of $YBa_2Cu_3O_{6+x}$.[47] These systems are exactly the ones where a BKT behavior could be expected: they have a very low in-plane superfluid density and a large anisotropy ($\eta \sim 10^{-4}$) of the coupling constants. However, the measured $J_s(T)$ goes smoothly across the $T_{BKT}$ estimated from the 2D relation (5.35). In view of the above discussion, such a behavior does not automatically rule out the possibility that any BKT physics is at play: indeed, if the vortex-core energy is larger than in the XY model, the transition can move away from the universal 2D case. Moreover, the effect of inhomogeneity can further round off the downturn induced by vortex proliferation (see the next section), making it hardly visible in the experiments. In analogy with the case of conventional superconductors discussed above, a hint on the realistic value of the vortex-core energy in cuprate systems can be inferred by the direct comparison between superfluid-density measurements and the theoretical prediction. We carried out such an analysis in the case of bilayer films of underdoped YBCO[13] for several doping values, showing that $\mu$ attains always values larger than in the XY model, with $1.5 < \mu/\mu_{XY} < 3$, see Fig. 5.2(b). Moreover, both $\mu$ and the inhomogeneity increase as the system gets underdoped, supporting further the interpretation that a similar effect can be at play also in the data of Ref. 47. Interestingly, a very similar trend has been reported recently in NbN systems as disorder increases,[15] and it has been interpreted by the authors as an effect of the increasing separation between the energy scales associated with local pairing and superfluid stiffness as disorder increases.[57,58] These results suggest that disorder and inhomogeneity also play a crucial role in the understanding of the BKT transition, as we shall discuss in the next section. Finally, a very recent report[71] on a heavy-fermion $CeCoIn_5/YbCoIn_5$ superlattice reached the conclusion that also in $CeCoIn_5$ the vortex-core energy is near to the BCS value. However, in this case the low value of the vortex-core energy has been attributed to a competing magnetic ordering within the vortex core, showing again how subtle is the identification of a "conventional" BKT mechanism in correlated systems.

## 5.4. Inhomogeneity

As we mentioned in the introduction, detailed measurements of superfluid density in thin films of superconductors became available only recently thanks to the efficient implementation of the two-coils mutual inductance technique, which gives access to the absolute value of the penetration depth of thin SC materials. Quite interestingly, in all the cases reported so far in the literature, concerning both conventional superconductors[14,15] and high-temperature superconductors[12,13,16,20,30] the BKT transition is never really sharp. At first sight, one could wonder if any finite-size effect is at play, as due to several factors: (i) the existence of a finite screening length $\Lambda \simeq 2\lambda^2/d$ due to the supercurrents in a charged superfluid;[6] (ii) the finite dimension of the system or (iii) the finite length $r_\omega$ intrinsically associated to the probe, which uses an ac field at a typical frequency $\omega$ of the order of the KHz. In all the above cases one should cut-off the RG equations (5.32) and (5.33) at a finite scale $\ell_{\max} = \ln L_{\max}/a_0$, leading to a rounding of the abrupt jump of the stiffness at $T_{BKT}$. However, in practice such rounding effects are hardly visible, since the decrease of $K(\ell)$ at $T > T_{BKT}$ is very fast, leading to visible rounding effect only for very short cut-off length scales of order of $\ell \simeq 2 - 3$. In real systems the cut-off length scales are usually much larger:

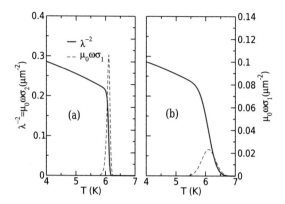

Fig. 5.3. Role of inhomogeneity, from Ref. 70. (a) $1/\lambda^2$ and $\mu_0\omega\sigma_1$ evaluated at $\omega = 50$ KHz for a single $\bar{J}(T)$ curve using parameter values appropriate for YBa$_2$Cu$_3$O$_{6+x}$ films from Ref. 13. Here $\mu = 3\mu_{XY}$. The finite frequency leads to a sharp but continuous decrease of $1/\lambda^2$ across $T_{BKT}$, along with a peak in $\sigma_1$. (b) $1/\lambda^2$ and $\mu_0\omega\sigma_1$ evaluated at finite frequency taking into account the presence of inhomogeneities, as explained in the text.

for example, both in the case of the NbN films of Ref. 14 and in the case of thin YBCO films from Ref. 13, $1/\lambda^2$ is or order of about 1 $\mu$m$^{-2}$ near the transition and $d$ is of the order of 1 nm, so that the Pearl length $\Lambda$ is of the order of 1 mm, i.e. comparable to the system size, leading to $\ell_{\max}$ around 10 ($a_o$ is of the order of the coherence length $\xi_0 \sim 1$ nm), which is practically an infinite cut-off for the RG. At the same time $r_\omega = \sqrt{14D/\omega}$ is the maximum length probed by the oscillating field, where $D \sim \hbar/m = 10^{16}$ Å$^2$/s is the diffusion constant of vortices and $\omega \simeq 50$ KHz is the frequency of the measurements,[13,14] giving again a large cut-off scale $r_\omega \sim 0.1$ mm and no visible rounding effect, see Fig. 5.3(a). It is worth noting that in the case of the experiments of Ref. 13, a second indication of the existence of pronounced rounding effects come from the experimental observation of a wide peak in the imaginary part of the conductivity around $T_{BKT}$. Following the dynamical BKT theory of Ambegaokar et al.[72] and Halperin and Nelson[5] such a peak is due to the bound- and free-vortex excitations contribution to the complex dielectric constant $\varepsilon(\omega) = \varepsilon^1 + i\varepsilon^2$ which appears in the complex conductivity $\sigma = \sigma_1 + i\sigma_2$ at a finite frequency $\omega$:

$$\sigma(\omega) = -\frac{1}{\lambda^2 e^2 \mu_0} \frac{1}{i\omega\varepsilon(\omega)}. \tag{5.43}$$

Here $\mu_0$ is the vacuum permittivity and we used MKS units as in Refs. 13 and 70. As in the case of the superfluid-density jump, we estimated[70] the width $\Delta T_\omega$ in $\sigma_1(\omega)$ due to the finite frequency $\omega$ of the measurements and we showed that it is expected to be much smaller than the experimental observations in Ref. 13, see Fig. 5.3(a).

On the light of the above discussion, we proposed in Ref. 70 that a more reasonable explanation for the rounding effects in the superfluid density and the large transient region in $\sigma_1(\omega)$ observed in the experiments is the sample inhomogeneity. In the case of cuprate systems such inhomogeneity is suggested by tunneling measurements in other families of cuprates,[48] where Gaussian-like fluctuations of the local gap value are observed. Recently similar local-gap distributions have been reported also in tunneling experiments in thin films of strongly-disordered conventional superconductors,[49–54] showing that an intrinsic tendency towards mesoscopic inhomogeneity appears even for systems with homogeneous intrinsic disorder. Even though the issue of the microscopic origin of this effect is beyond the scope of our work, nonetheless, we found that a Gaussian-like distribution of the superfluid-stiffness $J_0$ values around a given $\bar{J}$ can account very well for the data in Ref. 13 on two-unit-cell thick films of $YBa_2Cu_3O_{6+x}$. Thus, we compared with the experiments the quantity $J_{\text{inh}}(T) = \int dJ_0 P(J_0) J(T, J_0)$, where each $J(T, J_0)$ curve is obtained from the 2D RG equations using a bare superfluid stiffness $J = J_0 - \alpha T^2$. Each initial value $J_0$ has a probability $P(J_0) = \exp[-(J_0 - \bar{J}_0)^2 / 2\sigma^2] / (\sqrt{2\pi}\sigma)$ of being realized, where the bare average stiffness follows the typical temperature dependence due to quasiparticle excitations in a (disordered) $d$-wave superconductor, i.e. $\bar{J}_0(T) = \bar{J}_0 - \alpha T^2$, with $\bar{J}_0$ and $\alpha$ fixed by the experimental data at low $T$, where $J_{\exp}(T)$ is practically the same as $\bar{J}(T)$. Using a variance $\sigma = 0.05 \bar{J}_0$, we obtained a very good agreement with the experiments near the transition, as far as both the tail of $\lambda^{-2}$ and the position and width of $\sigma_1(\omega)$ are concerned, see Fig. 5.3(b).

Even though our approach to the issue of inhomogeneity is quite phenomenological, it has been shown more recently that it accounts very well also for the superfluid-density behavior in films on NbN.[14,15] In all these cases, indeed the rounding effect due to inhomogeneity is far more relevant than any finite-size effect due to screening or finite-frequency probes, explaining the lack of a very sharp jump even for relatively weakly-disordered films (see Fig. 5.1(b) above). Quite interestingly, a similar physical picture seems to be relevant also for a completely different class of materials, i.e. the SC metal-oxide interfaces, where experimental data for the superfluid density are not yet available, but the nature of the SC transition can be investigated in an indirect way by the analysis of the paraconductivity above $T_{\text{BKT}}$.[21] For these systems too, a BKT transition is likely to be expected, since the SC interfaces are very thin,[21] of the order of 15 nm, which is much lower than the SC coherence length $\xi_0 \simeq 70$ nm. Moreover, despite what occurs in thin films of conventional superconductors,[15] the drop of the resistivity above the transition is very smooth, suggesting that inhomogeneities on a mesoscopic scale broaden considerably the transition, as we discussed in detail in Ref. 32. On this aspect, it is worth recalling that any analysis of the BKT transition based on the paraconductivity data alone can suffer the unavoidable lack of knowledge about the exact extension of the BKT fluctuation regime. Indeed, as we mentioned in the introduction, the contribution of SC fluctuations to the paraconductivity is encoded in the temperature dependence of the correlation length $\xi(T)$ both within the BKT and GL theories, see Eq. (5.4). Thus, if the transition has BKT character, one should expect a crossover from the BKT exponential temperature dependence of $\xi(T)$,

$$\xi_{\text{BKT}} = A \exp(b/\sqrt{t}), \quad t \equiv \frac{T - T_{\text{BKT}}}{T_{\text{BKT}}}, \qquad (5.44)$$

to that of the power-law GL, $\xi_{\text{GL}}^2 \sim \xi_0 T_c/(T-T_c)$, where $T_c$ is the BCS mean-field critical temperature. In the case of thin films of superconductors, the BKT regime is, in practice, restricted to a very small range of temperatures near $T_{\text{BKT}}$, due to the fact that $T_c$ is at most twenty per cent larger than $T_{\text{BKT}}$. In the case of the SC interfaces, it has been proposed instead in the recent literature[21,22,31] that $T_c$ is far larger than $T_{\text{BKT}}$, so that the *whole* fluctuation regime above $T_{\text{BKT}}$ is dominated by BKT fluctuations alone, while the large tails observed experimentally near $T_{\text{BKT}}$ should be ascribed to finite-size effects, see Fig. 5.4(a). There is, however, a serious drawback in such interpretation, that originates from an incorrect application of the BKT relation (5.44) in the analysis of the experimental data. As we showed in detail in Ref. 32 by means of a RG analysis of the correlation length, the parameter $b$ which appears in the usual BKT formula (5.44) is directly connected to the distance $t_c = (T_c - T_{\text{BKT}})/T_{\text{BKT}}$ between the mean-field $T_c$ and the BKT transition temperature and to the value of the vortex-core energy with respect to the superfluid stiffness:

$$b_{\text{theo}} \sim \frac{4}{\pi^2} \frac{\mu}{J_s} \sqrt{t_c}, \quad t_c = \frac{T_c - T_{\text{BKT}}}{T_{\text{BKT}}}. \qquad (5.45)$$

As a consequence, $b$ increases when the BKT fluctuation regime extends in temperature. In the typical fits of the resistivity proposed in Refs. 22 and 31 one obtains values of the $b$ parameter that, according to Eq. (5.45) above, would imply a mean-field $T_c$ very near to $T_{\text{BKT}}$, see for instance Fig. 5.4(a). In other words, the fit leads to values of $b$ that contradict

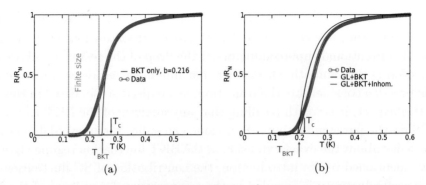

Fig. 5.4. Analysis of the paraconductivity effect on the resistivity of SC interfaces at the $T_{BKT}$ transition. The experimental data for the resistivity $R$ normalized to the normal-state value $R_N$ are taken from Ref. 22. Panel (a) shows an example of the approach proposed in Refs. 21, 22 and 31, where the whole range of temperatures above $T_{BKT}$ is dominated by SC fluctuations having BKT character, and finite-size effects are responsible for the resistive tail below $T_{BKT}$. However, the $b$ value obtained by the BKT fit based on Eq. (5.44) for the correlation length implies, according to Eq. (5.45), the $T_c$ value marked by the blue arrow, invalidating thus the assumption itself that the whole fluctuation regime has BKT character. Panel (b) is taken from Ref. 74 (Fig. 5.1 in that paper) and elucidates the approach developed in Refs. 32 and 74, based on the interpolation scheme (5.46) between BKT and GL fluctuations. The curve labeled "GL+BKT+inhom" has been computed by solving a random-resistor network problem with a Gaussian distribution ($\sigma = 0.035$ K) of the critical temperatures centered around $\bar{T}_{BKT} = 0.2$ K, marked by an arrow along with the corresponding $\bar{T}_c = 0.22$ K. The paraconductivity of each resistor follows Eq. (5.46) with $b = 0.25$, which according to Eq. (5.45) corresponds to $t_c = 0.1$ when $\mu/J_s \simeq 2$, an intermediate value between the XY model (5.17) and BCS (5.31) estimates. The curve labeled "GL+BKT" corresponds to the homogeneous case with a transition temperature $\bar{T}_{BKT}$.

the *a priori* assumption that the whole fluctuations regime is dominated by BKT fluctuations, as described by Eq. (5.44). Moreover, the interpretation of the tails extending below $T_{BKT}$ as due to finite-size effects[31] also leads to unphysical low values for the size of the homogeneous domains.[32] One would have to thus invoke a BKT transition of a completely differente nature (dislocations of a vortex crystal[73]), as it has been suggested in Ref. 21. However, even in this case, there is no theoretical understanding on how to reconcile such contradictory numbers obtained in the analysis of the resistive transition.

An alternative, and less far fetched, interpretation of the resistivity data can be based instead on the use of the interpolation formula for $\xi(T)$ proposed long ago by Halperin and Nelson[5]

$$\frac{R}{R_N} = \frac{1}{1 + (\xi/\xi_0)^2}, \quad \xi(T) = \xi_0 A \sinh \frac{b}{\sqrt{t}}. \tag{5.46}$$

One can easily see that Eq. (5.46) reduces to $\xi_{\text{BKT}}$ for small $t$ and to $\xi_{\text{GL}}$ for large $t$. Using the estimate (5.45) for the $b$ parameter, one realizes that the crossover occurs around $t \simeq b^2 \propto t_c$, so that for (realistic) small values of $t_c$, most of the fluctuation regime is dominated by GL-type fluctuations. Moreover, we attributed the broadening of the transition to the effect of inhomogeneity, that induces a distribution of possible realizations of local $R/R_N$ values corresponding to the local $J$ values discussed above. An example is shown in Fig. 5.4(b), where the average resistivity has been computed by solving a random-resistor-network model for a set of resistors undergoing a metal–superconductor transition, as discussed recently in Ref. 74. It must be emphasized that the issue of the physical value of the $b$ parameter in the BKT expression (5.44) of the correlation length has been often overlooked in the literature also in context of other materials,[24,28] leading to questionable conclusions concerning the existence of BKT physics based only on erroneous fits of the resistivity. In contrast, a recent analysis on NbN films,[15] based on the comparison between the estimate of $b$ deduced from the analysis of the superfluid density below $T_{\text{BKT}}$ and the resistivity above $T_{\text{BKT}}$ has confirmed the theoretical estimate (5.45) of the $b$ parameter, that must be used as an unavoidable constraint in any BKT fit of the paraconductivity.

## 5.5. The Case of a Finite Magnetic Field

As we discussed in the previous sections, the sine-Gordon model provided us with the most appropriate formalism to investigate not only the role of a vortex-core energy value different from that of the XY model, but also the occurrence of BKT physics in the presence of a relevant perturbation (in the RG sense), as the interlayer coupling. It is then worth asking the question of the possible relevance of such a mapping for another typical relevant perturbation, i.e. a finite external magnetic field. The nature of BKT physics in the presence of a magnetic field has been, of course, already investigated in the past,[4,8,75] with a renewed interest in the more recent literature[38–41] triggered once more by experiments in cuprate superconductors carried out at finite field, as the measurement of the Nerst effect or the magnetization. Unfortunately, contrary to the case of the **B** = 0 transition, the efforts have been partly unsatisfactory. In particular, most of the literature on the subject has rested on the use of the mapping into the Coulomb-gas problem, where the effects of the magnetic field can be incorporated as an excess of positive charges, in analogy with the finite population of vortices with a given vorticity due to the presence of the external field. There is, however, a fundamental drawback of this approach, since it gives the physical observables as a function of the magnetic induction **B** instead of the magnetic field **H**. This is, of course, not convenient at low applied field, since inside the superconductor **B** vanishes even for a finite external field **H**. Motivated by this observation and by the occurrence of an anomalous nonlinear regime for the magnetization measured in cuprate systems, we showed recently[76] that a suitable mapping into the sine-Gordon model provides a very simple and physically transparent way to deal with the finite magnetic field case, leading to a straightforward definition of the physical observables and clarifying the role of both **B** and **H**. Since the construction of the mapping has not been shown in Ref. 76 due to space limitation, we shall discuss here both the model derivation and its basic physical consequences. This will complete our overview of the sine-Gordon approach to BKT physics, and it will allow us to also clarify some aspects of the charged superfluid that have not been underlined yet in the previous sections.

As a starting point, we use again the XY model (5.1), where the coupling to $B$ is introduced via the minimal-coupling prescription for the vector potential **A**,

$$H = J \sum_{\langle i,j \rangle} [1 - \cos(\theta_i - \theta_j - F_{ij})], \quad (5.47)$$

where

$$F_{ij} = F_\mu(r) = \frac{2\pi}{\Phi_0} \int_r^{r+\mu} \mathbf{A} \cdot d\mathbf{l} \approx \frac{2\pi a}{\Phi_0} \mathbf{A}_\mu(r). \quad (5.48)$$

Here $\mu = \hat{x}, \hat{y}$ is the vector from one site to the neighboring one and $\Phi_0 = hc/2e$ is the flux quantum. As a first step, we make use of the Villain approximation, that amounts to replace the cosine in Eq. (5.47) with a function having the same minima for each multiple $2\pi m$ of the gauge-invariant phase difference, i.e.

$$\exp[-\beta J(1 - \cos(\theta_i - \theta_j - F_{ij}))] \Rightarrow \sum_{m=-\infty}^{m=\infty} \exp\left[-\frac{\beta J}{2}(\theta_i - \theta_j - F_{ij} - 2\pi m)^2\right]. \quad (5.49)$$

By following the analogous derivation given in Ref. 59 for the case **A** = 0, we make use of the Poisson summation formula

$$\sum_{m=-\infty}^{m=\infty} h(m) = \sum_{l=-\infty}^{l=\infty} \int dz \, h(z) e^{2\pi i l z}, \quad (5.50)$$

and by performing explicitly the integration over $z$ for each link $ij$ we are left with the following structure of the partition function:

$$Z = Z_J \sum_{\{l_{ij}\}} \int_0^{2\pi} d\theta_1 \cdots d\theta_N e^{-\sum_{\langle i,j \rangle} \frac{l_{ij}^2}{2\beta J} + i \sum_{\langle i,j \rangle} l_{ij}(\theta_i - \theta_j - F_{ij})}, \qquad (5.51)$$

where $N$ is the number of lattice sites and $l_{ij}$ are $2N$ integer variables defined for each link $ij$. The prefactor $Z_J = [(1/\beta J)^{1/2}]^{2N} = (1/\beta J)^N$ accounts for the $z$ integration above for each link. Since each $\theta$ variable is defined on a period, we have $\int_0^{2\pi} d\theta e^{i\theta\alpha} = \delta(\alpha)$. As a consequence, the integration over the $\theta_i$ variables in the above equation leads to $N$ constraint equations:

$$\sum_\mu l_\mu(\mathbf{r}) - l_\mu(\mathbf{r} - \mu) = 0, \qquad (5.52)$$

which are the discrete equivalent of $\nabla \cdot \mathbf{l} = 0$. These equations can be satisfied by defining for each site of the lattice a field $n$ such that:

$$l_x(\mathbf{r}) = n(\mathbf{r}) - n(\mathbf{r} - y), \qquad (5.53)$$
$$l_y(\mathbf{r}) = n(\mathbf{r} - x) - n(\mathbf{r}), \qquad (5.54)$$

i.e. the discrete equivalent of the relation $\mathbf{l} = \nabla \times (n\hat{z})$. From Eqs. (5.53) and (5.54), Eq. (5.51) can then be rewritten as:

$$Z = Z_J \sum_{\{n(\mathbf{r})\}} e^{\sum_{\mathbf{r},\mu} \frac{1}{2\beta J} (\Delta_\mu n(\mathbf{r}))^2} e^{-i\frac{2\pi a}{\Phi_0} \sum_\mathbf{r} \hat{z} \cdot (\mathbf{A} \times \Delta n)}, \qquad (5.55)$$

where $\Delta_\mu = n(\mathbf{r} + \mu) - n(\mathbf{r})$ is the discrete derivative. Finally, we can use again the Poisson summation formula to get:

$$Z = Z_J \int \mathcal{D}\phi(\mathbf{r}) \sum_{\{m(\mathbf{r})\}} e^{-\sum_{\mathbf{r},\mu} \frac{1}{2\beta J} (\Delta_\mu \phi(\mathbf{r}))^2 + 2\pi i \sum_\mathbf{r} \phi(\mathbf{r}) m(\mathbf{r}) - i \frac{2\pi a}{\Phi_0} \sum_\mathbf{r} \hat{z} \cdot (\mathbf{A} \times \Delta\phi)}, \qquad (5.56)$$

where the $m(\mathbf{r})$ variables assume arbitrary positive and negative integer values. Before completing the mapping into the sine-Gordon model, we notice that in the above equation, the $\phi$ variables can be integrated out exactly. By rewriting the last term in the action as $\sum_\mathbf{r} \hat{z} \cdot (\mathbf{A} \times \Delta\phi) = \sum_\mathbf{r} \phi \hat{z} \cdot (\Delta \times \mathbf{A}) = Ba^2 \sum_\mathbf{r} \phi$ one can easily see that the final result is given by

$$Z = Z_\| \sum_{\{m(\mathbf{r})\}} e^{-2\pi^2 \beta J \sum_{\mathbf{r},\mathbf{r}'}[m(\mathbf{r})-f]V(r-r')[m(\mathbf{r}')-f]}, \qquad f = \frac{Ba^2}{\Phi_0}. \qquad (5.57)$$

Here $Z_\| = Z_J \Pi_{q>0}(\beta J/\mathbf{q}^2) \equiv \Pi_{q>0}(1/\beta J \mathbf{q}^2)$ is the overall contribution due to the longitudinal excitations. In full analogy with the result of Sec. 5.2, the longitudinal modes decouple from the transverse ones, so their contribution

$Z_\parallel$ to the partition function can be discarded in what follows. The function $V(\mathbf{r}) = \sum_{\mathbf{k}} e^{i\mathbf{k}\cdot\mathbf{r}} V(\mathbf{k})$ is defined through the Fourier transform $V^{-1}(\mathbf{k}) = (4 - 2\cos k_x - 2\cos k_y)$ of the $\Delta_\mu$ operator on the square lattice, i.e.

$$V(\mathbf{r}) = \int \frac{d^2\mathbf{k}}{(2\pi)^2} \frac{e^{i\mathbf{k}\cdot\mathbf{r}}}{[4 - 2\cos k_x - 2\cos k_y]}. \qquad (5.58)$$

Equation (5.57) generalizes the Coulomb-gas formula (5.11) above to the case of a finite magnetic field. As we did in Sec. (5.2), we can separate in $V(\mathbf{r})$ the singular part in $\mathbf{r} = 0$ by defining the regular function $G(\mathbf{r})$, i.e. $V(\mathbf{r}) = V(0) + G(\mathbf{r})$. Then one sees that

$$\sum_{\mathbf{r},\mathbf{r}'}[m(\mathbf{r}) - f]V(r - r')[m(\mathbf{r}') - f]$$

$$= V(0)\left[\sum_\mathbf{r}(m(\mathbf{r}) - f)\right]^2 + \sum_{\mathbf{r},\mathbf{r}'}[m(\mathbf{r}) - f]G(r - r')[m(\mathbf{r}') - f].$$

It follows that also in this case only neutral configurations have a statistical weight different from zero. However, in the presence of a magnetic field the neutrality condition reads:

$$\sum_\mathbf{r}(m(\mathbf{r}) - f) = N_v - \frac{Ba^2N}{\phi_0} = 0 \qquad (5.59)$$

which means that the total flux $N_v\Phi_0$ carried out by the (unbalanced) vortices equals the total flux $Ba^2N$ of the magnetic field across the sample. The definition (5.58) allows us also to determine the value (5.17) of the chemical potential $\mu$ in the lattice XY model. Indeed, one can see that at the scale of the lattice spacing $V(\mathbf{r}) - V(0)$ gives:

$$V(\mathbf{r} = \hat{x}) - V(0) = \int \frac{d^2\mathbf{k}}{(2\pi)^2} \frac{\cos k_x - 1}{[4 - 2\cos k_x - 2\cos k_y]}$$

$$= \frac{1}{2}\int \frac{d^2\mathbf{k}}{(2\pi)^2} \frac{\cos k_x + \cos k_y - 2}{[4 - 2\cos k_x - 2\cos k_y]} = -\frac{1}{4}, \qquad (5.60)$$

so that from Eq. (5.57) it follows that the cost to put two vortices at distance $a$ apart is $\beta\mu = \beta\pi^2 J/2$, consistent with the value (5.17) that we quoted above.

Let us now go back to Eq. (5.56) and let us complete the mapping into the sine-Gordon model. We notice that in Eq. (5.56) the variable $\phi$ is still defined on the square lattice. We can, however, resort to a continuum approximation by taking into account the energetic cost $\mu$ of the vortex creation on the shortest length scale of the problem via a chemical-potential-like term $e^{\ln y \sum_\mathbf{r} m^2(\mathbf{r})}$ in Eq. (5.56):

$$Z = \int \mathcal{D}\phi(\mathbf{r}) \sum_{\{m(\mathbf{r})\}} e^{-\sum_{\mathbf{r},\mu} \frac{1}{2\beta J}(\Delta_\mu \phi(\mathbf{r}))^2 + 2\pi i \sum_\mathbf{r} \phi(\mathbf{r})m(\mathbf{r}) - i\frac{2\pi a}{\Phi_0} \sum_\mathbf{r} \hat{z} \cdot (\mathbf{A} \times \Delta\phi)} e^{\ln y \sum_\mathbf{r} m^2(\mathbf{r})}. \tag{5.61}$$

The vortex-core energy term favors the formation of vortices of smallest vorticity, i.e. $m(\mathbf{r}) = 0, \pm 1$. If we limit ourselves to this case, the sum over the integer variables $m(\mathbf{r})$ can be performed explicitly as:

$$\sum_{m(\mathbf{r})=0,\pm 1} e^{\ln y m^2 + 2\pi i m \phi} = 1 + 2y \cos(2\pi\phi) \approx e^{2y \cos(2\pi\phi)}. \tag{5.62}$$

Inserting this into Eq. (5.56), taking the limit of the continuum ($\sum_\mathbf{r} \to (1/a^2) \int d^2\mathbf{r}$), and rescaling $\phi \to \phi/\pi$, we finally obtain a partition function expressed in terms of the $\phi$ field only, that generalizes Eq. (5.23) above:

$$S_B = \int d\mathbf{r} dz \left[ \frac{(\nabla\phi)^2}{2\pi K} - \frac{g}{\pi a^2} \cos 2\phi + \frac{2i}{\Phi_0} \mathbf{A} \cdot (\nabla \times \hat{z}\phi) \right] \delta(z), \tag{5.63}$$

where we used again the definitions (5.28) and (5.29) for $K, g$. In Eq. (5.63) we added also the explicit $z$ dependence of the action, which is needed, since the $\mathbf{A}$ field depends, in general, also on the $z$ out-of-plane coordinate. The $\delta(z)$ function gives the proper boundary conditions for a truly 2D case (where there is no SC current outside the plane), while in the physical case of a SC film of finite thickness $d$, we shall assume that the sample quantities are averaged over $|z| < d/2$.

The action (5.63) and its corresponding partition function $Z = \int \mathcal{D}\phi e^{-S_B}$ allow for a straightforward definition of the physical observables at finite field. For example, the electric current follows as usual from the functional derivative of $F = -(1/\beta) \ln Z$ with respect to the gauge field $\mathbf{A}$, i.e.

$$\mathbf{J}_s(\mathbf{r}, z) = -c \frac{\partial F}{\partial \mathbf{A}(\mathbf{r}, z)} = -\frac{2ick_B T}{\Phi_0} \langle \nabla \times \hat{z}\phi(\mathbf{r}) \rangle \delta(z). \tag{5.64}$$

Equation (5.64) makes it evident that the current is purely transverse, as expected for vortex excitations, according to the discussion given in Sec. 5.2 above. A second quantity that can be easily obtained is the magnetization $\mathbf{M} = (\mathbf{B} - \mathbf{H})/4\pi$, defined as the functional derivative of $F$ with respect to $\mathbf{B}(\mathbf{r}, z) = \nabla \times \mathbf{A}$. By integrating by part and using the identity $\int (\mathbf{A} \times \nabla\phi) \cdot \hat{z} = \int \nabla\phi \cdot (\hat{z} \times \mathbf{A}) = -\int \phi \nabla \cdot (\hat{z} \times \mathbf{A}) = \int \phi \hat{z} \cdot (\nabla \times \mathbf{A})$, we can rewrite the last term of Eq. (5.63) as

$$S_B = \int d\mathbf{r} dz \left[ \frac{(\nabla\phi)^2}{2\pi K} - \frac{g}{\pi a^2} \cos 2\phi + \frac{2i}{\Phi_0} \mathbf{B}(\mathbf{r}, z) \cdot \hat{z}\phi(\mathbf{r}) \right] \delta(z). \tag{5.65}$$

As a consequence, the functional derivative with respect to **B** gives immediately

$$\mathbf{M}(\mathbf{r}) = -\frac{1}{d}\int dz \frac{\partial F}{\partial \mathbf{B}(\mathbf{r},z)} = -\hat{z}\frac{2ik_BT}{d\Phi_0}\langle\phi(\mathbf{r})\rangle, \quad (5.66)$$

and analogously the uniform magnetic susceptibility $\chi = \partial M_z/\partial B_z$ is:

$$\chi = -(4k_BT)/(d\phi_0^2)\int d\mathbf{r}[\langle\phi(\mathbf{r})\phi(0)\rangle - \langle\phi(\mathbf{r})\rangle\langle\phi(0)\rangle]. \quad (5.67)$$

Notice that the total magnetic moment $\mathcal{M}$ associated with the current $\mathbf{J}_s$ is $\mathcal{M} = (1/2c)\int d\mathbf{r}dz\,(\mathbf{r}\times\mathbf{J}_s)$.[77] Using the definition (5.64) of $\mathbf{J}_s$, one can easily verify that $\mathcal{M} = \int d^2\mathbf{r}dz\mathbf{M}$, i.e. the magnetization is the density of magnetic moment of the sample, as expected.[77] Finally, by exploiting the fact that $e^{-\beta\mu}e^{\pm i2\phi}$ is the operator which creates up and down vortices with density $n_\pm$ respectively, we have a straightforward definition of the average vortex number $n_F = a^2(\langle n_+\rangle+\langle n_-\rangle)$ and of the excess vortex number $n = a^2(\langle n_+\rangle - \langle n_-\rangle)$ per unit cell as a function of $\phi$ as:

$$n_F = 2e^{-\beta\mu}\langle\cos(2\phi)\rangle, \quad n = 2e^{-\beta\mu}\langle\sin(2\phi)\rangle. \quad (5.68)$$

In Eq. (5.66) above, the average value of $\phi$ is computed with the action (5.63), so that the magnetization is given as a function of the magnetic induction **B**. However, as we mentioned above, at low external field, it would be more convenient to compute **M** as a function of the applied field **H**. This can be achieved by using the Gibbs free energy $\mathcal{G} = -k_BT\ln Z_G$, where the partition function $Z_G$ includes also the contribution of the electromagnetic field $Z_G = \int \mathcal{D}\phi\mathcal{D}A e^{-S}$ and:

$$S = S_B + \int d\mathbf{r}dz\left\{\frac{(\nabla\times\mathbf{A})^2}{8\pi k_BT} - \frac{(\nabla\times\mathbf{A})\cdot\mathbf{H}}{4\pi k_BT}\right\}. \quad (5.69)$$

Before discussing explicitly the case of a finite external field **H**, we would like to stress how Eq. (5.69) gives also a very convenient description of the role of charged supercurrents in a 2D superconductor. Indeed, even when $\mathbf{H} = 0$, the electromagnetic field **A** in Eq. (5.69) above describes the magnetic field created by the current themselves in a charged superfluid. In this case, since the SC currents live in the plane, **A** also is a two-dimensional vector, so that $\nabla\times\mathbf{A} = (-\partial_z A_y, \partial_z A_x, \partial_x A_y - \partial_y A_x)$. Moreover, if we choose the Coulomb (or radial) gauge $\nabla\cdot\mathbf{A} = 0$, we have that in Fourier space $(\mathbf{k}_\parallel\times\mathbf{A})^2 = k_\parallel^2\mathbf{A}^2$. We can then rewrite in Fourier space the terms in **A** of Eq. (5.69) at $\mathbf{H} = 0$ as:

$$\int \frac{d^3k}{(2\pi)^3}\left\{-\frac{2}{\Phi_0}\phi(\mathbf{k}_\parallel)|\mathbf{k}_\parallel\times A(\mathbf{k}_\parallel,k_z)| - \frac{(k_z^2+k_\parallel^2)}{8\pi T}\mathbf{A}^2(\mathbf{k}_\parallel,k_z)\right\}. \quad (5.70)$$

By integrating $\mathbf{A}$ at Gaussian level, we obtain a $\phi^2$ contribution to the action of the form:

$$\int \frac{d^3\mathbf{k}}{(2\pi)^3} \frac{8\pi T}{\Phi_0^2} \frac{\mathbf{k}_\parallel^2}{(k_z^2 + \mathbf{k}_\parallel^2)} |\phi(\mathbf{k}_\parallel)|^2 \,. \tag{5.71}$$

Since $\phi$ depends on $\mathbf{k}_\parallel$ only, we can integrate $k_z$ and obtain that the overall Gaussian action for the $\phi$ field reads:

$$S_G = \int \frac{d^2\mathbf{k}_\parallel}{(2\pi)^2} \frac{1}{2\pi K} [k_\parallel^2 + k_\parallel \Lambda^{-1}] |\phi(\mathbf{k}_\parallel)|^2 \,, \tag{5.72}$$

where we defined:

$$\frac{1}{\Lambda} = \frac{8\pi^2 KT}{\Phi_0^2} \equiv \frac{d}{2\lambda^2} \,. \tag{5.73}$$

The last equality follows from the definitions (5.28) and (5.5) of $K = \pi J/T$ and $J = \Phi_0^2 d/16\pi^3 \lambda^2$, respectively, and it allows us to identify $\Lambda$ with the so-called Pearl screening length.[6] Indeed, due to the $k_\parallel \Lambda^{-1}$ terms in Eq. (5.72), one sees that the potential between vortices, which according to the above discussion is the Fourier transform of the Gaussian $\phi$ propagator, decays as $e^{-r/\Lambda}$ at scales $r \gg \Lambda$, instead of the usual $\log r$ dependence observed at all length scales in neutral superfluids.

Let us go back now to the case of a finite external field in Eq. (5.69) and let us integrate again the gauge field $\mathbf{A}$. Since now, an additional term $-(i/4\pi T)\mathbf{A}(\mathbf{k}) \cdot (\mathbf{k} \times \mathbf{H}(-\mathbf{k}))$ is present in $S$, one obtains the following contribution to the action:

$$\frac{8\pi T}{k^2} \left| \frac{\phi(-\mathbf{k}_\parallel)}{\phi_0} (\hat{z} \times \mathbf{k}_\parallel) + \frac{i}{8\pi T}(\mathbf{k} \times \mathbf{H}(-\mathbf{k})) \right| \times [\mathbf{k} \to -\mathbf{k}] \,. \tag{5.74}$$

The quadratic term in $\phi$ corresponds to Eq. (5.71) above, leading to the screening of the vortex potential. The remaining terms can be written as:

$$S_{\phi-H} = \int \frac{d^3\mathbf{k}}{(2\pi)^3} \frac{2i}{\Phi_0 k^2} (\hat{z} \times \mathbf{k}_\parallel \phi(\mathbf{k}_\parallel)) \cdot (\mathbf{k} \times \mathbf{H}(-\mathbf{k})) \,, \tag{5.75}$$

and

$$S_{H-H} = \int \frac{d^3\mathbf{k}}{(2\pi)^3} \frac{1}{8\pi T} \frac{(\mathbf{k} \times \mathbf{H})^2}{k^2} \,. \tag{5.76}$$

Using the identity

$$\int \frac{d^3\mathbf{k}}{(2\pi)^3} F_1(\mathbf{k}) F_2(-\mathbf{k}) \frac{1}{\mathbf{k}^2} = \int d^3\mathbf{r} d^3\mathbf{r}' F_1(\mathbf{r}) F_2(\mathbf{r}') \frac{1}{4\pi |\mathbf{r} - \mathbf{r}'|} \,, \tag{5.77}$$

and the Maxwell equation relating the magnetic field $\mathbf{H}$ to the distribution of the external current $\mathbf{J}_{\text{ext}}$ producing the field itself

$$\nabla \times \mathbf{H} = \frac{4\pi}{c} \mathbf{J}_{\text{ext}}, \tag{5.78}$$

one can easily see that Eqs. (5.75) and (5.76) can be written in real space as:

$$S_{\phi-H} = 2i \int d^3r d^3r' \frac{[\hat{z} \times \nabla \phi(\mathbf{r})] \cdot [\nabla' \times \mathbf{H}(\mathbf{r}')]}{4\pi |\mathbf{r} - \mathbf{r}'|}, \tag{5.79}$$

and

$$S_{H-H} = -\frac{1}{8\pi T} \int d^3r d^3r' \frac{\mathbf{J}_{\text{ext}}(\mathbf{r}) \cdot \mathbf{J}_{\text{ext}}(\mathbf{r}')}{4\pi |\mathbf{r} - \mathbf{r}'|}, \tag{5.80}$$

respectively. Integrating by part, Eq. (5.79) can be rewritten as:

$$\begin{aligned}
S_{\phi-H} &= \frac{2i}{c} \int d^3r d^3r' \frac{[\hat{z} \times \nabla \phi(\mathbf{r})] \cdot \mathbf{J}_{\text{ext}}(\mathbf{r}')}{|\mathbf{r} - \mathbf{r}'|} \\
&= -\frac{2i}{c} \int d^3r d^3r' \phi(\mathbf{r}) \nabla_\mathbf{r} \cdot \frac{[\mathbf{J}_{\text{ext}}(\mathbf{r}') \times \hat{z}]}{|\mathbf{r} - \mathbf{r}'|} \\
&= \frac{2i}{c} \int d^3r d^3r' \phi(\mathbf{r}) \hat{z} \cdot \frac{[\mathbf{J}_{\text{ext}}(\mathbf{r}') \times (\mathbf{r} - \mathbf{r}')]}{|\mathbf{r} - \mathbf{r}'|^3} \\
&= \frac{2i}{\Phi_0} \int d^3r \phi(\mathbf{r}) \hat{z} \cdot \mathbf{H}^0(\mathbf{r}).
\end{aligned} \tag{5.81}$$

In the last equality of Eq. (5.81), we introduced the reference field $\mathbf{H}^0$, which corresponds to the magnetic field generated by the same distribution of currents $\mathbf{J}_{\text{ext}}$ *in the vacuum*. According to the Laplace formula, $\mathbf{H}^0$ is given exactly by

$$\mathbf{H}^0(\mathbf{r}) = \frac{1}{c} \int d^3r' \frac{\mathbf{J}_{\text{ext}}(\mathbf{r}') \times (\mathbf{r} - \mathbf{r}')}{|\mathbf{r} - \mathbf{r}'|^3}, \tag{5.82}$$

leading to Eq. (5.81) above. One also recognizes in the term (5.80) the magnetic energy density associated to the reference field $\mathbf{H}^0$,

$$S_{H-H} = -\frac{1}{8\pi T} \int d^3r (\mathbf{H}^0)^2. \tag{5.83}$$

Indeed, since $\mathbf{H}^0$ is the field created by the currents $\mathbf{J}_{\text{ext}}$ in the vacuum, it satisfies $\nabla \cdot \mathbf{H}^0 = 0$, so that $(\mathbf{k} \times \mathbf{H}^0)^2 = (\mathbf{H}^0)^2 k^2$. Thus $\mathbf{H}^0(\mathbf{k})^2 = (\nabla \times \mathbf{H}^0)/k^2 = \mathbf{J}_{\text{ext}}(\mathbf{k})^2/k^2$ that is the Fourier transform of Eq. (5.80). In summary, the full sine-Gordon action after integration of the gauge field can be rewritten as:

$$S = \int \frac{d^2\mathbf{k}_\parallel}{(2\pi)^2} \frac{k_\parallel^2 + k\Lambda^{-1}}{2\pi K} |\phi(\mathbf{k}_\parallel)|^2 - \frac{g}{\pi a^2} \int d\mathbf{r} \cos 2\phi$$
$$+ \frac{2i}{\Phi_0} \int d\mathbf{r}\, \phi\, \hat{z} \cdot \mathbf{H}^0(\mathbf{r}, z=0) - \int dr dz \frac{(\mathbf{H}^0)^2}{8\pi k_B T}. \quad (5.84)$$

Equation (5.84) is the desired result to be used to evaluate the physical observable as a function of the reference field $\mathbf{H}^0$. Once more, the sine-Gordon mapping facilitates a quite powerful framework for the investigation of BKT physics of a superconductor embedded in an external field. Indeed, apart from the fact that it includes automatically the screening effect of the supercurrents discussed above, the action (5.84) expressed in terms of $\mathbf{H}^0$ has two main advantages. First of all, $\mathbf{H}^0$ is the field quoted in the experimental measurements, since what is known *a priori* are only the generating currents $\mathbf{J}_{\text{ext}}$. Indeed, $\mathbf{H}^0$ does not coincide, in general, with the real field $\mathbf{H}$ even outside the sample, since the $\mathbf{H}$ configuration takes into account also the field exclusion from the SC sample, the so-called demagnetization effects. For simple sample geometries, one can include these effects in a demagnetization coefficient $\eta$, and write in general the following relation between $\mathbf{B} = \mathbf{H} + 4\pi\mathbf{M}$ and $\mathbf{H}^0$:[77]

$$(1-\eta)\mathbf{H} + \eta\mathbf{B} = \mathbf{H}^0 \Rightarrow \mathbf{B} = \mathbf{H}^0 + 4\pi(1-\eta)\mathbf{M}. \quad (5.85)$$

In the complete Meissner phase one has $\mathbf{B} = 0$, which implies $-4\pi\mathbf{M} = \mathbf{H}$. However, from Eq. (5.85) it follows that $\mathbf{H} = \mathbf{H}^0/(1-\eta)$ so that:

$$\mathbf{M} = -\frac{1}{4\pi}\frac{\mathbf{H}^0}{1-\eta}. \quad (5.86)$$

While for a cylinder $\eta = 0$ and $\mathbf{H} = \mathbf{H}^0$, for a film of thickness $d$ and transversal dimension $R$, $\eta \sim 1 - d/R$, so one expects to find $\mathbf{M} \sim (R/d)\mathbf{H}^0$ below $H_{c1}$, i.e. a much smaller critical field with respect to the same system in the 3D geometry.[77,78] Since the magnetization $M$ calculated from Eq. (5.84) is already a function of $\mathbf{H}^0$, it will include automatically all the demagnetization effects and the complications of the thin-film geometry.

These properties have been derived in Ref. 76, where the magnetization has been computed by a variational approximation for the cosine term in the model (5.84). While we refer the reader to Ref. 76 for more details concerning these calculations, we would like to mention here one particular result, that is related to the discussion of the previous sections. It concerns the behavior of the field-induced magnetization above $T_{\text{BKT}}$, that is expected[5] to be proportional to $H$ with a coefficient depending on the SC correlation length:

$$M = -\frac{k_B T}{d\Phi_0^2}\xi^2 H. \tag{5.87}$$

In full analogy with the paraconductivity discussed in the previous sections, the functional dependence of the low-field magnetization $M$ on the BKT correlation length $\xi$ in Eq. (5.87) is the same as in the GL theory. While this result was already known in the literature,[5,38] our calculations based on the model (5.84) allowed to establish an upper limit $H_l$ for the validity of the linear regime (5.87)

$$H \lesssim H_l = 0.1\frac{\Phi_0}{\xi^2}\sqrt{\frac{T - T_{\text{BKT}}}{T}}. \tag{5.88}$$

Notice that the above relation can be approximately expressed as the condition $\xi \gg \ell_B$ for the low-field limit to be applied, where $\ell_B^2 = \Phi_0/H$ is the magnetic length scale.[44] As $T$ approaches $T_{\text{BKT}}$, $\xi$ increases rapidly and the field $H_l$ becomes rapidly smaller than the lowest field accessible in the standard experimental set-up. This effect can explain, for example, the nonlinear magnetization effects reported recently in several measurements in cuprate superconductors.[34-36] Indeed, the persistence of a nonlinear magnetization up to $H \sim 0.01$ T in a wide range of temperatures above $T_{\text{BKT}}$ can be a signature of the rapid decrease of $H_l$ as $T \to T_{\text{BKT}}$, which does not contradict, but eventually supports the BKT nature of the SC fluctuations in these systems. Moreover, since $\xi$ increases as $\mu$ increases, the extremely low values of $H_l$ measured in Ref. 36 suggest a value of $\mu$ larger than $\mu_{\text{XY}}$, in agreement with the result discussed in Sec. 5.3 based on the analysis of the superfluid density. On the other hand, the existence of inhomogeneities can also alter the straightforward manifestation of a linear magnetization above $T_{\text{BKT}}$, an issue that has not been explored yet neither in the context of cuprates nor in the case of conventional superconductors. Finally, we would like to mention that even though some theoretical work exists[38] on the RG approach to the BKT transition at finite magnetic field based on the Coulomb-gas analogy, a full analysis of the more general model (5.84) is still lacking. Such an approach could eventually improve the estimate (5.88) of the linear regime, based on a variational calculation that is not expected to capture the correct critical behavior as the transition is approached.

## 5.6. Conclusions

It is clear that BKT theory has profoundly changed our understanding of quasi-2D superconductors and given us a tool to tackle such challenging and interesting problems. However, more than 40 years after the original discovery, the occurrence of the BKT transition in several quasi-2D superconducting materials remains partly controversial. One can, in general, identify two possible sources of discrepancies between theoretical predictions and the current experimental scenario. On the one hand, the original formulation was based on the paradigmatic case of the XY model, that is, only one possible model where the BKT transition occurs. Even though it correctly reproduces the critical behavior of all the systems belonging to the same universality class, quantitative discrepancies away from criticality can be observed in different models. This is the case of the strong superfluid-stiffness renormalization *below* the transition temperature $T_{BKT}$ in the case of superconducting films of conventional superconductors, where the vortex-core energy attains values significantly different from the XY-model prediction. On the other hand, emerging new materials and improved experimental techniques offer new scenarios for the occurrence of the BKT transition, which coexists with several other phenomena. An example is provided by the case of cuprate superconductors, that are layered systems formed by strongly-correlated 2D SC layers. In this case, the deviations of the vortex-core energy from the XY model value can eventually lead to a qualitatively different behavior of the superfluid-density jump at the transition or to strong nonlinear field-induced magnetization effects above $T_{BKT}$. In the present article, we have reviewed a possible approach to all these issues based on the sine-Gordon model. Even though this is certainly not a new approach for the pure 2D case, in the presence of additional relevant perturbations, it provides a very convenient framework to investigate BKT physics. Indeed, it allows not only to incorporate easily the effects of a vortex-core energy value different from the XY-model, but also to describe the coupling to the electromagnetic field in a clear way, giving a straightforward and elegant description of the charged superfluid. Finally, we would like to emphasize once more that an interesting issue, that applies equally well to conventional and unconventional superconductors, is posed by the role of the intrinsic sample inhomogeneity. Even though we outlined here a kind of mesoscopic approach to the emergence of spatially inhomogenous SC properties, a more microscopic approach to the effect of disorder on the BKT transition would be required, as suggested by some recent numerical works.[79,80] The theoretical and experimental investigation of this issue will certainly offer another perspective on BKT transition in low-dimensional superconductors.

## Acknowledgments

We thank M. Cazalilla, S. Caprara, A. Caviglia, A. Ho, M. Gabay, S. Gariglio, M. Grilli, J. Lesueur, N. Reyren, P. Raychaudhuri, and J. M. Triscone for enjoyable collaborations and discussions. This work was supported in part by the Swiss NSF under MaNEP and Division II.

## References

1. V. L. Berezinsky, *Sov. Phys. JETP* **34**, 610 (1972).
2. J. M. Kosterlitz and D. J. Thouless, *J. Phys. C* **6**, 1181 (1973).
3. J. M. Kosterlitz, *J. Phys. C* **7**, 1046 (1974).
4. S. Doniach and B. A. Huberman, *Phys. Rev. Lett.* **42**, 1169 (1979).
5. B. I. Halperin and D. R. Nelson, *J. Low. Temp. Phys.* **36**, 599 (1979).
6. J. Pearl, *Appl. Phys. Lett.* **5**, 65 (1964).
7. D. R. Nelson and J. M. Kosterlitz, *Phys. Rev. Lett.* **39**, 1201 (1977).
8. P. Minnaghen, *Rev. Mod. Phys.* **59**, 1001 (1987).
9. T. Giamarchi, *Quantum Physics in One Dimension* (Oxford University Press, Oxford, 2004).
10. A. Larkin and A. A. Varlamov, *Theory of Fluctuations in Superconductors* (Oxford University Press, Oxford, 2005).
11. S. J. Turneaure, T. R. Lemberger and J. M. Graybeal, *Phys. Rev. B* **63**, 174505 (2001).
12. A. Rüfenacht, J.-P. Locquet, J. Fompeyrine, D. Caimi and P. Martinoli, *Phys. Rev. Lett.* **96**, 227002 (2006).
13. I. Hetel, T. R. Lemberger and M. Randeria, *Nat. Phys.* **3**, 700 (2007).
14. A. Kamlapure, M. Mondal, M. Chand, A. Mishra, J. Jesudasan, V. Bagwe, L. Benfatto, V. Tripathi and P. Raychaudhuri, *Appl. Phys. Lett.* **96**, 072509 (2010).
15. M. Mondal, S. Kumar, M. Chand, A. Kamlapure, G. Saraswat, G. Seibold, L. Benfatto and P. Raychaudhuri, *Phys. Rev. Lett.* **106**, 047001 (2011).
16. J. Yong, M. Hinton, A. McCray, M. Randeria, M. Naamneh, A. Kanigel and T. Lemberger, *Phys. Rev. B* **85**, 180507 (2012).
17. S. J. Turneaure, E. R. Ulm and T. R. Lemberger, *J. Appl. Phys.* **79**, 4221 (1996).
18. R. W. Crane, N. P. Armitage, A. Johansson, G. Sambandamurthy, D. Shahar and G. Grüner, *Phys. Rev. B* **75**, 094506 (2007).
19. W. Liu, M. Kim, G. Sambandamurthy and N. Armitage, *Phys. Rev. B* **84**, 024511 (2011).
20. L. S. Bilbro, R. V. Aguilar, G. Logvenov, O. Pelleg, I. Bozović and N. P. Armitage, *Nat. Phys.* **7**, 298 (2011).
21. N. Reyren, S. Thiel, A. D. Caviglia, L. F. Kourkoutis, G. Hammerl, C. Richter, C. W. Schneider, T. Kopp, A.-S. Retschi, D. Jaccard, M. Gabay, D. A. Muller, J.-M. Triscone and J. Mannhart, *Science* **317**, 1196 (2007).
22. A. D. Caviglia, S. Gariglio, N. Reyren, D. Jaccard, T. Schneider, M. Gabay,

S. Thiel, G. Hammerl, J. Mannhart and J.-M. Triscone, *Nature (London)* **456**, 624 (2008).
23. J. Biscaras, N. Bergeal, A. Kushwaha, T. Wolf, A. Rastogi, R. Budhani and J. Lesueur, *Nat. Comm.* **1**, 89 (2010).
24. J. T. Ye, S. Inoue, K. Kobayashi, Y. Kasahara, H. T. Yuan, H. Shimotani and Y. Iwasa, *Nature Mater.* **9**, 125 (2010).
25. V. Emery and S. Kivelson, *Nature (London)* **374**, 434 (1995).
26. P. A. Lee, N. Nagaosa and X.-G. Wen, *Rev. Mod. Phys.* **78**, 17 (2006).
27. J. Corson, R. Mallozzi, J. Orenstein, J. Eckstein and J. N. Bozovic, *Nature* **398**, 221 (1999).
28. D. Matthey, N. Reyren, T. Schneider and J.-M. Triscone, *Phys. Rev. Lett.* **98**, 057002 (2007).
29. O. Yuli, I. Asulin, O. Millo, D. Orgad, L. Iomin and G. Koren, *Phys. Rev. Lett.* **101**, 057005 (2008).
30. G. Logvenov, A. Gozar and I. Bozovic, *Science* **326**, 699 (2009).
31. T. Schneider, A. Caviglia, S. Gariglio, N. Reyren and J.-M. Triscone, *Phys. Rev. B* **79**, 184502 (2009).
32. L. Benfatto, C. Castellani and T. Giamarchi, *Phys. Rev. B* **80**, 214506 (2009).
33. Y. Wang, L. Li and N. P. Ong, *Phys. Rev. B* **73**, 024510 (2006).
34. L. Li, J. G. Checkelsky, S. Komiya, Y. Ando and N. P. Ong, *Nat. Phys.* **3**, 311 (2007).
35. L. Li, Y. Wang, S. Komiya, S. Ono, Y. A. G. D. Gu and N. P. Ong, *Phys. Rev. B* **81**, 054510 (2010).
36. L. Li, Y. Wang, M. J. Naughton, S. Ono, Y. Ando and N. P. Ong, *Europhys. Lett.* **72**, 451 (2005).
37. E. Bernardi, A. Lascialfari, A. Rigamonti, L. Romanò, M. Scavini and C. Oliva, *Phys. Rev. B* **81**, 064502 (2010).
38. D. A. H. V. Oganesyan and S. L. Sondhi, *Phys. Rev. B* **73**, 094503 (2006).
39. S. Raghu, D. Podolsky, A. Vishwanath and D. A. Huse, *Phys. Rev. B* **78**, 184520 (2008).
40. D. Podolsky, S. Raghu and A. Vishwanath, *Phys. Rev. Lett.* **99**, 117004 (2007).
41. A. Andersson and J. Lidmar, *Phys. Rev. B* **83**, 174502 (2011).
42. O. Cyr-Choinière, R. Daou, F. Laliberté, D. LeBoeuf, N. Doiron-Leyraud, J. C. J.-Q. Yan, J.-G. Cheng, J.-S. Zhou, J. B. Goodenough, S. Pyon, T. Takayama, H. Takagi, Y. Tanaka and L. Taillefer, *Nature (London)* **458**, 743 (2009).
43. A. Levchenko, M. R. Norman and A. A. Varlamov, *Phys. Rev. B* **83**, 020506 (2011).
44. A. Pourret, H. Aubin, J. Lesueur, C. A. Marrache-Kikuchi, L. Berge, L. Dumoulin and K. Behnia, *Nat. Phys.* **2**, 683 (2006).
45. B. Leridon, J. Vanacken, T. Wambecq and V. V. Moshchalkov, *Phys. Rev. B* **76**, 012503 (2007).
46. F. Rullier-Albenque, H. Alloul and G. Rikken, *Phys. Rev. B* **84**, 014522 (2011).
47. D. M. Broun, W. A. Huttema, P. J. Turner, S. Ozcan, B. Morgan, R. Liang, W. N. Hardy and D. A. Bonn, *Phys. Rev. Lett.* **99**, 237003 (2007).
48. K. K. Gomes, A. N. Pasupathy, A. Pushp, S. Ono, Y. Ando and A. Yazdani, *Nature* **447**, 569 (2007).

49. B. Sacépé, C. Chapelier, T. I. Baturina, V. M. Vinokur, M. R. Baklanov and M. Sanquer, *Phys. Rev. Lett.* **101**, 157006 (2008).
50. B. Sacépé, C. Chapelier, T. I. Baturina, V. M. Vinokur, M. R. Baklanov and M. Sanquer, *Nature Communications* **1**, 140 (2010).
51. B. Sacépé, T. Dubouchet, C. Chapelier, M. Sanquer, M. Ovadia, D. Shahar, M. Feigel'man and L. Ioffe, *Nature Phys.* **7**, 239 (2011).
52. M. Mondal, A. Kamlapure, M. Chand, G. Saraswat, S. Kumar, J. Jesudasan, L. Benfatto, V. Tripathi and P. Raychaudhuri, *Phys. Rev. Lett.* **106**, 047001 (2011).
53. M. Chand, G. Saraswat, A. Kamlapure, M. Mondal, S. Kumar, J. Jesudasan, V. Bagwe, L. Benfatto, V. Tripathi and P. Raychaudhuri, *Phys. Rev. B* **85**, 014508 (2012).
54. Y. Noat, T. Cren, C. Brun, F. Debontridder, V. Cherkez, K. Ilin, M. Siegel, A. Semenov, H.-W. Hubers and D. Roditchev, arXiv:1205.3408 (2012).
55. M. V. Feigel'man, L. B. Ioffe, V. E. Kravtsov and E. Cuevas, *Ann. Phys.* **325**, 1390 (2010).
56. Y. Dubi, Y. Meir and Y. Avishai, *Nature* **449**, 876 (2008).
57. K. Bouadim, Y. L. Loh, M. Randeria and N. Trivedi, *Nature Phys.* **7**, 884 (2011).
58. G. Seibold, L. Benfatto, C. Castellani and J. Lorenzana, *Phys. Rev. Lett.* **108**, 207004 (2012).
59. N. Nagaosa, *Quantum Field Theory in Condensed Matter Physics* (Springer, New York, 1999).
60. D. J. Bishop and J. D. Reppy, *Phys. Rev. B* **22**, 5171 (1980).
61. L. Benfatto, C. Castellani and T. Giamarchi, *Phys. Rev. Lett.* **98**, 117008 (2007).
62. V. Cataudella and P. Minnaghen, *Physica C* **166**, 442 (1990).
63. B. Chattopadhyay and S. R. Shenoy, *Phys. Rev. Lett.* **72**, 400 (1994).
64. M. Friesen, *Phys. Rev. B* **51**, 632 (1995).
65. S. W. Pierson, *Phys. Rev. B* **51**, 6663 (1995).
66. P. Minnaghen and P. Olsson, *Phys. Rev. B* **44**, 4503 (1991).
67. I. Nandori, *J. Phys. Cond. Matt.* **19**, 236226 (2007).
68. A. F. Ho, M. A. Cazalilla and T. Giamarchi, *Phys. Rev. Lett.* **92**, 130405 (2004).
69. M. A. Cazalilla, A. F. Ho and T. Giamarchi, *New. J. of Phys.* **8**, 158 (2006).
70. L. Benfatto, C. Castellani and T. Giamarchi, *Phys. Rev. B* **77**, 100506(R) (2008).
71. J. H. She and A. V. Balatsky, *Phys. Rev. Lett.* **109**, 077002 (2012).
72. V. Ambegaokar, B. I. Halperin, D. R. Nelson and E. D. Siggia, *Phys. Rev. B* **21**, 1806 (1979).
73. M. Gabay and A. Kapitulnik, *Phys. Rev. Lett.* **71**, 2138 (1993).
74. S. Caprara, M. Grilli, L. Benfatto and C. Castellani, *Phys. Rev. B* **84**, 014514 (2011).
75. P. Minnaghen, *Phys. Rev. B* **23**, 5745 (1981).
76. L. Benfatto, C. Castellani and T. Giamarchi, *Phys. Rev. Lett.* **99**, 207002 (2007).

77. L. D. Landau and E. M. Lifchitz, *Electrodynamics of Continuous Media* (Pergamon, Oxford, 1984).
78. A. L. Fetter and P. C. Hohenberg, *Phys. Rev.* **159**, 330 (1967).
79. A. Erez and Y. Meir, *Europhys. Lett.* **91**, 47003 (2010).
80. G. Y. Conduit and Y. Meir, *Phys. Rev. B* **84**, 064513 (2011).

Chapter 6

# The Two-Dimensional Fully Frustrated XY Model

Stephen Teitel

*Department of Physics and Astronomy, University of Rochester, Rochester, NY 14627, USA*

In this chapter, we discuss the two-dimensional uniformly frustrated XY model, which arises as a model for a periodic array of Josephson junctions in an applied magnetic field. We will focus primarily on the fully frustrated model, which exhibits both phase angle ordering of the XY spins as well as a discrete $Z_2$ ordering corresponding to the spatial structure of vortices in the ground state of the model.

## 6.1. Introduction

The seminal works of Berezinskii[1,2] and of Kosterlitz and Thouless[3,4] on the role of topological excitations in mediating continuous phase transitions have found wide application to numerous physical systems. In this chapter, we discuss a class of classical two-dimensional statistical models, known as the uniformly frustrated XY model, which serves as a model for describing behavior in a planar periodic array of Josephson junctions in a perpendicular applied magnetic field. We will focus primarily on the specific case known as the fully frustrated XY model (FFXY). We will show how the ideas of Berezinskii, and of Kosterlitz and Thouless are crucial for an understanding of behavior in the FFXY, and we will make use of them in two different contexts: first, in arguing about the loss of phase angle coherence in the model due to a vortex–antivortex unbinding transition, and second, in discussing a kink–antikink transition that takes place along the domain walls separating regions of different discrete symmetry in the system. We will discuss the intimate connection that exists between these two different phenomena.

## 6.2. The Uniformly Frustrated XY Model

The Hamiltonian for the ordinary two-dimensional (2D) XY model on a square lattice is,

$$\mathcal{H} = -J \sum_{i\mu} \mathbf{S}(\mathbf{r}_i) \cdot \mathbf{S}(\mathbf{r}_i + \hat{\mu}). \tag{6.1}$$

Here $\mathbf{S}(\mathbf{r}_i)$ is a planar spin of unit magnitude on site $\mathbf{r}_i = (x, y)$ of a square lattice, the sum is over nearest neighbor bonds in directions $\hat{\mu} = \hat{x}, \hat{y}$, and $J > 0$ is the ferromagnetic coupling constant. We will take the coordinates $x$ and $y$ as integers.

If one represents the spin $\mathbf{S}(\mathbf{r}_i)$ in terms the angle $\theta(\mathbf{r}_i)$ it makes with respect to some fixed direction, then the Hamiltonian (6.1) can be rewritten as,

$$\mathcal{H} = -J \sum_{i\mu} \cos(\theta(\mathbf{r}_i + \hat{\mu}) - \theta(\mathbf{r}_i)). \tag{6.2}$$

The latter expression for the Hamiltonian suggests a particular physical application of the model. If we regard the nodes of our lattice as superconducting islands, with $\theta(\mathbf{r}_i)$ the phase angle of the superconducting wavefunction on island $i$, $\psi(\mathbf{r}_i) = |\psi|e^{i\theta(\mathbf{r}_i)}$, then Eq. (6.2) becomes the Hamiltonian for an array of Josephson junctions, with one junction on each bond of the lattice.[5–7]

This mapping to a Josephson junction array then suggests an interesting generalization of the XY model. The Hamiltonian (6.2) represents a Josephson junction array in the absence of any applied magnetic field. If a magnetic field is applied, we must modify Eq. (6.2) so that the phase angle difference becomes the *gauge invariant* phase angle difference between neighboring nodes,

$$\mathcal{H} = -J \sum_{i\mu} \cos(\theta(\mathbf{r}_i + \hat{\mu}) - \theta(\mathbf{r}_i) - A_\mu(\mathbf{r}_i)), \tag{6.3}$$

where

$$A_\mu(\mathbf{r}_i) \equiv \frac{2\pi}{\Phi_0} \int_{\mathbf{r}_i}^{\mathbf{r}_i + \hat{\mu}} \mathbf{A} \cdot \ell \tag{6.4}$$

is proportional to the line integral of the magnetic vector potential $\mathbf{A}$ across the bond at note $i$ in direction $\hat{\mu}$, and $\Phi_0 = hc/2e$ is the flux quantum. The sum of the $A_\mu(\mathbf{r}_i)$ going counterclockwise around any closed path $C$ of bonds on the lattice is $2\pi$ times the number of magnetic flux quanta $f_C$ penetrating

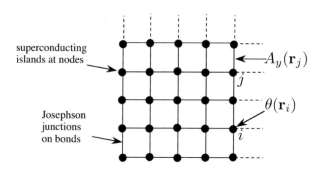

Fig. 6.1. Schematic geometry of the square lattice Josephson junction array with Hamiltonian as in Eq. (6.3).

the path,

$$\sum_C A_\mu(\mathbf{r}_i) = \frac{2\pi}{\Phi_0} \oint_C \mathbf{A} \cdot \boldsymbol{\ell} = 2\pi \frac{\Phi_C}{\Phi_0} \equiv 2\pi f_C \quad (6.5)$$

where $\Phi_C$ is the total magnetic flux through the path $C$. In Fig. 6.1 the geometry of such a Josephson junction array is illustrated.

The addition of the phase factor $A_\mu(\mathbf{r}_i)$ to the argument of the cosine in Eq. (6.3) adds *frustration* to the system; in general, the ground state will no longer be ferromagnetic with all $\theta(\mathbf{r}_i)$ equal, but rather the $\theta(\mathbf{r}_i)$ will vary from node to node so as to try and minimize the gauge invariant phase angle difference $\theta(\mathbf{r}_i + \hat{\mu}) - \theta(\mathbf{r}_i) - A_\mu(\mathbf{r}_i)$ across the bonds. Depending on the values of the $A_\mu(\mathbf{r}_i)$, the ground state can develop inhomogeneous spatial structure.

For the case of a uniform magnetic field $H$ applied perpendicular to the plane of the array, the number $f$ of flux quanta per unit cell of the square array is constant, and Eq. (6.3) is known as the *uniformly frustrated* XY model with uniform frustration $f$.[8,9] The ground state of the system now consists of a periodic configuration of vortices in the phase angle $\theta(\mathbf{r}_i)$, similar to the Abrikosov vortex lattice in a type-II superconductor. However, whereas in a homogeneous continuous superconductor the Abrikosov vortex lattice is always triangular, in the Josephson array vortices are constrained to sit at the centers of the unit cells of the array lattice. The result, in general, is a nontrivial spatial structure for the vortices that results from a competition between vortex–vortex repulsion and commensurability with the geometry of the Josephson array. In Fig. 6.2 are shown the ground state vortex structures for several simple rational fractions $f$.

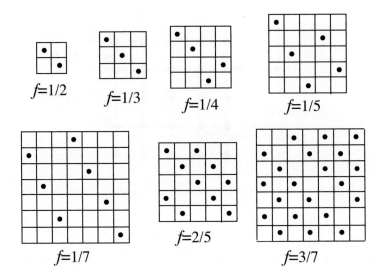

Fig. 6.2. Ground state vortex structures for several simple rational fractions of the uniform frustration $f$. The ground state is given by the periodic tiling of space with the shown structure. Solid dots indicate the location of vortices in the phase angle $\theta(\mathbf{r}_i)$.

A convenient choice of gauge for representing the uniform frustration $f$ is to take,

$$A_\mu(\mathbf{r}_i) = \begin{cases} 2\pi f y & \mu = x, \text{ i.e. on horizontal bonds} \\ 0 & \mu = y, \text{ i.e. on vertical bonds.} \end{cases} \quad (6.6)$$

If we now take $f \to f + 1$, we see that the $A_\mu(\mathbf{r}_i)$ on the horizontal bonds change by integral multiples of $2\pi$, leaving the Hamiltonian (6.3) invariant. It is thus only necessary to consider $f$ in the range of $-1/2$ to $+1/2$.

## 6.3. The Fully Frustrated XY Model

The main focus of this article will be the special case $f = 1/2$. In this case, using the gauge choice of Eq. (6.6), the horizontal bonds take values $0$ and $\pi$ (modulus $2\pi$) on alternating rows of bonds. For the Hamiltonian (6.3), bonds with $A_\mu(\mathbf{r}_i) = \pi$ become antiferromagnetic bonds, as $J\cos(\varphi - \pi) = -J\cos(\varphi)$. As the product of bonds around any unit cell of the array is always negative, this model is known as the *fully frustrated* XY model (FFXY).[10,11]

In Fig. 6.3 we show ground state configurations for the FFXY model. The gauge invariant phase angle difference across all bonds is $\pi/4$, as shown in the top row of Fig. 6.3. As in the ordinary XY model, rotating all spins,

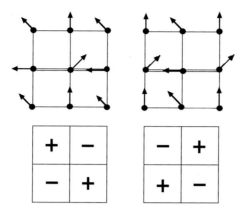

Fig. 6.3. Ground states of the fully frustrated XY model. Top: arrows denote spin directions, giving the phase angles $\theta(\mathbf{r}_i)$; double horizontal line denotes the antiferromagnetic bonds, where $A_\mu(\mathbf{r}_i) = \pi$. Bottom: corresponding charge configurations with $q_i = \pm 1/2$.

$\theta(\mathbf{r}_i) \to \theta(\mathbf{r}_i) + \theta_0$ with $\theta_0$ a constant, leaves the Hamiltonian (6.3) invariant. The ground state breaks this continuous symmetry $U(1)$ by picking out a particular direction for the spins. However the particular spatial structure of the ground state also leads to the breaking of a discrete two-fold symmetry $Z_2$. This is most readily seen in the "charge" representation.

Let us denote the gauge invariant phase angle difference across the bond leaving node $i$ in direction $\hat{\mu}$ by,

$$\varphi_\mu(\mathbf{r}_i) \equiv [\theta(\mathbf{r}_i + \hat{\mu}) - \theta(\mathbf{r}_i) - A_\mu(\mathbf{r}_i)]_{-\pi}^{\pi} \quad (6.7)$$

where the notation $[\ldots]_{-\pi}^{\pi}$ means that we take the value modulus $2\pi$ so that it lies in the interval $(-\pi, \pi]$. For simplicity of notation, we will denote a unit cell (plaquette) of the array by the position $\mathbf{r}_i$ of the node at the cell's lower left corner. We can then compute the circulation of the gauge invariant phase going counterclockwise around the unit cell at $\mathbf{r}_i$,

$$\varphi_x(\mathbf{r}_i) + \varphi_y(\mathbf{r}_i + \hat{x}) - \varphi_x(\mathbf{r}_i + \hat{y}) - \varphi_y(\mathbf{r}_i) = 2\pi(n_i - f) \equiv 2\pi q_i, \quad (6.8)$$

where $n_i$ is the integer vorticity in the phase angle $\theta(\mathbf{r}_i)$ going around the unit cell at $\mathbf{r}_i$, and $f$ is the uniform frustration, with $f = 1/2$ for the fully frustrated model. At low temperatures, the vorticity will take only the values $n_i = 0, +1$, leading to charges $q_i = \pm 1/2$. This charge analogy will be pursued in greater detail in a following section. Here we just note that at low temperature one finds only configurations with equal numbers of $+1/2$ and $-1/2$ charges. In the ground state, these charges are arranged in a checkerboard pattern, as shown in the bottom row of Fig. 6.3, with two possibilities for the sublattice on which to put the positive charges. Taking $q_i \to -q_i$ is then a discrete symmetry of the Hamiltonian (6.3) that is broken in the ground state.

The ground state of the FFXY model thus breaks both continuous $U(1)$ and discrete $Z_2$ symmetries. A main question of interest is whether, upon cooling, these two different symmetries are broken at the same or different temperatures.

### 6.3.1. *Phase angle ordering*

Just as in the ordinary XY model, one can consider smooth perturbations about the ground state to describe the low lying excitations of the system; this is the *spinwave approximation*. Writing $\theta(\mathbf{r}_i) = \theta_0(\mathbf{r}_i) + \delta\theta(\mathbf{r}_i)$, where $\theta_0(\mathbf{r}_i)$ is the value of the phase angle in the ground state and $\delta\theta(\mathbf{r}_i)$ is a smooth deviation, we can expand the Hamiltonian (6.3) for small $\delta\theta(\mathbf{r}_i)$,

$$\mathcal{H} = -J \sum_{i\mu} \cos(\theta_0(\mathbf{r}_i + \hat{\mu}) - \theta_0(\mathbf{r}_i) - A_\mu(\mathbf{r}_i) + \delta\theta(\mathbf{r}_i + \hat{\mu}) - \delta\theta(\mathbf{r}_i))$$

$$\approx \frac{2NJ}{\sqrt{2}} + \frac{J}{2\sqrt{2}} \sum_{i\mu} (\delta\theta(\mathbf{r}_i + \hat{\mu}) - \delta\theta(\mathbf{r}_i))^2, \qquad (6.9)$$

where we use $\cos(\theta_0(\mathbf{r}_i + \hat{\mu}) - \theta_0(\mathbf{r}_i) - A_\mu(\mathbf{r}_i)) = \cos(\pi/4) = 1/\sqrt{2}$ for all bonds in the ground state, and note that the linear term in $\delta\theta(\mathbf{r}_i)$ vanishes because we are expanding about the ground state. Assuming a smoothly varying $\delta\theta(\mathbf{r}_i)$, one can then compute the spin–spin correlation within this Gaussian spinwave approximation. The form of the Hamiltonian (6.9) is exactly the same as in the ordinary unfrustrated ($f = 0$) XY model, except for the factor $1/\sqrt{2}$. Proceeding with the same steps as in the case of the ordinary XY model[12] one finds,

$$\langle \mathbf{S}(\mathbf{r}_i) \cdot \mathbf{S}(\mathbf{r}_j) \rangle = e^{i(\theta_0(\mathbf{r}_j) - \theta_0(\mathbf{r}_i))} (\pi|\mathbf{r}_i - \mathbf{r}_j|)^{-\sqrt{2}T/2\pi J}, \qquad (6.10)$$

where here, and throughout this chapter, we take $k_B = 1$.

Thus, within this spinwave approximation, spin correlations decay algebraically at all temperatures. Since the correlation vanishes as $r \to \infty$ there is no long-range phase angle ordering at low temperatures. Yet the algebraic decay, denoted *quasi-long-range order*, is still more ordered than the exponential decay one expects at sufficiently large temperatures. Thus the spinwave approximation must leave out essential excitations that convert the algebraic decay to exponential as temperature increases. These excitations are the fluctuation of vortices in the phase angles $\theta(\mathbf{r}_i)$ away from their ground state positions shown in Fig. 6.3.

A convenient measure of phase coherence in the system is given by the *helicity modulus*, $\Upsilon(T)$.[13,14] To define the helicity modulus, we need to consider the boundary conditions applied to a finite sample. We take our array to be a finite square of length $L$ in the $x$ and $y$ directions, with a total of $N = L^2$ sites. Rather than apply the usual periodic boundary conditions, we can apply *twisted boundary conditions*, requiring,

$$\theta(\mathbf{r}_i + L\hat{\mu}) = \theta(\mathbf{r}_i) + \Delta_\mu, \qquad (6.11)$$

where $\Delta_\mu$ is the total phase angle twist applied across the system in direction $\hat{\mu} = \hat{x}, \hat{y}$.

We can now transform to a new set of variables,

$$\theta'(\mathbf{r}_i) \equiv \theta(\mathbf{r}_i) - \mathbf{r}_i \cdot \mathbf{\Delta}/L \qquad (6.12)$$

so that the $\theta'(\mathbf{r}_i)$ obey periodic boundary conditions, $\theta'(\mathbf{r}_i + L\hat{\mu}) = \theta'(\mathbf{r}_i)$. The Hamiltonian then becomes,

$$\mathcal{H} = -J \sum_{i\mu} \cos(\theta'(\mathbf{r}_i + \hat{\mu}) - \theta'(\mathbf{r}_i) - A_\mu(\mathbf{r}_i) - \Delta_\mu/L). \qquad (6.13)$$

We can now ask if the free energy $\mathcal{F}$ of the system depends on the boundary twist $\mathbf{\Delta}$ that is applied. If $\mathcal{F}(\mathbf{\Delta})$ varies with $\mathbf{\Delta}$ then the system is sensitive to the boundary conditions; what happens at the boundary effects the bulk, hence the system has phase angle coherence. If $\mathcal{F}(\mathbf{\Delta})$ is independent of $\mathbf{\Delta}$, then what happens at the boundary has no effect on the bulk; the system has no phase angle ordering. In terms of the Josephson junction array analogy, a dependence of $\mathcal{F}(\mathbf{\Delta})$ on $\mathbf{\Delta}$ means that an applied phase angle twist drives a net supercurrent through the system; the array is superconducting with superconducting phase angle coherence. When $\mathcal{F}(\mathbf{\Delta})$ is independent of $\mathbf{\Delta}$, applying a phase angle twist causes no net supercurrent to flow; the array is in the normal state.

For the FFXY model, the ground state energy is a minimum when the twist $\mathbf{\Delta} = 0$. We can thus measure the dependence of $\mathcal{F}$ on $\mathbf{\Delta}$ by measuring the stiffness of the free energy about this minimum. For small $\Delta_\mu$ we can approximate,

$$\mathcal{F}(\mathbf{\Delta}) \simeq \mathcal{F}(0) + \frac{1}{2}\Upsilon|\mathbf{\Delta}|^2, \qquad (6.14)$$

where the helicity modulus $\Upsilon$ is defined by,

$$\Upsilon_\mu \equiv \left.\frac{\partial^2 \mathcal{F}}{\partial \Delta_\mu^2}\right|_{\mathbf{\Delta}=0}. \qquad (6.15)$$

Using,
$$\mathcal{F}(\mathbf{\Delta}) = -T \ln Z(\mathbf{\Delta}) \tag{6.16}$$

with
$$Z(\mathbf{\Delta}) = \left(\prod_i \int_0^{2\pi} d\theta'(\mathbf{r}_i)\right) \exp(-\mathcal{H}[\theta'(\mathbf{r}_i); \mathbf{\Delta}]/T), \tag{6.17}$$

we arrive at,
$$\Upsilon_\mu = \frac{J}{N} \left\langle \sum_i \cos(\theta(\mathbf{r}_i + \hat{\mu}) - \theta(\mathbf{r}_i) - A_\mu(\mathbf{r}_i)) \right\rangle$$
$$- \frac{J^2}{TN} \left\langle \left[\sum_i \sin(\theta(\mathbf{r}_i + \hat{\mu}) - \theta(\mathbf{r}_i) - A_\mu(\mathbf{r}_i))\right]^2 \right\rangle \tag{6.18}$$

where the averages are evaluated in the untwisted ensemble $\mathbf{\Delta} = 0$. For the FFXY model one has $\Upsilon_x = \Upsilon_y$, hence we denote these simply as $\Upsilon$.

At low temperatures one can use the spinwave approximation to evaluate the helicity modulus. The calculation proceeds analogously to the case of the ordinary XY model,[14] with the substitution $J \to J/\sqrt{2}$. To quadratic order in $\delta\theta(\mathbf{r}_i)$ only the first term in Eq. (6.18) contributes and one finds,

$$\Upsilon \approx \frac{J}{\sqrt{2}} - \frac{T}{4}. \tag{6.19}$$

Thus the spinwaves cause $\Upsilon$ to decrease as $T$ increases.

But the vanishing of the helicity modulus, marking the loss of quasi-long-range phase ordering, is due not to these spinwave fluctuations, but rather to vortex fluctuations, via the same mechanism proposed by Berezinskii[2] and by Kosterlitz and Thouless[3] for the ordinary XY model. We now sketch the Kosterlitz–Thouless argument for the instability of the system to vortex excitations.

Consider inserting a free (unpaired) vortex $n = +1$ superimposed on the ground state of the system. Summing phase angle differences going counterclockwise around any closed path containing the vortex yields, by definition, $2\pi$. Such a vortex would lead to a phase angle shift $\delta\theta(\mathbf{r}_i)$ at site $\mathbf{r}_i = (x, y)$, given by,

$$\delta\theta(\mathbf{r}_i) = \arctan\left[\frac{y - y_0}{x - x_0}\right] \tag{6.20}$$

where the vortex is located at position $\mathbf{r}_0 = (x_0, y_0)$ at the center of one of the cells of the array. Far from the center of the vortex, the phase angle shifts across any given bond are small, and so one can approximate $\delta\theta(\mathbf{r}_i + \hat{\mu}) - \delta\theta(\mathbf{r}_i) \approx \hat{\mu} \cdot \nabla\delta\theta(\mathbf{r}_i)$ and use the Hamiltonian of Eq. (6.9) to compute the total energy $E_v$ of adding the vortex. We get,

$$E_v \approx \frac{J}{2\sqrt{2}} \sum_{i\mu} |\hat{\mu} \cdot \nabla\delta\theta(\mathbf{r}_i)|^2 = \frac{J}{2\sqrt{2}} \sum_i \frac{1}{|\mathbf{r}_i - \mathbf{r}_0|^2} \qquad (6.21)$$

$$\approx \frac{J}{2\sqrt{2}} 2\pi \int_1^R r dr \frac{1}{r^2} \approx \frac{\pi J}{\sqrt{2}} \ln R = \pi \Upsilon(0) \ln R, \qquad (6.22)$$

where $R$ is the radial size of the array, and we used $\Upsilon(0) = J/\sqrt{2}$ for the zero temperature value of the helicity modulus. Note that the corrections to the result in Eq. (6.22) come from bonds close to the vortex center $\mathbf{r}_0$, where $\delta\theta(\mathbf{r}_i+\hat{\mu}) - \delta\theta(\mathbf{r}_i)$ is not necessarily small. However these corrections remain finite, and hence Eq. (6.22) gives the correct leading asymptotic behavior in the thermodynamic limit, $R \to \infty$.

At finite temperature, one should consider the *free energy* $F_v$ to add the vortex. This involves two important changes to Eq. (6.22). Firstly, one must consider the entropy $S$ associated with adding the vortex. In a system of radius $R$, there are $\pi R^2$ cells on which to center the vortex, and so $S = \ln \pi R^2$. Secondly, at finite $T$, one should replace $\Upsilon(0)$ in Eq. (6.22) by $\Upsilon(T)$, as can be argued as follows. The $\ln R$ term in Eq. (6.22) comes from putting a $2\pi$ phase angle twist around the vortex at large distances $r$ from the vortex center. This corresponds to a slowly varying phase angle gradient $1/r$. At $T \to 0$, when all thermal phase angle fluctuations are frozen out, the energy increase for inserting this phase angle twist is determined by the *bare spinwave stiffness* $J/\sqrt{2} = \Upsilon(0)$, as in the spinwave Hamiltonian (6.9). At finite $T$, however, the $2\pi$ twist around the vortex occurs in the presence of thermal fluctuations in the phase angles. One should thus replace the bare spinwave stiffness by a *renormalized spinwave stiffness* that measures the increase in free energy from applying slowly varying phase angle gradients in the presence of such thermal fluctuations. This is just the helicity modulus $\Upsilon(T)$. Thus the increase in free energy to add a vortex somewhere in the system is,

$$F_v = \pi \Upsilon(T) \ln R - TS = \pi \Upsilon(T) \ln R - T \ln \pi R^2$$
$$\approx [\pi \Upsilon(T) - 2T] \ln R, \qquad (6.23)$$

to leading order as $R \to \infty$. For $\Upsilon(T) > 2T/\pi$, $F_v \to +\infty$ as $R \to \infty$. Adding a vortex then costs infinite free energy and so is suppressed. But for $\Upsilon(T) < 2T/\pi$, $F_v \to -\infty$ as $R \to \infty$, and such vortices are free to enter the system. Since $\Upsilon(T)$ is a monotonic decreasing function of $T$, we necessarily cross from the first case to the second as $T$ increases.

Once such a free vortex is thermally excited, it is free to diffuse throughout the system. Since the phase angle change going completely around a vortex must be $2\pi$, each time such a vortex moves a distance $\Delta y$ in the $\hat{y}$ direction, the net phase angle twist across the system in the $\hat{x}$ direction changes by $2\pi\Delta y/L$ (assuming no other vortices have moved). Diffusion of free vortices can thus unwind any applied boundary twist $\Delta$ and cause the system to lose all phase angle coherence, driving the helicity modulus $\Upsilon \to 0$. We thus have the famous Kosterlitz–Thouless stability criterion for a finite helicity modulus, and hence for phase angle coherence, which continues to hold for the fully frustrated model. If we denote $T_{\rm KT}$ as the Kosterlitz–Thouless temperature above which free vortices may be thermally excited and the system loses phase coherence, then

$$\Upsilon(T) \geq \frac{2}{\pi}T \qquad T \leq T_{\rm KT} \tag{6.24}$$

$$\Upsilon(T) = 0 \qquad T > T_{\rm TK}. \tag{6.25}$$

We thus conclude that $\Upsilon(T)$ takes a discontinuous drop to zero as $T$ increases above $T_{\rm KT}$. If Eq. (6.24) holds as an equality at $T_{\rm KT}$, then $\Upsilon(T_{\rm KT})/T_{\rm KT} = 2/\pi$ has the same *universal jump* to zero found in the ordinary XY model.[15,16] It is, however, possible that a first order transition or other mechanism could cause the loss of phase coherence to occur at a lower $T$, so that jump in $\Upsilon/T$ at the transition is larger than this universal value. The issue of whether this jump is the universal value or larger remains one of the often disputed questions concerning the FFXY model.

We return now to the spin–spin correlation function, considered previously in Eq. (6.10). The same argument concerning the replacement of the *bare* with the *renormalized* spin stiffness constant, that was used in going from Eq. (6.22) to (6.23), can be applied here to include the effects on the spin–spin correlation function of thermal excitations that go beyond the spinwave approximation. One thus replaces $\Upsilon(0) = J/\sqrt{2}$ in Eq. (6.10) by $\Upsilon(T)$ to arrive at,

$$\langle \mathbf{S}(\mathbf{r}_i) \cdot \mathbf{S}(\mathbf{r}_j) \rangle \sim |\mathbf{r}_i - \mathbf{r}_j|^{-\eta(T)}, \quad \text{with} \quad \eta(T) = \frac{T}{2\pi\Upsilon(T)}. \tag{6.26}$$

This result was first derived for the ordinary XY model by Berezinskii,[2] and later by José et al.[17] The condition of Eq. (6.24) then implies $\eta(T) \leq 1/4$, with $\eta(T_{\rm KT}) = 1/4$ if the universal jump in $\Upsilon/T$ holds. Above $T_{\rm KT}$, where

$\Upsilon = 0$, Eq. (6.26) is consistent with the change in the spin–spin correlation from algebraic to exponential decay.

### 6.3.2. Charge lattice ordering

The ground state structure, shown in Fig. 6.3, consists of a checkerboard pattern of alternating "charges" $q_i \pm 1/2$ at sites in the centers of the square unit cells of the array. If we identify $q = +1/2$ with an up "spin" and $q = -1/2$ with a down "spin" ("spin" here having nothing to do with the real planar spins $\mathbf{S}(\mathbf{r}_i)$), then the ground state charge structure has the same symmetry as an antiferromagnetic Ising model. Accordingly, we can define a charge ordering parameter corresponding to the "staggered magnetization",

$$M \equiv \sum_i (-1)^{x_i+y_i} q_i \qquad (6.27)$$

where $q_i$ is the charge at the center of the cell located at position $\mathbf{r}_i = (x_i, y_i)$. The prefactor $(-1)^{x_i+y_i}$ thus alternates in sign on neighboring sites.

We expect that $\langle M \rangle$ is finite at sufficiently low temperatures, but that it vanishes above a well defined critical temperature $T_\mathrm{I}$. Since $\langle M \rangle$ is a scalar order parameter, we naively expect that this transition is in the Ising universality class, with critical exponents characteristic of the two-dimensional Ising model. In this case, $\langle M \rangle$ should vanish continuously as $\langle M \rangle \sim (T_\mathrm{I} - T)^\beta$, with $\beta = 1/8$. Furthermore, we would expect there to be a logarithmically diverging specific heat ($\alpha = 0$) at $T_\mathrm{I}$ and a diverging correlation length $\xi_\mathrm{I} \sim |T - T_\mathrm{I}|^{-\nu}$ with $\nu = 1$. Whether or not the transition at $T_\mathrm{I}$ is indeed in the Ising universality class has been another of the main questions concerning the FFXY model.

### 6.3.3. Summary

To conclude this section we summarize the above main points. In the two-dimensional FFXY model, we expect upon cooling to find a spontaneous breaking of the $U(1)$ symmetry at $T_\mathrm{KT}$ associated with the onset of phase angle coherence, and a breaking of the discrete $Z_2$ symmetry at $T_\mathrm{I}$ associated with the formation of an ordered charge lattice. The questions of interest are: (i) does $T_\mathrm{KT} = T_\mathrm{I}$, and if not, which is larger; (ii) is the phase angle ordering transition at $T_\mathrm{KT}$ in the universality class of the Berezinskii–Kosterlitz–Thouless transition of the ordinary 2D XY model; and (iii) is the charge lattice ordering transition at $T_\mathrm{I}$ in the Ising universality class?

## 6.4. Mapping to the Coulomb Gas

### 6.4.1. *Duality transformation*

Considerable insight into the nature of the transitions in the 2D FFXY model can be obtained via a duality mapping onto an equivalent problem of two-dimensional Coulomb interacting charges, as first shown by Kosterlitz and Thouless for the ordinary 2D XY model.[3] To simplify this duality transformation of the lattice FFXY model, it is customary to replace the cosine interaction of Eq. (6.3) with the Villain function[11] (sometimes referred to as the *periodic Gaussian* function) $V(\varphi)$, defined by,

$$e^{-V(\varphi)/T} \equiv \sum_{m=-\infty}^{\infty} e^{-J(\varphi+2\pi m)^2/2T}. \tag{6.28}$$

The partition function, with twisted boundary conditions, then becomes,

$$Z(\boldsymbol{\Delta}) = \left(\prod_i \int_0^{2\pi} d\theta'(\mathbf{r}_i)\right) \left(\prod_{i\mu} \sum_{m_\mu(\mathbf{r}_i)=-\infty}^{\infty}\right) e^{-\mathcal{H}[\theta'(\mathbf{r}_i),m_\mu(\mathbf{r}_i);\boldsymbol{\Delta}]/T} \tag{6.29}$$

where $m_\mu(\mathbf{r}_i)$ is an integer valued variable on each bond of the array and

$$\mathcal{H}[\theta'(\mathbf{r}_i), m_\mu(\mathbf{r}_i); \boldsymbol{\Delta}]$$
$$= -\frac{J}{2} \sum_{i\mu} \left[\theta'(\mathbf{r}_i + \hat{\mu}) - \theta'(\mathbf{r}_i) - A_\mu(\mathbf{r}_i) - \Delta_\mu/L + 2\pi m_\mu(\mathbf{r}_i)\right]^2. \tag{6.30}$$

We can think of the integers $m_\mu(\mathbf{r}_i)$ as representing the number of $2\pi$ twists in the phase angle $\theta$ as one crosses the bond from site $\mathbf{r}_i$ to site $\mathbf{r}_i + \hat{\mu}$.

Several different approaches to the duality transformation have been given in the literature.[17-20] Here we follow an approach due to Vallat and Beck,[20] and derive the mapping for the case of the general uniformly frustrated XY model. We start by defining the "current" on each bond of the array as,

$$v_\mu(\mathbf{r}_i) \equiv \theta'(\mathbf{r}_i + \hat{\mu}) - \theta'(\mathbf{r}_i) - A_\mu(\mathbf{r}_i) - \Delta_\mu/L + 2\pi m_\mu(\mathbf{r}_i) \tag{6.31}$$

in terms of which

$$\mathcal{H} = \frac{J}{2} \sum_{i\mu} v_\mu^2(\mathbf{r}_i). \tag{6.32}$$

Next we decompose $v_\mu(\mathbf{r}_i)$ into three pieces,

$$v_\mu(\mathbf{r}_i) = v_\mu^0 + v_\mu^T(\mathbf{r}_i) + v_\mu^L(\mathbf{r}_i) \tag{6.33}$$

where
$$v_\mu^0 \equiv \frac{1}{L^2} \sum_i v_\mu(\mathbf{r}_i) \tag{6.34}$$

is the average current flowing in the array in direction $\hat{\mu}$, and $v_\mu^T(\mathbf{r}_i)$ and $v_\mu^L(\mathbf{r}_i)$ are the transverse and longitudinal parts of $v_\mu(\mathbf{r}_i)$, defined by,

$$\mathbf{D} \cdot \mathbf{v}^T(\mathbf{r}_i) \equiv \sum_{\mu=x,y} [v_\mu^T(\mathbf{r}_i) - v_\mu^T(\mathbf{r}_i - \hat{\mu})] = 0 \tag{6.35}$$

$$\hat{z} \cdot [\mathbf{D} \times \mathbf{v}^L(\mathbf{r}_i)] \equiv v_x^L(\mathbf{r}_i) + v_y^L(\mathbf{r}_i + \hat{x}) - v_x^L(\mathbf{r}_i + \hat{y}) - v_y^L(\mathbf{r}_i) = 0 \tag{6.36}$$

where $\mathbf{D} \cdot \mathbf{v}$ is the discrete analog of divergence, and $\hat{z} \cdot [\mathbf{D} \times \mathbf{v}]$ is the discrete analog of the two-dimensional curl, which is just the sum of $v_\mu(\mathbf{r}_i)$ going counterclockwise around the unit cell of the array with lower left corner at $\mathbf{r}_i$. By definition, the spatial averages of $v_\mu^T(\mathbf{r}_i)$ and $v_\mu^L(\mathbf{r}_i)$ are zero.

Substituting the decomposition of Eq. (6.34) into the Hamiltonian (6.22), one can show that all the cross terms vanish and one gets,

$$\mathcal{H} = \frac{J}{2} \sum_{i\mu} \{[v_\mu^0]^2 + [v_\mu^T(\mathbf{r}_i)]^2 + [v_\mu^L(\mathbf{r}_i)]^2\} = \mathcal{H}^0 + \mathcal{H}^T + \mathcal{H}^L. \tag{6.37}$$

One can now solve for $v_\mu^T(\mathbf{r}_i)$ and $v_\mu^L(\mathbf{r}_i)$. The discrete difference operators behave just as do their continuum differential counterparts. Since the curl of $v_\mu^L(\mathbf{r}_i)$ vanishes, one can always write $v_\mu^L(\mathbf{r}_i)$ as the gradient of a scalar field $\chi(\mathbf{r}_i)$,

$$v_\mu^L(\mathbf{r}_i) = \chi(\mathbf{r}_i + \hat{\mu}) - \chi(\mathbf{r}_i), \tag{6.38}$$

and since the spatial average of $v_\mu^L(\mathbf{r}_i)$ vanishes, the scalar field $\chi(\mathbf{r}_i)$ must satisfy periodic boundary conditions. The contribution of $v_\mu^L(\mathbf{r}_i)$ to the Hamiltonian is then just,

$$\mathcal{H}^L = \frac{J}{2} \sum_{i\mu} [\chi(\mathbf{r}_i + \hat{\mu}) - \chi(\mathbf{r}_i)]^2, \tag{6.39}$$

and has the same form as the spinwave approximation.

Since the divergence of $v_\mu^T(\mathbf{r}_i)$ vanishes, one can always write it in terms of a scalar field $\lambda(\mathbf{r}_i)$ such that

$$\mathbf{v}^T(\mathbf{r}_i) = \hat{z} \times \mathbf{D}\lambda(\mathbf{r}_i), \quad \text{with} \quad D_\mu \lambda(\mathbf{r}_i) \equiv [\lambda(\mathbf{r}_i + \hat{\mu}) - \lambda(\mathbf{r}_i)], \tag{6.40}$$

and since the spatial average of $v_\mu^T(\mathbf{r}_i)$ vanishes, the scalar field $\lambda(\mathbf{r}_i)$ must satisfy periodic boundary conditions. The contribution of $v_\mu^T(\mathbf{r}_i)$ to the

Hamiltonian is then,

$$\mathcal{H}^T = \frac{J}{2}\sum_i |\hat{z} \times \mathbf{D}\lambda(\mathbf{r}_i)|^2 = \frac{J}{2}\sum_i |\mathbf{D}\lambda(\mathbf{r}_i)|^2 = -\frac{J}{2}\sum_i \lambda(\mathbf{r}_i)D^2\lambda(\mathbf{r}_i) \tag{6.41}$$

where

$$D^2\lambda(\mathbf{r}_i) \equiv \sum_{\mu=x,y}[\lambda(\mathbf{r}_i+\hat{\mu}) - 2\lambda(\mathbf{r}_i) + \lambda(\mathbf{r}_i-\hat{\mu})] \tag{6.42}$$

is the discrete Laplacian operator, and we have "integrated" by parts in the last step of Eq. (6.41).

Unlike $\chi(\mathbf{r}_i)$, which make take any value, $\lambda(\mathbf{r}_i)$ is constrained as follows. The curl of $v_\mu^T(\mathbf{r}_i)$ must satisfy,

$$\hat{z} \cdot [\mathbf{D}\times\mathbf{v}^T(\mathbf{r}_i)] = \hat{z} \cdot [\mathbf{D}\times\mathbf{v}(\mathbf{r}_i)] = \hat{z} \cdot [\mathbf{D}\times 2\pi\mathbf{m}(\mathbf{r}_i)] - \hat{z} \cdot [\mathbf{D}\times\mathbf{A}(\mathbf{r}_i)] \tag{6.43}$$

as the contribution to the curl of $v_\mu(\mathbf{r}_i)$ from $[\theta'(\mathbf{r}_i+\hat{\mu}) - \theta'(\mathbf{r}_i)]$ and $\Delta_\mu/L$ vanishes. We then have,

$$\hat{z} \cdot [\mathbf{D}\times\mathbf{A}(\mathbf{r}_i)] \equiv 2\pi f, \tag{6.44}$$

with $f$ the uniform frustration arising from the circulation of the magnetic vector potential around the unit cell at $\mathbf{r}_i$, and we can define,

$$\hat{z} \cdot [\mathbf{D}\times 2\pi\mathbf{m}(\mathbf{r}_i)] \equiv 2\pi n(\mathbf{r}_i), \tag{6.45}$$

with $n(\mathbf{r}_i)$ the integer vorticity at unit cell $i$, arising from the circulation of the phase angle twists $m_\mu(\mathbf{r}_i)$ around the cell.

We then have,

$$\hat{z} \cdot [\mathbf{D}\times\mathbf{v}^T(\mathbf{r}_i)] = 2\pi[n(\mathbf{r}_i) - f] \equiv 2\pi q(\mathbf{r}_i), \tag{6.46}$$

with $q(\mathbf{r}_i)$ the "charge" centered on cell $i$.

We now write the curl of $v_\mu^T(\mathbf{r}_i)$ in terms of the scalar field $\lambda(\mathbf{r}_i)$,

$$\hat{z}\cdot[\mathbf{D}\times\mathbf{v}^T(\mathbf{r}_i)] = -\mathbf{D}\cdot[\hat{z}\times\mathbf{v}^T(\mathbf{r}_i)] = -\mathbf{D}\cdot[\hat{z}\times[\hat{z}\times\mathbf{D}\lambda(\mathbf{r}_i)]] = D^2\lambda(\mathbf{r}_i), \tag{6.47}$$

which combined with Eq. (6.46) gives,

$$D^2\lambda(\mathbf{r}_i) = 2\pi q(\mathbf{r}_i). \tag{6.48}$$

Hence, $\lambda(\mathbf{r}_i)$ must solve the discrete Poisson equation with $q(\mathbf{r}_i)$ as the sources,

$$\lambda(\mathbf{r}_i) = -\sum_j G(\mathbf{r}_i - \mathbf{r}_j)q(\mathbf{r}_j) \tag{6.49}$$

where $G(\mathbf{r}_i - \mathbf{r}_j)$ is the lattice Green's function with periodic boundary conditions,[17]

$$G(\mathbf{r}_i) = \frac{2\pi}{L^2} \sum_{\mathbf{k}} \frac{e^{i\mathbf{q}\cdot\mathbf{r}_i}}{4 - 2\cos k_x - 2\cos k_y}, \qquad (6.50)$$

where one sums over all wavevectors $\mathbf{k}$ consistent with periodic boundary conditions, i.e. $k_\mu = 2\pi\ell_\mu/L$ with $\ell_\mu = 0, \ldots L-1$. One has $G(\mathbf{r}) \sim -\ln|\mathbf{r}|$ for large $|\mathbf{r}|$. The contribution of $v_\mu^T(\mathbf{r}_i)$ to the Hamiltonian then becomes,

$$\mathcal{H}^T = -\frac{J}{2}\sum_i \lambda(\mathbf{r}_i) D^2 \lambda(\mathbf{r}_i) = \pi J \sum_{i,j} q(\mathbf{r}_i) G(\mathbf{r}_i - \mathbf{r}_j) q(\mathbf{r}_j), \qquad (6.51)$$

which has precisely the form of a gas of charges $q(\mathbf{r}_i)$ interacting via the two-dimensional Coulomb potential. Requiring $\mathcal{H}^T$ to be finite in the thermodynamic limit, $L \to \infty$, imposes the constraint of charge neutrality,

$$\sum_i q(\mathbf{r}_i) = 0, \quad \text{or} \quad \frac{1}{L^2}\sum_i n(\mathbf{r}_i) = f. \qquad (6.52)$$

Finally we consider the spatial average $v_\mu^0$. Doing a discrete "integration" by parts, we have,

$$v_x^0 \equiv \frac{1}{L^2}\sum_{x=0}^{L-1}\sum_{y=0}^{L-1} v_x(x,y) = -\frac{1}{L^2}\sum_{x=0}^{L-1}\sum_{y=0}^{L-1} y D_y v_x(x,y) + \frac{1}{L}\sum_{x=0}^{L-1} v_x(x,0), \qquad (6.53)$$

where in the last term we made use of periodic boundary conditions $v_x(x,L) = v_x(x,0)$. The first term on the right-hand side can be rewritten as,

$$-\frac{1}{L^2}\sum_{y=0}^{L-1} y \sum_{x=0}^{L-1} D_y v_x(x,y) = \frac{1}{L^2}\sum_y y \sum_x 2\pi q(x,y) = \frac{2\pi P_y}{L^2}, \qquad (6.54)$$

where $P_y = \sum_i y_i q(\mathbf{r}_i)$ is the $y$ component of the total dipole moment in the system. The second term can be rewritten as,

$$\frac{1}{L}\sum_{x=0}^{L-1} v_x(x,0) = \frac{1}{L}(2\pi m_x^0 - \Delta_x), \qquad (6.55)$$

where we made use of the gauge choice of Eq. (6.6) to set $A_x(x,0) = 0$, used periodic boundary conditions on $\theta'(\mathbf{r}_i)$ to conclude $\sum_x [\theta'(\mathbf{r}_i+\hat{x}) - \theta'(\mathbf{r}_i)] = 0$, and defined $m_x^0 \equiv \sum_x m_x(x,0)$ an integer.

Combining the above results gives,

$$v_x^0 = \frac{1}{L}\left(\frac{2\pi P_y}{L} - \Delta_x + 2\pi m_x^0\right). \qquad (6.56)$$

A similar calculation gives,

$$v_y^0 = \frac{1}{L}\left(-\frac{2\pi P_x}{L} - \Delta_y + 2\pi m_y^0\right), \qquad (6.57)$$

so that

$$\mathcal{H}^0 = \frac{J}{2}\left[\left(\frac{2\pi P_y}{L} - \Delta_x + 2\pi m_x^0\right)^2 + \left(-\frac{2\pi P_x}{L} - \Delta_y + 2\pi m_y^0\right)^2\right]. \qquad (6.58)$$

The above transformations have mapped the degrees of freedom from $\{\theta(\mathbf{r}_i), m_\mu(\mathbf{r}_i)\}$ to $\{\chi(\mathbf{r}_i), n(\mathbf{r}_i), m_x^0, m_y^0\}$. We can now evaluate the partition function summing over these new degrees of freedom. We group these sums as follows,

$$Z(\boldsymbol{\Delta}) = \sum_{\{n(\mathbf{r}_i)\}} e^{-\mathcal{H}^T[n(\mathbf{r}_i)]/T} \sum_{m_x^0, m_y^0 = -\infty}^{\infty} e^{-\mathcal{H}^0[n(\mathbf{r}_i), m_x^0, m_y^0]/T}$$

$$\times \left(\prod_i \int_{-\infty}^{\infty} d\chi(\mathbf{r}_i)\right) e^{-\mathcal{H}^L[\chi(\mathbf{r}_i)]/T}. \qquad (6.59)$$

The Gaussian integrals in the last spinwave-like factor are easily done, and give an analytic multiplicative factor $Z_0(T)$ to the partition function. Since this term is independent of both the vortex degrees of freedom $n(\mathbf{r}_i)$ and the twist $\boldsymbol{\Delta}$, it plays no role in either the phase angle ordering or the vortex lattice ordering transition. In the XY model with Villain interaction, the spinwave excitations are thus completely decoupled from the vortex excitations, and we henceforth drop this term. Using Eq. (6.58) for $\mathcal{H}^0$, the sums over $m_x^0$ and $m_y^0$ can be done and they result in the Villain function $V(\varphi)$ of Eq. (6.28). We thus get the partition function of the dual Coulomb gas,

$$Z(\boldsymbol{\Delta}) = \sum_{\{n(\mathbf{r}_i)\}} e^{-\mathcal{H}_{\text{CG}}[n(\mathbf{r}_i)]/T} \qquad (6.60)$$

where the sum over vortex configurations $\{n(\mathbf{r}_i)\}$ is constrained by the condition of charge neutrality, Eq. (6.52), and the Coulomb gas Hamiltonian is,

$$\mathcal{H}_{\text{CG}} = V\left(\frac{2\pi P_y}{L} - \Delta_x\right) + V\left(-\frac{2\pi P_x}{L} - \Delta_y\right) + \pi J \sum_{i,j} q(\mathbf{r}_i) G(\mathbf{r}_i - \mathbf{r}_j) q(\mathbf{r}_j)$$

$$(6.61)$$

### 6.4.2. *Mechanisms for destruction of order*

Note, the terms in the Coulomb gas Hamiltonian involving the dipole moment $\mathbf{P}$ are the consequence of imposing the boundary condition with fixed twist $\boldsymbol{\Delta}$. In many derivations of the duality transformation, these terms were missing. Leaving off these terms corresponds to working in an ensemble in which the twist $\boldsymbol{\Delta}$ is free to fluctuate as a thermal degree of freedom. However, to get an expression for the helicity modulus in the dual Coulomb gas representation, it is important to keep them. From Eq. (6.15) we have,

$$\Upsilon_x = \left.\frac{\partial^2 \mathcal{F}}{\partial \Delta_x^2}\right|_{\boldsymbol{\Delta}=0} = \left\langle V''\left(\frac{2\pi P_y}{L}\right)\right\rangle - \frac{1}{T}\left\langle V'\left(\frac{2\pi P_y}{L}\right) V'\left(\frac{2\pi P_y}{L}\right)\right\rangle, \tag{6.62}$$

where $V'(\varphi)$ and $V''(\varphi)$ are the first and second derivatives of the Villain function. Note, at very low $T$ where the Villian function is approximately parabolic $V(\varphi) \approx \frac{1}{2}J\varphi^2$ for most of the interval $(-\pi, \pi]$, one has $V'' \approx J$ and $V'(\varphi) \approx J\varphi$, yielding,

$$\frac{\Upsilon_x}{J} \approx 1 - \frac{4\pi^2 J}{TL^2}\langle P_y^2\rangle \quad \text{at very low } T. \tag{6.63}$$

Equation (6.63) was first dervied by Berezinskii[2] for the case of the ordinary XY model. Thus the helicity modulus is reduced by the fluctuations in the total dipole moment of the system, and the above leads to the identification $\Upsilon/J \equiv \epsilon^{-1}$, where $\epsilon$ is the dielectric function of the Coulomb gas.

The broken $U(1)$ and $Z_2$ symmetries of the low temperature ordered phase then suggest two different types of dipole excitations that cause $\Upsilon(T)$ to decrease from its $T = 0$ value as the system is heated. These are illustrated in Fig. 6.4. The excitation in Fig. 6.4(a) is obtained by displacing one of the vortices $n = +1$ from its position $\mathbf{r}_i$ in the ground state, and moving it a distance $\mathbf{d}$ to a site $\mathbf{r}_j$ which previously had no vortex. The result is that $q(\mathbf{r}_i)$ changes from $+1/2$ to $-1/2$, while $q(\mathbf{r}_j)$ changes from $-1/2$ to $1/2$, creating a net dipole moment $\mathbf{P} = \mathbf{d}$. Alternatively, we can view this excitation as the creation of a new vortex–antivortex pair with $n = +1$ at site $\mathbf{r}_j$ and $n = -1$ at site $\mathbf{r}_i$. The unbinding of such vortex–antivortex pairs, with $\mathbf{d}$ diffusing to large values, corresponds to the Kosterlitz–Thouless mechanism for the destruction of the $U(1)$ phase angle coherence at $T_{\mathrm{KT}}$ in the ordinary 2D XY model.[3,4]

The excitations in Figs. 6.4(b) and 6.4(c) are obtained by taking a closed domain of sites and flipping the sign of all the charges in the domain; to keep the overall system charge neutrality, the domain itself must be charge neutral. Such flipped domains are Ising-like excitations and the growth of such domains leads to the vanishing of the $Z_2$ Ising-like order parameter $\langle M\rangle$ at $T_{\mathrm{I}}$. Depending on the precise shape of such domains, they may also create dipole moments as large as the area of the domain, as in the

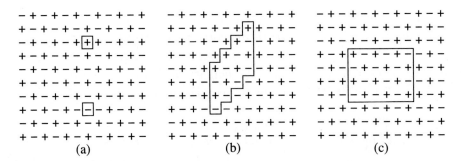

Fig. 6.4. Vortex excitations leading to a net dipole moment $P_y$. (a) A vortex–antivortex pair excitation; (b) An Ising-like domain excitation creating a net dipole moment; (c) An Ising-like domain excitation with no net dipole moment. A (+) denotes a charge $q = +1/2$ while a (−) denotes a charge $-1/2$. Solid lines are bonds separating charges of equal sign.

example shown in Fig. 6.4(b), and thus also serve to decrease the helicity modulus $\Upsilon$.

It was further noticed by Halsey,[21] that Ising-like excitations can be thought of as having fractional $\pm 1/4$ charges at the corners of the domain, as can be seen by coarse graining the charges over a $2 \times 2$ block of cells. The sum of these corner charges, going completely around the enclosing domain wall, must vanish for a neutral domain. It was suggested[21-24] that such corner charges played an important role: when the Ising-like transition occurs at $T_\mathrm{I}$, the domain wall tension will vanish, allowing paired $+1/4$ and $-1/4$ corner charges to unbind, thus destroying phase angle coherence by a similar pair unbinding mechanism as in Fig. 6.4(a). Thus the Ising-like transition would necessarily trigger the KT-like transition. However, once the domain wall energy vanishes, domains of the type in Fig. 6.4(b) should also proliferate and become large, driving $\Upsilon \to 0$ through the large dipole moments created by such domains; domains with no net dipole moment, as in Fig 6.4(c), could not lead to a reduction in $\Upsilon$ even if they became arbitrarily large with unbound corner charges, since the dipole moments of these corner charges must always sum up to zero. Indeed, the coarse graining that gives rise to the corner charge picture is just an equivalent way to represent the net dipole moment that may appear on Ising-like domains. It is thus not clear that the transitions at $T_\mathrm{KT}$ or $T_\mathrm{I}$ are better understood in terms of a corner charge unbinding scenario. Nevertheless, we will see in a later section that a corner charge unbinding transition does indeed take place at a temperature well below $T_\mathrm{I}$, where the domain wall tension is still finite, and that this lower transition has crucial ramifications for the question as to whether $T_\mathrm{KT}$ is equal to, or less than, $T_\mathrm{I}$.

## 6.5. Numerical Results

The first numerical study of the 2D FFXY model was by Teitel and Jayaprakash[8] in 1983. Using ordinary Metropolis Monte Carlo simulations, they used Eq. (6.18) to compute the helicity modulus $\Upsilon(T)$, to investigate the vanishing of phase angle coherence at $T_{\text{KT}}$. They computed the specific heat $C$, using energy fluctuations, to look for the charge ordering transition at $T_{\text{I}}$. If, as naively expected, the transition at $T_{\text{I}}$ is in the universality class of the 2D Ising model, one expects a logarithmically diverging specific heat at $T_{\text{I}}$. Their numerical results for $\Upsilon(T)$ and $C(T)$, for system sizes $L = 8 - 32$, are reprinted in Fig. 6.5. $\Upsilon$ appears to be taking a sharp drop to zero, that becomes steeper as $L$ increases, at a $T_{\text{KT}}$ not inconsistent with the Kosterlitz–Thouless universal jump. The specific heat peak steadily increases with $L$, consistent with a $\ln L$ dependence, with the peak location approaching a $T_{\text{I}}$ that is very close to $T_{\text{KT}}$.

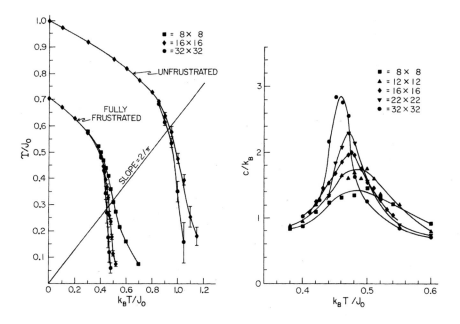

Fig. 6.5. (Left) Helicity modulus $\Upsilon$ versus temperature $T$ for the unfrustrated ($f = 0$) and fully frustrated ($f = 1/2$) cases, and various lattice sizes $L \times L$. A line of slope $2/\pi$ indicates the universal jump in $\Upsilon(T_{\text{KT}})/T_{\text{KT}}$ of a Kosterlitz–Thouless transition. (Right) Specific heat $C$ of the fully frustrated ($f = 1/2$) model for various lattice sizes $L \times L$. The smooth curves through the data are drawn as guides to the eye and are not the result of any theoretical computation. [Reprinted from Ref. 8, © (1983) by The American Physical Society]

It is worthwhile to quote from Teitel and Jayaprakash's conclusions: "Two possible scenarios seem likely: (i) As $T_\text{I}$ is approached from below, the Ising excitations [cf. Fig. 6.4(b)] result in a steep drop in $\Upsilon(T)$ from its low-$T$ value. As $\Upsilon/k_\text{B}T$ approaches $2/\pi$, however, the KT excitations [cf. Fig. 6.4(a)] become important, producing a universal jump $2/\pi$ in $\Upsilon(T)/k_\text{B}T$ at some temperature $T_\text{KT} \leq T_\text{I}$. (ii) As $T_\text{I}$ is approached from below, the Ising excitations result in a nonuniversal jump in $\Upsilon/k_\text{B}T > 2/\pi$ at the same temperature as the specific-heat peak $T_\text{I}$. Our numerical simulations cannot adequately distinguish between these two possibilities."

In the following years, numerous theoretical and numerical works were carried out to try and resolve the three main questions: (1) is the jump in $\Upsilon/T$ at $T_\text{KT}$ the universal $2/\pi$ or is it larger; (2) is the transition at $T_\text{I}$ characterized by the critical exponents of the 2D Ising model; (3) is there only a single phase transition in the model with $T_\text{KT} = T_\text{I}$, or are there two separate transitions with $T_\text{KT} < T_\text{I}$? Theoretical analyses,[24–30] based on symmetry arguments and renormalization group calculations, generally suggested a single phase transition $T_\text{KT} = T_\text{I}$ and nonuniversal jump in $\Upsilon$ at $T_\text{KT}$. Numerical investigations were carried out on the original square lattice model of Eq. (6.3),[31–42] on the equivalent antiferromagnetic XY model on a triangular lattice (also fully frustrated),[32,42–50] and on the dual lattice Coulomb gas model.[51–54] These numerical works generally led to conflicting conclusions on all of the above three questions.

Several extensions to the model were introduced in order to tune the separation between $T_\text{KT}$ and $T_\text{I}$ and so explore the fully frustrated model within the context of a larger parameter space. Berg et al.[55] introduced an asymmetric model in which the strength of the antiferromagnetic bond was set to $\eta J$, in comparison with the ferromagnetic bonds $-J$. Granato and Kosterlitz[30,56] mapped this onto two coupled unfrustrated XY models with unequal couplings. Symmetry arguments[27,28] led to a proposed coupled "XY-Ising" model believed to be in the same universality class as the FFXY model, and extensive theoretical and numerical analysis of this model was carried out by Granato, Kosterlitz, and co-workers.[24,57–59] Thijssen and Knops[52,53] introduced an additional term in the Coulomb gas model to tune the Ising-like domain wall energy. Cristofano et al.[60] related the FFXY model to behavior in more general twisted conformal field theories. Minnhagen and co-workers[61–63] modified the cosine interaction, $-J\cos(\varphi)$, to $(2J/p^2)[1 - \cos^{2p^2}(\varphi/2)]$; $p = 1$ is the usual fully frustrated model, but for $p$ sufficiently large, the nature of the ground state changes. While these extended models led to interesting phase diagrams, in which $T_\text{KT}$ and $T_\text{I}$

could be clearly separated over large regions of the parameter space, numerical simulations at the specific point in the parameter space corresponding to the FFXY model generally lacked the accuracy to conclusively determine the behavior, and conflicting results remained.

The simplest and cleanest numerical demonstration that the FFXY model has two very close but distinct phase transitions, with $T_{\rm KT} < T_{\rm I}$, was given by Olsson[38] in 1996. Olsson utilized the Kosterlitz–Thouless stability criterion Eq. (6.24), $\Upsilon(T)/T \geq 2\pi$, to define the set of temperatures $T_L$ such that $\Upsilon(T_L, L)/T_L = 2\pi$, for finite systems with length up to $L = 128$. It was observed that the $T_L$ were monotonically decreasing with $L$, with an apparent finite limit as $L \to \infty$, thus providing an upper bound on $T_{\rm KT}$. Olsson then measured the staggered magnetization at these points, $M_L \equiv \langle M(T_L, L) \rangle$, and observed that the $M_L$ monotonically *increased* with increasing $L$, thus demonstrating that $M(T_{\rm KT}) \geq \lim_{L \to \infty} M_L$ must be finite as $L \to \infty$. In Fig. 6.6, we show Olsson's results from Ref. 38, plotting $\langle M \rangle$ versus $\Upsilon \pi/(2T)$ for various system sizes $L$. The dashed vertical line defines the temperatures $T_L$.

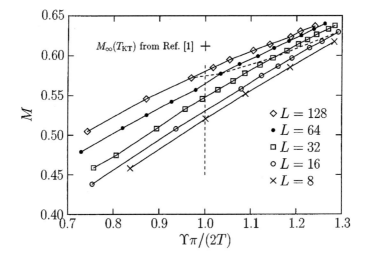

Fig. 6.6. Plot of staggered magnetization $\langle M \rangle$ versus $\Upsilon \pi/(2T)$ for systems of different length $L$, for the FFXY model with cosine interaction on a square lattice. For $L = 64$ and 128, the temperature difference between the neighboring points is $0.001J$. The vertical dashed line determines $T_L$, the size dependent upper bound on $T_{\rm KT}$, as obtained via the Kosterlitz–Thouless stability condition, $\Upsilon(T_{\rm KT}) \geq (2\pi)/T_{\rm KT}$. That $M_L \equiv \langle M(T_L, L) \rangle$ increases monotonically with increasing $L$ along this dashed vertical line indicates that $M(T_{\rm KT})$ is finite in the thermodynamic limit, and hence $T_{\rm KT} < T_{\rm I}$. [Reprinted from Ref. 38, © (1996) by The American Physical Society]

For two separated transitions, $T_{\rm KT} < T_{\rm I}$, the general expectation has been that the loss of $Z_2$ symmetry at the transition $T_{\rm I}$ should be in the usual 2D Ising universality class, while the loss of $U(1)$ symmetry at $T_{\rm KT}$ should be in the Kosterlitz–Thouless universality class, with a universal jump in the helicity modulus, $\Upsilon(T_{\rm KT})/T_{\rm KT} = 2\pi$. Olsson[37–39] has provided numerical support for this scenario and noted that the non-Ising critical exponents at $T_{\rm I}$ often cited in the literature are most likely due to crossover effects resulting from the very close proximity of $T_{\rm KT}$ to $T_{\rm I}$ (see also Ref. 62 for a related point of view). This scenario has been confirmed in very recent high precision numerical simulations by Hasenbusch et al.,[64,65] who carried out detailed finite size scaling analyses using the largest system sizes to date, $L = 1000$. Among their findings is that the spin correlation length $\xi_s$, obtained from the exponential decay of the spin–spin correlation function Eq. (6.10) above $T_{\rm KT}$, is $\xi_s \simeq 120$ at $T = T_{\rm I}$, thus confirming that the non-Ising exponents reported in earlier works are due to crossover effects dominating the results for too small system sizes $L$. A nice, detailed, review of earlier numerical results is presented by these authors in Ref. 65. The most recent numerical simulations of the FFXY model by Okumura et al.,[66] also using system sizes up to $L = 1000$, confirm the results of Hasenbusch et al. that there are two close but distinct transitions $T_{\rm KT} < T_{\rm I}$, that the transition at $T_{\rm I}$ is characterized by the usual 2D Ising critical exponents, and that $\xi_s(T_{\rm I}) \simeq 120$. However, their analysis finds a slightly lower value of $T_{\rm KT}$ and that, while the phase angle disordering transition is still continuous, the jump in $\Upsilon(T_{\rm KT})/T_{\rm KT}$ is larger than the expected universal value $2/\pi$.

### 6.6. Kink–Antikink Unbinding Transition

While Olsson[38] convincingly demonstrated in 1996 that the FFXY model did indeed have two distinct transitions $T_{\rm KT} < T_{\rm I}$, it remained until 2002 for Korshunov[67] to provide the theoretical understanding behind this phenomenon, in terms of the unbinding of kink–antikink pairs that occur along the domain walls of Ising-like excitations, such as shown in Figs. 6.4(b) and 6.4(c).

Korshunov's argument can be rephrased in terms of the dual Coulomb gas model introduced in Sec. 6.4. As $T$ increases to $T_{\rm I}$, increasingly large Ising-like domains get excited in the system. If such domains can carry a net dipole moment that scales with the size of the domain $\xi_{\rm I}$, such as in Fig. 6.4(b), and if $\xi_{\rm I}$ diverges continuously upon approaching $T_{\rm I}$, $\xi_{\rm I} \sim |T - T_{\rm I}|^{-\nu}$, then by Eqs. (6.62)–(6.63) diverging dipole fluctuations would drive the inverse dielectric constant $\epsilon^{-1}$, and hence the helicity modulus $\Upsilon \equiv$

$J\epsilon^{-1}$, *continuously* to zero at $T_{\rm I}$. However the Kosterlitz–Thouless stability criterion of Eq. (6.24) ensures that $\Upsilon$ may not go continuously to zero; once $\Upsilon/T$ falls below $2/\pi$, vortex–antivortex pair excitations as in Fig. 6.4(a) unbind to produce freely diffusing vortices that drive $\Upsilon$ discontinuously to zero. This necessarily happens at a $T_{\rm KT}$ that must be lower than $T_{\rm I}$, unless a first order phase transition preempts the continuous Ising-like transition and drives both $\Upsilon$ and $\langle M \rangle$ discontinously to zero at a common temperature.

However, for a domain excitation to be considered "Ising-like", the free energy to create the domain should scale proportional to the domain wall length. Because the charges in the Coulomb gas interact with a long range logarithmic potential, one finds that at $T = 0$, and presumably also at sufficiently low $T$, the only domains for which the excitation energy scales proportional to wall length are those which have *zero* total dipole moment! Such domains, as in Fig. 6.4(c), cannot contribute to any reduction in $\Upsilon$. If this remained true up to high temperatures, it would suggest that the loss of Ising-like order with the vanishing of $\langle M \rangle$ might be completely decoupled from the loss of phase angle coherence with the vanishing of $\Upsilon$. Korshunov demonstrated, however, that this does not remain true due to an unbinding of kink–antikink pairs along the domain wall, that takes place at a $T_{\rm w}$ well below $T_{\rm I}$.

We are interested in the behavior of domains that become large on the scale of the system length $L$. Let us therefore imagine a system containing a single domain wall running the length of the system. One can introduce such a domain wall by choosing a system of size $L \times (L+1)$, with $L$ even. The $2 \times 2$ repeated cell structure of the ground state, as shown in Fig. 6.3, then ensures that the ground state necessarily contains such a system spanning domain wall. The lowest energy configuration for such a domain wall would be perfectly flat, as illustrated in Fig. 6.7(a). The simplest excitation of the domain wall would consist of a single *kink* of unit step height, as illustrated in Fig. 6.7(b). However, as discussed at the end of Sec. 6.4.2, each corner of a domain wall can be thought of as consisting of a net localized charge of $q = \pm 1/4$. As the two corner charges of a kink have the same sign, the net charge of a kink is $q_{\rm kink} = \pm 1/2$ (for the kink of Fig. 6.7(b), $q_{\rm kink} = +1/2$). An isolated kink, as in Fig. 6.7(b), would therefore destroy charge neutrality and have a Coulomb energy that grows proportional to $\ln L$.

To keep the excitation energy finite, such kinks must therefore appear only as *kink–antikink pairs*, separated by a length $\ell$, where $\ell$ must be even to preserve charge neutrality. The energy of the pair is then finite and proportional to $\ln \ell$. Such a kink–antikink pair, as illustrated in Fig. 6.7(c),

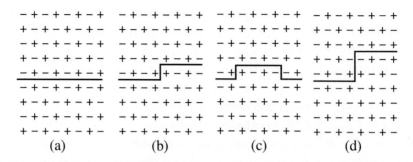

Fig. 6.7. Various configurations of the domain wall in a $L \times (L+1)$ system: (a) ground state, (b) isolated kink of unit height, (b) finite width step of unit height (kink–antikink pair), (d) isolated kink of height two. A $(+)$ indicates the presence of a vortex in the XY model, or a charge $q_i = +1/2$ in the dual Coulomb gas; a $(-)$ indicates the absence of a vortex, or a charge $q_i = -1/2$.

creates a net dipole moment in the system, $p_x = q_{\text{kink}}\ell$. At low temperatures, the $\ln \ell$ energy keeps kink–antikink pairs confined to small separations $\ell \leq \ell^*(T)$. However, as $T$ increases, the entropy associated with the placement of the kinks will cause a reduction in the free energy of such excitations, leading to a *kink–antikink unbinding transition* at a $T_\text{w}$ where $\ell^*(T_\text{w}) \to \infty$. Above $T_\text{w}$, large dipole moments can thus form on the Ising-like domains with no increase in free energy, except for a contribution that is proportional to the length of the domain wall. If $T_\text{w} < T_\text{I}$, we then recover the argument for $T_\text{KT} < T_\text{I}$ outlined at the start of this section.

So far we have discussed only kinks of unit height. One may also have kinks of general height $h$. A kink of $h = 2$ is illustrated in Fig. 6.7(d). For $h$ odd, the two corner charges that comprise the kink have the same sign, and so such kinks must again be paired with antikink(s) of equal opposite total charge. For $h$ even, however, the two corner charges of the kink have the *opposite* sign, and the energy of an isolated kink is therefore finite and proportional to $\ln h$. Such kinks may thus be excited at any finite temperature. Note, however, that kinks with even $h$ do not create any net dipole moment in the system. Such even-step kinks can thus act to roughen domain walls, however they cannot contribute to a reduction in $\Upsilon$.

The preceding discussion can also be cast in terms of the phase angle coherence in the system, as originally presented by Korshunov. When a unit-step kink–antikink pair unbinds, and the domain wall of Fig. 6.7(a) shifts by one lattice spacing, the net phase angle twist $\Delta_y$ across the system changes by $\pi$ (a shift in the wall by two lattice spacing induces a twist of $2\pi$, which is equivalent to zero). Above $T_\text{w}$, such domain wall fluctuations therefore lead to fluctuating phase shifts between opposite sides of the domain wall, leading to the destruction of phase coherence across the domain wall. As $T_\text{I}$ is approached from below, and the size of the largest thermally excited domain diverges continuously, phase coherence across the system would vanish continuously. As $\Upsilon(T)$ decreases due to the domain excitations, the Kosterlitz–Thouless vortex instability is triggered at a $T_\text{KT} < T_\text{I}$, causing $\Upsilon$ to drop discontinuously to zero and destroying phase coherence.

We conclude by presenting the Kosterlitz–Thouless-like argument for the kink–antikink unbinding transition.[68] From Eq. (6.61), the energy of an isolated unit-step kink along a system spanning domain wall of length $L$ is,

$$E_\text{kink} = \pi \Upsilon q_\text{kink}^2 G(L) \approx \frac{\pi}{4} \Upsilon \ln L. \tag{6.64}$$

In the above expression, we have replaced the bare coupling constant $J$ of Eq. (6.61) with $\Upsilon = J\epsilon^{-1}$, so as to allow for the screening effects of other charge excitations elsewhere in the system. The entropy associated with the position of the kink along the domain wall is,

$$S_\text{kink} = \ln L. \tag{6.65}$$

The free energy for the isolated unit-step kink is therefore,

$$F_\text{kink} = E_\text{kink} - TS_\text{kink} = \left[\frac{\pi}{4}\Upsilon - T\right] \ln L. \tag{6.66}$$

The domain thus becomes unstable to the appearance of free unit-step kinks (due to the unbinding of kink–antikink pairs) when,

$$T_\text{w} = \frac{\pi}{4}\Upsilon(T_\text{w}). \tag{6.67}$$

Comparing to the similar criterion Eq. (6.23) for the instability of the system to the appearance of free vortices (due to the unbinding of vortex–antivortex pairs),

$$T_\text{KT} = \frac{\pi}{2}\Upsilon(T_\text{KT}), \tag{6.68}$$

we can thus, ignoring the temperature dependence of $\Upsilon$, estimate,

$$T_\text{w} \approx \frac{1}{2}T_\text{KT}. \tag{6.69}$$

Thus, as desired, the kink–antikink unbinding transition takes place well below the bulk transitions of the system. Numerical evidence for this kink–antikink unbinding scenario was given by Olsson and Teitel in Ref. 69.

## 6.7. FFXY on a Honeycomb Lattice

In the fully frustrated XY model on a triangular lattice, the charges of the dual Coulomb gas sit on the sites of a honeycomb lattice, and have a ground state structure as shown in Fig. 6.8. The ground state is thus doubly degenerate and breaks the same $Z_2$ symmetry as does the FFXY on a square lattice. As discussed above, we therefore expect the transitions of the FFXY on the triangular lattice to be qualitatively the same as those on the square lattice. However a FFXY on a honeycomb lattice, in which the charges $q_i = \pm 1/2$ sit on the sites of the dual triangular lattice, is a more complex problem with a higher degeneracy of ground states. This case was first numerically simulated in the XY formulation by Shih and Stroud,[46] then discussed theoretically by Korshunov.[23] Numerical study of the dual Coulomb gas on a triangular lattice was carried out by Lee and Teitel.[70]

Since charges of the same sign repel each other, the ground state configuration would like to have charges of opposite sign in nearest neighbor cells. For square or triangular lattices, a state is easily constructed which satisfies this condition simultaneously for each pair of neighboring cells, as in Figs. 6.3 and 6.8. However, for the XY honeycomb lattice, the geometry frustrates the construction of any such state; no matter how one tries to put down equal numbers of $+1/2$ and $-1/2$ charges, one inevitably must wind up with neighboring cells which are occupied by charges of the same sign. This leads to a high degeneracy of ground states.

Recall, the ordinary short range antiferromagnetic Ising model on a triangular lattice is fully frustrated and has a ground state degeneracy that grows exponentially with the number of sites $N$, and thus has a finite $T = 0$ entropy density. In the $f = 1/2$ Coulomb gas on the triangular lattice, dual

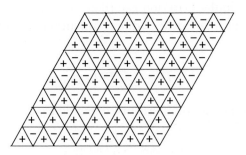

Fig. 6.8. Configuration of charges $q_i = \pm 1/2$ in the ground state of the FFXY on a triangular lattice. The ground state has a double discrete degeneracy just as is the case for the FFXY on a square lattice.

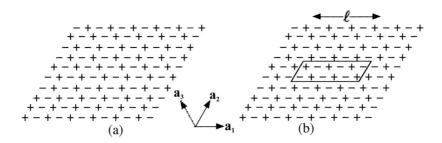

Fig. 6.9. (a) An example of a ground state of the $f = 1/2$ Coulomb Gas on a triangular lattice, which is dual to the FFXY model on a honeycomb lattice. A (+) is a charge $q_i = +1/2$, while a (−) is a charge $q_i = -1/2$. In a given direction, here $\mathbf{a}_1$, charges in each row alternate in sign, while in alternative directions the charges are sequenced randomly. The three lattice directions $\mathbf{a}_1$, $\mathbf{a}_2$ and $\mathbf{a}_3$ are indicated. (b) An example of a $2 \times \ell$ domain excitation that, as $\ell \to \infty$, leads to transitions among the different grounds states with the same ordering direction.

to the FFXY on the honeycomb lattice, the long range logarithmic interactions lift much of the degeneracy found in the corresponding Ising model, and the degeneracy of the ground state is $3(2^L)$, for a system of length $L$, thus giving a vanishing ground state entropy density.[70] A sample ground state is shown in Fig. 6.9(a).

The ground states may be described as follows. Pick one of the three directions $\mathbf{a}_i$ that define the triangular lattice. In each row oriented in this direction, the charges alternate in sign; we will refer to this as the *ordering direction*. In either of the two other directions $\mathbf{a}_j$, $j \neq i$ (which we will call the *complementary directions*), the charges may be sequenced completely at random. The reason for this degeneracy is easily seen in the dual Coulomb gas, where the Coulomb interaction on the triangular lattice is given by,[71]

$$G(\mathbf{r}) = \frac{\sqrt{3}\pi}{N} \sum_{\mathbf{k}} e^{i\mathbf{k}\cdot\mathbf{r}} G_{\mathbf{k}}, \quad G_{\mathbf{k}} \equiv \frac{1}{3 - \cos(\mathbf{k} \cdot \mathbf{a}_1) - \cos(\mathbf{k} \cdot \mathbf{a}_2) - \cos(\mathbf{k} \cdot \mathbf{a}_3)}.$$
(6.70)

If the charges alternate in sign along the direction $\mathbf{a}_1$, as for example in Fig. 6.9(a), then the only wavevectors $\mathbf{k}$ that appear in the Fourier transform of the charge distribution $q(\mathbf{r}_i)$ must satisfy $\mathbf{k} \cdot \mathbf{a}_1 = \pi$. Since $\mathbf{a}_3 = \mathbf{a}_2 - \mathbf{a}_1$, we then get $\cos(\mathbf{k} \cdot \mathbf{a}_3) = \cos(\mathbf{k} \cdot [\mathbf{a}_2 - \mathbf{a}_1]) = \cos(\mathbf{k} \cdot \mathbf{a}_2 - \pi) = -\cos(\mathbf{k} \cdot \mathbf{a}_2)$, and hence $G_{\mathbf{k}}$ is independent of the component of $\mathbf{k}$ perpendicular to $\mathbf{a}_1$. Since the Coulomb interaction part of the energy can be written in terms of Fourier transforms as $\propto \sum_{\mathbf{k}} q_{\mathbf{k}} G_{\mathbf{k}} q_{-\mathbf{k}}$, the energy of such a configuration is independent of the sequence of charges in the complementary directions.

Korshunov[23] has noted that this degeneracy holds for any $2\pi$-periodic interaction potential $V(\varphi)$. For the specific case of the Villain interaction of Eq. (6.28), in which spinwave excitations completely decouple from the vortex excitations (see Sec. 6.4.1), Lee and Teitel[70] have shown that domain excitations of size $2 \times \ell$, such as shown in Fig. 6.9(b), have an energy that saturates to a finite value as $\ell \to \infty$. Thus the free energy barrier for transitions between the different possible random sequences for a given ordering direction $\mathbf{a}_i$ remains finite, and transitions may occur at any finite $T$. However, this argument still leaves open the question of whether the system can have a sharp transition at a finite temperature into the *class* of states specified by a given ordering direction. Similarly, since the excitations illustrated in Fig. 6.8(b) create no net dipole moment, the possibility remains that there exists a finite temperature phase angle ordering transition below which $\Upsilon$ is finite.

Korshunov[23] proposed that for interactions other than Villain's, such as the physical one for Josephson junctions $V(\varphi) = -J\cos\varphi$, in which spinwave–vortex interactions are present,[14] thermal fluctuations will give contributions to the free energy that may lift the degeneracy among the ground states at finite low temperatures. In a recent work Korshunov and Douçot[72] have shown that such a fluctuation effect indeed occurs, but only at the anharmonic order. In this case, domain walls as in Fig. 6.9(b) acquire a finite free energy per unit length, $F_\mathrm{d}/\ell \sim \gamma T^2/J$, with $\gamma \sim 10^{-4}$, and the free energy of different ground states is similarly shifted by an amount $\sim \gamma T^2/J$ per unit cell; the lowest energy state is one in which the charges all have the same sign when looking in one of the complimentary directions. The smallness of the coefficient $\gamma$, however, implies that one needs to have very large lattice sizes (Korshunov estimates $L > 10^5$), in order for the total free energy of a system spanning domain wall to be greater than the proposed critical temperature of an ordering transition, and hence to observe the fluctuation induced ordering. Very similar effects have been reported by Korshunov[73,74] for the FFXY on a dice lattice[75] (with dual Coulomb gas on a kagome lattice).

Recently, new works on the FFXY model on the honeycomb lattice have attempted to address some of these issues. Numerical simulations and a scaling analysis by Granato[76] argue that both phase angle and vortex ordering transitions occur only at $T = 0$. Korshunov,[77] however, has argued for three distinct ordered phases at low temperatures. The existence of the two ordered phases at the lowest temperatures relies on the small but finite domain wall free energy induced by the above spinwave–vortex interaction.

However the third of these ordered phases persists even in the absence of this effect, and consists of an ordering in the *orientation* of zero-energy domain walls. This phase is disordered via a first order phase transition mediated by the appearance of free fractional vortices.

## 6.8. Conclusions

In this chapter we have endeavored to describe the rich phenomena associated with the fully frustrated XY model on different lattices. Many of the key concepts in understanding this phenomena have their clear origin in the ideas of Berezinskii and of Kosterlitz and Thouless. The broader class of uniformly frustrated XY models offers an even richer set of systems in which to explore the relation between continuous and discrete symmetries, and complex spatial ground states. Numerous works have explored specific cases. Denniston and Tang[78,79] have shown that for the $f = 1/3$ model on a square lattice, and the $f = 2/5$ on a square lattice with quenched bond disorder, the vortex pattern ordering transition is in the same universality class as the ordinary 2D Ising model, whereas the pure $f = 2/5$ model has a first-order transition. Kolahchi and Straley have shown that for $f = 5/11$ on a square lattice, the ground state consists of a periodic superlattice of vacancies on the otherwise checkerboard pattern of $f = 1/2$, with the basic periodic cell of that structure having size $22 \times 22$ in contrast to earlier suggestions[9] that for $f = p/q$ the ground state would always be compatible with a $q \times q$ periodic cell. Franz and Teitel[81] found for this $f = 5/11$ case that, upon heating, the superlattice of vacancies first melted to a liquid while preserving the background checkerboard $f = 1/2$-like structure, and only at a higher temperature did the $f = 1/2$-like structure itself melt. Franz and Teitel,[81] and later Gotcheva and Teitel[82] studied dilute frustrations $f = 1/q$ on triangular and square lattices, finding, for $q$ large enough, distinct vortex lattice unpinning and vortex lattice melting transitions upon heating the ground state. Korshunov et al.,[83] have studied the cases $f = 1/4$ and $f = 1/3$ on a triangular lattice, and found large ground state degeneracies similar to those discussed above for $f = 1/2$ on the honeycomb lattice. For the $f = 1/3$ model on a dice lattice, Korshunov[84] has argued that the vortex pattern is disordered at any finite temperature, yet there remains a sharp transition at finite temperature to a low temperature phase in which there exists phase angle ordering with a finite helicity modulus $\Upsilon$; numerical simulations by Cataudella and Fazio,[75] however, seem to show an ordered vortex structure at low temperature.

Considerable effort has gone into attempts to understand the behavior at *irrational* values of the frustration, in particular the frustration $f^* = (3 - \sqrt{5})/2$, which is related to the golden mean. In the first study of this model, Halsey[85] proposed that the system had a finite temperature transition into a disordered but frozen vortex glass state. However subsequent study of the Coulomb gas version of the model by Gupta et al.,[86] and of the XY version of the model by Denniston and Tang[87] and by Kolahchi and Fazli,[88] found evidence for a finite temperature transition $T_c$ to an ordered vortex pattern. Although it appears that the ground states of the system at $f^*$ may depend in detail upon the precise form of the interaction potential $V(\varphi)$, for both the Villain interaction (i.e. the Coulomb gas model) and the cosine interaction, the ordered vortex structure below $T_c$ appears to possess anisotropic phase angle coherence, with the helicity modulus becoming finite in one direction, while remaining zero in the orthogonal direction until a much lower temperature where all vortices become pinned. Other recent works,[89-91] however, have argued for a zero temperature glass transition.

Determining the ground state vortex pattern for a general $f = p/q$ has received considerable attention. Straley and Barnett[92] determined ground states for all cases with $q \leq 20$ for the cosine XY model, and found that the periodicity of the ground state can in general be larger than $q$. Denniston and Tang[87] have found that ground state periodicity may be as large as $q^2$. Kolahchi[93,94] has proposed schemes to generate potential ground states for rational and irrational frustrations of the cosine XY model. Lee et al.[95] have proposed a scheme to construct ground states for all $1/3 < f < 1/2$ in the dual Coulomb gas. A comprehensive review of the many facets of the general uniformly frustrated XY model lies outside the scope of this short chapter. We therefore conclude with the observation that much rich phenomena remains to be explored.

## Acknowledgments

I am grateful to the many colleagues from whom I learned about the XY model, and the many colleagues and students who joined me in working on related problems. These include, Vinay Ambegaokar, Petter Minnhagen, C. Jayaprakash, David Stroud, Eytan Domany, K. K. Mon, Ying-hong Li, Jong-Rim Lee, John Chiu, Peter Olsson, Marcel Franz, Michel Gingras, Sergey Korshunov, Pramod Gupta, Tao Chen, and Violeta Gotcheva.

## References

1. V. L. Berezinskii, Violation of long range order in one-dimensional and two-dimensional systems with a continuous symmetry group. I. Classical systems, *Zh. Eksp. Teor. Fiz.* **59**, 907 (1970); [*Sov. Phys.–JETP* **32**, 493 (1971)].
2. V. L. Berezinskii, Violation of long range order in one-dimensional and two-dimensional systems with a continuous symmetry group. II. Quantum systems, *Zh. Eksp. Teor. Fiz.* **61**, 1144 (1971); [*Sov. Phys.–JETP* **34**, 610 (1972)].
3. J. M. Kosterlitz and D. J. Thouless, Ordering, metastability and phase transitions in two-dimensional systems, *J. Phys. C* **6**, 1181 (1973).
4. J. M. Kosterlitz, The critical properties of the two-dimensional XY model, *J. Phys. C* **7**, 1046 (1974).
5. D. J. Resnick, J. C. Garland, J. T. Boyd, S. Shoemaker and R. S. Newrock, Kosterlitz–Thouless transition in proximity-coupled superconducting arrays, *Phys. Rev. Lett.* **47**, 1542 (1981).
6. R. F. Voss and R. W. Webb, Phase coherence in a weekly coupled array of 20 000 NB Josephson junctions, *Phys. Rev. B* **25**, 3446 (1982).
7. C. J. Lobb, Josephson junction arrays and superconducting wire networks, *Hel. Phys. Acta* **65**, 219 (1992).
8. S. Teitel and C. Jayaprakash, Phase transitions in frustrated two-dimensional XY models, *Phys. Rev. B* **27**, 598 (1983).
9. S. Teitel and C. Jayaprakash, Josephson-junction arrays in transverse magnetic fields, *Phys. Rev. Lett.* **51**, 1999 (1983).
10. J. Villain, Spin glass with non-random interactions, *J. Phys. C* **10**, 1717 (1977).
11. J. Villain, Two-level systems in a spin-glass model: I. General formalism and two-dimensional model, *J. Phys. C* **10**, 4793 (1977).
12. See, for example, M. Plischke and B. Bergersen, *Equilibrium Statistical Physics*, 3rd ed. (World Scientific, Singapore, 2006), Sec. 6.6.
13. M. E. Fisher, M. N. Barber and D. Jasnow, Helicity modulus, superfluidity, and scaling in isotropic systems, *Phys. Rev. A* **8**, 1111 (1973).
14. T. Ohta and D. Jasnow, XY model and the superfluid density in two dimensions, *Phys. Rev. B* **20**, 139 (1979).
15. D. R. Nelson and J. M. Kosterlitz, Universal jump in the superfluid density of two-dimensional superfluids, *Phys. Rev. Lett.* **39**, 1201 (1977).
16. P. Minnhagen and G. G. Warren, Superfluid density of a two-dimensional fluid, *Phys. Rev. B* **24**, 2526 (1981).
17. J. José, L. P. Kadanoff, S. Kirkpatrick and D. R. Nelson, Renormalization, vortices, and symmetry-breaking perturbations in the two-dimensional planar model, *Phys. Rev. B* **16**, 1217 (1977).
18. E. Fradkin, B. Huberman and S. H. Shenker, Gauge symmetries in random magnetic systems, *Phys. Rev. B* **18**, 4789 (1978).
19. R. Savit, Vortices and the low-temperature structure of the X-Y model, *Phys. Rev. B* **17**, 1340 (1978).
20. A. Vallat and H. Beck, Coulomb-gas representation of the two-dimensional XY model on a torus, *Phys. Rev. B* **50**, 4015 (1994).

21. T. C. Halsey, Topological defects in the fully frustrated XY model and in $^3$He-A films, *J. Phys. C* **18**, 2437 (1985).
22. S. E. Korshunov and G. V. Uimin, Phase transitions in two-dimensional uniformly frustrated XY models. I. Antiferromagnetic model on a triangular lattice, *J. Stat. Phys.* **43**, 1 (1986).
23. S. E. Korshunov, Phase transitions in two-dimensional uniformly frustrated XY models. II. General scheme, *J. Stat. Phys.* **43**, 17 (1986).
24. E. Granato, Domain wall induced XY disorder in the fully frustrated XY model, *J. Phys. C: Solid State Phys.* **20**, L215 (1987).
25. P. Minnhagen, Nonuniversal jumps and the Kosterlitz–Thouless transition, *Phys. Rev. Lett.* **54**, 2351 (1985).
26. P. Minnhagen, Empirical evidence of a nonuniversal Kosterlitz–Thouless jump for frustrated two-dimensional XY models, *Phys. Rev. B* **32**, 7548 (1985).
27. M. Y. Choi and S. Doniach, Phase transitions in uniformly frustrated XY models, *Phys. Rev. B* **31**, 4516 (1985).
28. M. Yosefin and E. Domany, Phase transitions in fully frustrated spin systems, *Phys. Rev. B* **32**, 1778 (1985).
29. M. Y. Choi and D. Stroud, Critical behavior of pure and diluted XY models with uniform frustration, *Phys. Rev. B* **32**, 5773 (1985).
30. E. Granato and J. M. Kosterlitz, Critical behavior of coupled XY models, *Phys. Rev. B* **33**, 4767 (1986).
31. J. M. Thijssen and H. J. F. Knops, Monte Carlo transfer-matrix study of the frustrated XY model, *Phys. Rev. B* **42**, 2438 (1990).
32. J. Y. Lee, J. M. Kosterlitz and E. Granato, Monte Carlo study of frustrated XY models on a triangular and square lattice, *Phys. Rev. B* **43**, 11531 (1991).
33. G. Ramirez-Santiago and J. V. José, Correlation functions in the fully frustrated 2D XY model, *Phys. Rev. Lett.* **68**, 1224 (1992).
34. G. Ramirez-Sanitago and J. V. José, Critical exponents of the fully frustrated two-dimensional XY model, *Phys. Rev. B* **49**, 9567 (1994).
35. E. Granato and M. P. Nightingale, Chiral exponents of the square-lattice frustrated XY model: A Monte Carlo transfer-matrix calculation, *Phys. Rev. B* **48**, 7438 (1993).
36. S. Y. Lee and K. C. Lee, Phase transitions in the fully frustrated XY model studied with use of the microcanonical Monte Carlo technique, *Phys. Rev. B* **49**, 15184 (1994).
37. P. Olsson, Two phase transitions in the fully frustrated XY model, *Phys. Rev. Lett.* **75**, 2758 (1995).
38. P. Olsson, Olsson replies: Two phase transitions in the fully frustrated XY model, *Phys. Rev. Lett.* **77**, 4850 (1996).
39. P. Olsson, Monte Carlo study of the villain version of the fully frustrated XY model, *Phys. Rev. B* **55**, 3585 (1997).
40. V. Cataudella and M. Nicodemi, Efficient cluster dynamics for the fully frustrated XY model, *Physica A* **233**, 293 (1996).
41. E. H. Boubcheur and H. T. Diep, Critical behavior of the two-dimensional fully frustrated XY model, *Phys. Rev. B* **58**, 5173 (1998).

42. Y. Ozeki and N. Ito, Nonequilibrium relaxation analysis of fully frustrated XY models in two dimensions, *Phys. Rev. B* **68**, 054414 (2003).
43. D. H. Lee, J. D. Joannopoulos, J. W. Negele and D. P. Landau, Discrete symmetry breaking and novel critical phenomena in an antiferromagnetic planar (XY) model in two dimensions, *Phys. Rev. Lett.* **52**, 433 (1984).
44. D. H. Lee, J. D. Joannopoulos, J. W. Negele and D. P. Landau, Symmetry analysis and Monte Carlo study of a frustrated antiferromagnetic planar (XY) model in two dimensions, *Phys. Rev. B* **33**, 450 (1986).
45. S. Miyashita and H. Shiba, Nature of the phase transition of the two-dimensional antiferromagnetic planar rotator model on the triangular lattice, *J. Phys. Soc. Jpn.* **53**, 1145 (1984).
46. W. Y. Shih and D. Stroud, Superconducting arrays in a magnetic field: Effects of lattice structure and a possible double transition, *Phys. Rev. B* **30**, 6774 (1984).
47. J. E. Van Himbergen, Monte Carlo study of a generalized-planar-model antiferromagnet with frustration, *Phys. Rev. B* **33**, 7857 (1986).
48. H.-J. Xu and B. W. Southern, Phase transitions in the classical XY antiferromagnet on the triangular lattice, *J. Phys. A: Math. Gen.* **29** L133 (1996).
49. S. Lee and K. C. Lee, Phase transitions in the fully frustrated triangular XY model, *Phys. Rev. B* **57**, 8472 (1998).
50. J. D. Noh, H. Rieger, M. Enderie and K. Knorr, Critical behavior of the frustrated antiferromagnetic six-state clock model on a triangular lattice, *Phys. Rev. E* **66**, 026111 (2002).
51. G. S. Grest, Critical behavior of the two-dimensional uniformly frustrated charged Coulomb gas, *Phys. Rev. B* **39**, 9267 (1989).
52. J. M. Thijssen and H. J. F. Knops, Monte Carlo study of the Coulomb-gas representation of frustrated XY models, *Phys. Rev. B* **37**, 7738 (1988).
53. J. M. Thijssen, Phase diagram of the frustrated XY model on a triangular lattice, *Phys. Rev. B* **40**, 5211 (1989).
54. J.-R. Lee, Phase transitions in the two-dimensional classical lattice Coulomb gas of half-integer charges, *Phys. Rev. B* **49**, 3317 (1994).
55. B. Berg, H. T. Diep, A. Ghazali and P. Lallemand, Phase transitions in two-dimensional uniformly frustrated XY spin systems, *Phys. Rev. B* **34**, 3177 (1986).
56. E. Granato and J. M. Kosterlitz, Frustrated XY model with unequal ferromagnetic and antiferromagnetic bonds, *J. Phys. C: Solid State Phys.* **19**, L59 (1986).
57. E. Granato, J. M. Kosterlitz, J. Y. Lee and M. P. Nightingale, Phase transitions in coupled XY-Ising systems, *Phys. Rev. Lett.* **42**, 1090 (1991)
58. J. Y. Lee, E. Granato and J. M. Kosterlitz, Nonuniversal critical behavior and first-order transitions in a coupled XY-Ising model, *Phys. Rev. B* **44**, 4819 (1991).
59. M. P. Nightingale, E. Granato and J. M. Kosterlitz, Conformal anomaly and critical exponents of the XY Ising model, *Phys. Rev. B* **52**, 7402 (1994).
60. G. Cristofano, V. Marotta, P. Minnhagen, A. Naddeo and G. Niccoli, CFT

description of the fully frustrated XY model and phase diagram analysis, *J. Stat. Mech. Theo. Expt.*, P11009 (2006).
61. P. Minnhagen, B. J. Kim, S. Bernhardsson and G. Cristofano, Phase diagram of generalized fully frustrated XY model in two dimensions, *Phys. Rev. B* **76**, 224403 (2007).
62. P. Minnhagen, B. J. Kim, S. Bernhardsson and G. Cristofano, Symmetry-allowed phase transitions realized by the two-dimensional fully frustrated XY class, *Phys. Rev. B* **78**, 184432 (2008).
63. P. Minnhagen, S. Bernhardsson and B. J. Kim, The groundstates and phases of the two-dimensional fully frustrated XY model, *Int. J. Mod. Phys. B* **23**, 2929 (2009).
64. M. Hasenbusch, A. Pelissetto and E. Vicari, Transitions and crossover phenomena in fully frustrated XY systems, *Phys. Rev. B* **72**, 184502 (2005).
65. M. Hasenbusch, A. Pelissetto and E. Vicari, Multicritical behavior in the fully frustrated XY model and related systems, *J. Stat. Mech: Theo. Expt.* P12002 (2005).
66. S. Okumura, H. Yoshino and H. Kawamura, Spin-chirality decoupling and critical properties of a two-dimensional fully frustrated XY model, *Phys. Rev. B* **83**, 094429 (2011).
67. S. E. Korshunov, Kink pairs unbinding on domain walls and the sequence of phase transitions in the fully frustrated XY models, *Phys. Rev. Lett.* **88**, 167007 (2002).
68. The unbinding transition of logarithmically interacting charges in one dimension was first discussed in: S. A. Bulgadaev, Phase Transitions in Gases with Generalized Charges, *Phys. Lett. A* **86**, 213 (1981); Phase Transitions in Gases with Generalized Charges Interacting Through a Logartithmic Law. 2. $D = 1$, Isotropic Case, *Teor. Mat. Fiz.* **51**, 424 (1982) [*Theor. Math. Phys.* **51**, 593 (1982)].
69. P. Olsson and S. Teitel, Kink–antikink unbinding transition in the two-dimensional fully frustrated XY model, *Phys. Rev. Lett.* **71**, 104423 (2005).
70. J.-R. Lee and S. Teitel, Phase transitions in classical two-dimensional lattice Coulomb gases, *Phys. Rev. B* **46**, 3247 (1992).
71. V. Gotcheva, Y. Wang, A. T. J. Wang and S. Teitel, Continuous-time Monte Carlo and spatial ordering in driven lattice gases: Application to driven vortices in periodic superconducting networks, *Phys. Rev. B* **72**, 064505 (2005).
72. S. E. Korshunov and B. Douçot, Fluctuations and vortex-pattern ordering in the fully frustrated XY model on a honeycomb lattice, *Phys. Rev. Lett.* **93**, 097003 (2004)
73. S. E. Korshunov, Vortex ordering in fully frustrated superconducting systems with a dice lattice, *Phys. Rev. B* **63**, 134503 (2001).
74. S. E. Korshunov, Fluctuation-induced vortex pattern and its disordering in the fully frustrated XY model on a dice lattice, *Phys. Rev. B* **71**, 174501 (2005).
75. V. Cataudella and R. Fazio, Glassy dynamics of Josephson arrays on a dice lattice, *Europhys. Lett.* **61**, 341 (2003).
76. E. Granato, Zero-temperature transition and correlation-length exponent of the frustrated XY model on a honeycomb lattice, *Phys. Rev. B* **85**, 054508 (2012).

77. S. E. Korshunov, Ordered phases and phase transitions in the fully frustrated XY model on a honeycomb lattice, *Phys. Rev. B* **85**, 134526 (2012).
78. C. Denniston and C. Tang, Domain walls and phase transitions in the frustrated two-dimensional XY model, *Phys. Rev. Lett.* **79**, 451 (1997).
79. C. Denniston and C. Tang, Low-energy excitations and phase transitions in the frustrated two-dimensional XY model, *Phys. Rev. B* **58**, 6591 (1998).
80. M. Kolahchi and J. P. Straley, Ground state of the uniformly frustrated two-dimensional XY model near $f = 1/2$, *Phys. Rev. B* **43**, 7651 (1991).
81. M. Franz and S. Teitel, Vortex-lattice melting in two-dimensional superconducting networks and films, *Phys. Rev. B* **51**, 6551 (1995).
82. V. Gotcheva and S. Teitel, Depinning transition of a two-dimensional vortex lattice in a commensurate periodic potential, *Phys. Rev. Lett.* **86**, 2126 (2001).
83. S. E. Korshunov, A. Vallat and H. Beck, Frustrated XY models with accidental degeneracy of the ground state, *Phys. Rev. B* **51**, 3071 (1995).
84. S. E. Korshunov, Uniformly frustrated XY model without a vortex-pattern ordering, *Phys. Rev. Lett.* **94**, 087001 (2005).
85. T. C. Halsey, Josephson-junction array in an irrational magnetic field: A superconducting glass?, *Phys. Rev. Lett.* **55**, 1018 (1985).
86. P. Gupta, S. Teitel and M. J. P. Gingras, Glassiness versus order in densely frustrated Josephson arrays, *Phys. Rev. Lett.* **80**, 105 (1998).
87. C. Denniston and C. Tang, Incommensurability in the frustrated two-dimensional XY model, *Phys. Rev. B* **60**, 3163 (1999).
88. M. R. Kolahchi and H. Fazli, Behavior of a Josephson-junction array with golden mean frustration, *Phys. Rev. B* **62**, 9089 (2000).
89. S. Y. Park, M. Y. Choi, B. J. Kim, G. S. Jeon and J. S. Chung, Intrinsic finite-size effects in the two-dimensional XY model with irrational frustration, *Phys. Rev. Lett.* **85**, 3484 (2000).
90. E. Granato, Zero-temperature resistive transition in Josephson-junction arrays at irrational frustration, *Phys. Rev. B* **75**, 184527 (2007).
91. E. Granato, Phase and vortex correlations in superconducting Josephson-junction arrays at irrational magnetic frustration, *Phys. Rev. Lett.* **101**, 027002 (2008).
92. J. P. Straley and G. M. Barnett, Phase diagram for a Josephson network in a magnetic field, *Phys. Rev. B* **48**, 3309 (1993).
93. M. R. Kolahchi, Finding local minimum states of Josephson-junction arrays in a magnetic field, *Phys. Rev. B* **56**, 95 (1997).
94. M. R. Kolahchi, Ground state of uniformly frustrated Josephson-junction arrays at irrational frustration, *Phys. Rev. B* **59**, 9569 (1999).
95. S. J. Lee, J.-R. Lee and B. Kim, Patterns of striped order in the classical lattice Coulomb gas, *Phys. Rev. Lett.* **88**, 025701 (2002).

Chapter 7

# Charges and Vortices in Josephson Junction Arrays

Rosario Fazio

*NEST, Scuola Normale Superiore and Istituto Nanoscienze – CNR,
56126 Pisa, Italy*

Gerd Schön

*Institut für Theoretische Festkörperphysik and DFG-Center for
Functional Nanostructures (CFN), Karlsruhe Institute of Technology,
76128 Karlsruhe, Germany*

> Classical Josephson junction arrays have been studied as model systems demonstrating the properties of the Berezinskii–Kosterlitz–Thouless transition between a superconducting and a resisitive state, driven by the unbinding of vortices. In junction arrays with small capacitance charge degrees of freedom and quantum fluctuations gain importance. Vortices acquire a quantum dynamics, under certain conditions charges and vortices are dual to each other, and a quantum phase transition may occur between a Mott insulating and a superconducting phase. Here, we review the description of Josephson junction arrays in the quantum regime and some of their intriguing properties.

## 7.1. Introduction

A Josephson junction array (JJA) consists of a network of superconducting islands which are weakly coupled by tunnel junctions (Fig. 7.1). Below the BCS critical temperature each island is superconducting with a well defined superconducting gap. The whole array, however, may still be in a resistive state due to the effect of phase fluctuations. A transition to the zero-resistance state, where global coherence is established across the whole sample, occurs at a lower temperature of the order of the Josephson coupling energy $E_J$, which is the energy scale associated with Cooper pair tunneling

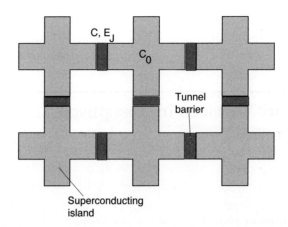

Fig. 7.1. A schematic view of a Josephson array. A regular network of superconducting islands (here a square lattice) is weakly coupled by Josephson tunnel junctions. Each junction is characterized by the Josephson coupling energy and the junction capacitance $C$. In addition, each island has a capacitance $C_0$ to the ground.

between neighboring islands. Classical JJAs, such as those fabricated in the early eighties,[1] constitute a physical realization of the XY-model. In two-dimensional arrays the transition to the phase coherent state due to the binding of vortex–antivortex pairs is of the Berezinskii–Kosterlitz–Thouless (BKT) type.[2,3] The properties of this transition were demonstrated in experiments.[4] Further investigations, both experimental and theoretical, identified Josephson arrays as ideal model systems in which also frustration effects, classical vortex dynamics, as well as nonlinear dynamics and chaos can be studied in a controlled way.[5–7]

When the size of the superconducting islands is reduced to the submicron scale, the charging energy scale $E_C$, i.e. the energy associated with adding single electron charges, may become comparable or even larger than the Josephson coupling. In this regime, quantum effects are observable due to the number-phase uncertainty relation between fluctuations of the charge on each island and those of the phase of the order parameter. Quantum fluctuations are responsible for a variety of zero-temperature (quantum) phase transitions.[8] A quantum JJA may be insulating at zero temperature even though each island is still superconducting. In the classical limit $E_J \gg E_C$, the system turns superconducting at low temperatures since the fluctuations of the phases are weak and the system is globally phase coherent. In the opposite limit, $E_C \gg E_J$, the array becomes a Mott insulator since the charge on each island is localized, and an activation energy of the order of

$E_C$ is required to transport charges through the system (Coulomb blockade of Cooper pairs). Quantum JJAs were first realized in Delft, and a direct superconductor-insulator transition has been observed in transport experiments.[9]

Also the quantum *dynamics* of JJAs is characterized by the properties of the charges and vortices. In the classical limit, $E_J \gg E_C$, vortices are the topological excitations that determine the (thermo)dynamic properties of JJAs. In quantum JJAs vortices behave as particles with a mass moving in a periodic potential and subject to forces and dissipation. They can move ballistically, and under appropriate conditions they can display quantum properties, such as quantum tunneling and quantum coherence, at a macroscopic level. In the opposite situation, $E_J \ll E_C$, the charges on each island are the relevant degrees of freedom. Vortices and charges play a dual role, and many features of JJAs in the two limits can be related to each other with the roles of charges and vortices being interchanged.

In this review, we consider the quantum model of a Josephson junction array. We analyze the transitions between different phases. We then demonstrate the duality between charges and vortices by presenting the effective action for these coupled degrees of freedom. In the regimes $E_J \gg E_C$ and $E_J \ll E_C$ we will study the dynamics of vortices and charges, respectively. We will not be able to give an exhaustive overview of the physics of JJAs. Our aim here is to present some aspects of quantum Josephson arrays which are related to the properties of BKT transition. For a detailed discussion of the properties of JJAs in the quantum regime we refer to the reviews.[10,11]

## 7.2. Model of a Josephson Junction Array in the Quantum Regime

Quantum effects in JJAs come into play whenever the scale of the charging energy (associated with non-neutral charge configurations of the islands) is comparable with the Josephson coupling. The electrostatic energy of charges at the array sites $i$ and $j$ is determined by the capacitance matrix $C_{ij}$, with dominant contributions coming from the junction capacitances $C$ and the capacitance to the ground $C_0$. Hence $C_{ij} = (C_0 + zC)\delta_{ij} - C\delta_{j,nn(i)}$, where $z$ is the coordination number of the lattice and the second delta-function involves only nearest neighbor sites. For $C \gg C_0$, which is the limit considered throughout this paper, the interaction decays logarithmically up to distances of the order of the screening length $\lambda \sim \sqrt{C/C_0}$ (in units of the lattice constant), while for larger distance, it decays exponentially.

In the following, we restrict our attention to low temperatures, where quasiparticles are frozen out (or contribute only to a static disorder, see below). Hence the relevant charges are Cooper pair charges $2e$. The electrostatic energy is further influenced by applied gate voltages. Adding the Josephson coupling energy in the presence of an external magnetic field, we thus arrive at the Hamiltonian

$$H = \frac{1}{2} \sum_{i,j} (q_i - q_{x,i}) U_{ij} (q_j - q_{x,j}) - E_J \sum_{\langle i,j \rangle} \cos\left(\phi_i - \phi_j - \frac{2\pi}{\Phi_0} A_{ij}\right). \quad (7.1)$$

The first term is the electrostatic energy, with $U_{ij} = \frac{1}{2}(2e)^2 C_{ij}^{-1}$, $2eq_i$ is the net charge on the $i$th island and $q_{x,i} = \sum_j C_{ij} V_{x,j}/2e$ is an external charge induced by a gate voltage $V_{x,j}$. The second term is the Josephson coupling energy, $\phi_i$ is the phase of the superconducting order parameter, and $A_{ij}$ the line integral of the vector potential defined according to the Peierls substitution. The sum in the Josephson coupling is constrained to nearest neighbor only. Finally $\Phi_0 = hc/2e$ is the superconducting flux quantum. The commutation relation between charge number and phase operators $[q_i, e^{i\phi_j}] = \delta_{ij} e^{i\phi_j}$ make the model a quantum model.

In the Hamiltonian (7.1) the Josephson energy tends to establish global phase coherence, while the electrostatic term favors charge localization on each island and therefore tends to suppress superconducting coherence. Indeed, due to the commutation relation between charge and phase operators the two terms in the Hamiltonian cannot be minimized simultaneously, and a zero-temperature (quantum) phase transition will take place. The existence of this zero-temperature superconductor-insulator transition leads to dramatically different behavior of the resistance as a function of temperature for different values of the ratio $E_J/E_C$. A sketch of the phase diagram is shown in Fig. 7.2(b). Samples for which the Josephson coupling dominates undergo a finite-temperature transition to a superconducting state, which belongs to the Berezinskii–Kosterlitz–Thouless universality class. In the absence of charging effects, the critical temperature is $T_c^{(0)} = (\pi/2) E_J$. With increasing charging energy the critical temperature gets reduced until, above some critical value, only the resistive state survives. The conductance in the disordered regime decreases when lowering the temperature and vanishes at zero temperature. Thus at $T = 0$ the array shows a superconductor–insulator transition despite the fact that each island is superconducting. A rough sketch of the dependence of the resistance of the array as a function of temperature (as it is typically observed in the experiments) is shown in Fig. 7.2(a). The sketch of the phase diagram shown in Fig. 7.2(b) applies to

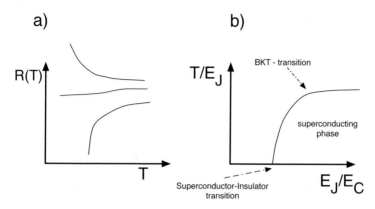

Fig. 7.2. (a) The superconductor-insulator transition in Josephson junction arrays is displayed in the temperature dependence of the electrical resistance. A sketch of what may be observed experimentally is shown for three arrays with different ratios of the Josephson coupling and charging energy, larger (superconducting), close to, and smaller (insulating) than the critical value. Close to the critical value the experiments may be further complicated by fluctuation and reentrance effects. (b) The phase diagram of a JJA with short range interaction of the charges shows two regions: a low-temperature coherent phase and a disordered phase separated by a BKT transition. The disordered phase at finite temperatures is resistive. At zero temperature, a transition between a superconducting and an insulating phase occurs.

the case of short-range charge interaction in the absence of external gates and magnetic fields. The case in which the electrostatic interaction is long-range will be discussed in the next section.

A very rich scenario emerges in the presence of external charges and/or magnetic frustration, which are controlled by gate voltages and applied magnetic fields. Theoretically, the phase diagram of JJAs has been studied by a variety of methods (mean-field, renormalization group, quantum Monte Carlo,...) and under different situations (dimensionality, frustration effects, tuning of the gate voltages,...). Here we cannot refer to all the important papers that addressed this problem. The first papers date back to the 80's,[12-15] and many more followed in the 90's. But still some controversy remains. Even the order of the superconductor–insulator transition is still debated, with very recent work[16] indicating that, under certain conditions, it may be of first order.

## 7.3. Charge-Vortex Duality

In classical Josephson arrays, described by the XY-model, the vortex unbinding transition can be made evident by mapping the phase-model onto

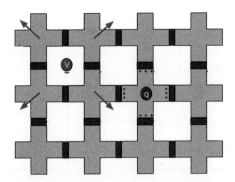

Fig. 7.3. An extra charge $q$ in a given island polarizes the neighboring ones up to a distance of the order of the screening length (of the order of $\sqrt{C/C_0}$). The figure shows also a vortex configuration with arrows indicating the phase of each island with respect to some reference direction.

a two-dimensional gas of logarithmically interacting vortices. They are the relevant topological excitations of the system and drive the BKT-transition. They repel each other if they have the same sign of vorticity, otherwise they attract each other. The mapping of the XY-model to a description in terms of its topological defects can be preformed by means of the Villain transformation.[17–19]

Also in the quantum regime, the properties of a Josephson array are to a large extent characterized by its topological excitations, which in this case are the vortices and the charges (see Fig. 7.3). By using the Villain transformation, in an extension of the procedure in the classical case, we could map the model of Eq. (7.1) onto a model expressed in terms of integer-valued charge degrees of freedom $q_i$ on the islands and vortices $v_i$ in the plaquettes.[20,21] The partition function of the JJA can be expressed as a sum over these fields,

$$Z = \sum_{[q,v]} e^{-S\{q,v\}}, \qquad (7.2)$$

where the effective action $S\{q,v\}$ reads

$$S\{q,v\} = \int_0^\beta d\tau \sum_{i,j} \left\{ \frac{1}{2} q_i(\tau) U_{ij} q_j(\tau) + \pi E_J v_i(\tau) G_{ij} v_j(\tau) \right.$$
$$\left. + i\, q_i(\tau) \Theta_{ij} \dot{v}_j(\tau) + \frac{1}{4\pi E_J} \dot{q}_i(\tau) G_{ij} \dot{q}_j(\tau) \right\}. \qquad (7.3)$$

(For transparency, we set here gate voltages and magnetic fields to zero.[22]) This action describes two coupled Coulomb gases, $q$ and $v$. The charges interact via the electrostatic interaction. The interaction among the vortices (second term) is described by the kernel $G_{ij}$, which is the Fourier transform of $k^{-2}$. At large distances it depends logarithmically on the distance, $G_{ij} \sim -\frac{1}{2}\ln r_{ij}$. The third term describes the coupling between charges and vortices. The function

$$\Theta_{ij} = \arctan\left[\frac{y_i - y_j}{x_i - x_j}\right] \quad (7.4)$$

represents the phase configuration at site $i$ when a vortex is centered at the site $j$. This coupling has a simple physical interpretation: a change of vorticity at site $j$ produces a voltage which is felt by the charge at site $i$. The last term in the action is due to the spin-wave contribution to the charge-correlation function.

### 7.3.1. Self-duality and phase diagram

The action (7.3) explicitly displays the duality between both excitations. It demonstrates that there is a dual transformation relating the classical vortex limit, $E_J \gg E_C$, to the opposite charge limit, $E_J \ll E_C$. An interesting case, with a particularly high degree of symmetry between vortex and charge degrees of freedom is realized in the limit $C_0 \ll C$. In this limit, the screening is suppressed and the charges interact logarithmically over large distances. That is, the inverse capacitance matrix has the same functional form as the kernel describing vortex interactions,

$$U_{ij} = \frac{2}{\pi} E_C G_{ij}. \quad (7.5)$$

In this case, the transformation

$$v_i \leftrightarrow q_i$$
$$\pi E_J \leftrightarrow 2E_C/\pi$$

produces an action of the same form as before, which implies that charges and vortices are dual to each other. It allows relating the classical vortex limit, $E_J \gg E_C$, to the opposite charge limit, $E_J \ll E_C$. In addition, there is a symmetry point for which the system is self-dual, i.e. invariant under the transformation, namely for

$$E_J/E_C = 2/\pi^2. \quad (7.6)$$

More precisely, the duality is strict for vanishing self-capacitance and in the absence of the spin-wave (last) term in Eq. (7.3). The latter term is irrelevant at the critical point; it merely shifts the transition point. However, it has important implications for the dynamical behavior, to be discussed below.

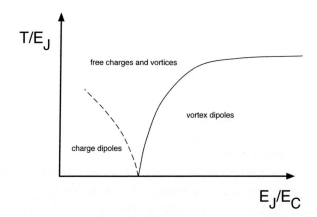

Fig. 7.4. Phase diagram of a Josephson array in the case in which the charges interact logarithmically, i.e. when $C_0 = 0$. In addition to the vortex unbinding transition there is also a charge unbinding transition set by the ratio between the temperature and the charging energy. Below the charge-unbinding transition the array is in an insulating state. The zero-temperature superconducting to insulating transition is expected at the self-dual point $E_J/E_C \sim 2/\pi^2$.

The self-duality in the limit of long-range Coulomb interaction has further interesting consequences for the phase diagram. Since in this case the interaction between charges on islands is logarithmic, analogous to vortex interactions in the limit $E_C \gg E_J$, charges will bind/unbind at a charge-BKT transition[23] with transition temperature $T_{ch}^{(0)} = E_C/\pi$. A finite Josephson coupling reduces the transition temperature. Thus, in the limit of logarithmically interacting charges, the phase diagram, displays three phases, as shown in Fig. 7.4. In addition to the global coherent phase, where vortices are bound in pairs, there is a finite temperature insulating phase where charges are bound in pairs. Only in the phase where vortices and charges are free the array is resistive. Experimental indication of this diagram has been reported.[24,25]

It should be noted, however, that an experimental demonstration of the charge-KBT transition is difficult for several reasons. One is that a non-vanishing capacitance to ground leads to a finite screening length, and hence the logarithmic interaction is valid only for a limited range of distances. Furthermore, while a uniformly applied gate voltage leads to a richly structured phase diagram with various charge ordered superstructures,[26] disorder in the offset charges, which is difficult to avoid in experiments, washes out much of this structure. Similarly, as pointed out by Feigel'man et al.[27] parity effects may contribute to mask the charge-BKT transition. Not withstanding further experimental confirmations of the charge-BKT transition, it is without doubt an interesting property that may appear in quantum Josephson junction arrays. Very recently, the charge-BKT transition has been also advocated to explain the properties of ultra-thin amorphous superconducting films.[28]

### 7.3.2. Magnetic and charge frustration

The charge-vortex duality manifests itself also in the way in which frustration can be induced in a quantum JJA. A frustrated JJA can be realized either by applying a uniform magnetic field, leading to a magnetic frustration, or by applying a uniform gate voltage with respect to the ground plane, leading to charge frustration. In the coupled-Coulomb gas action introduced in Eq. (7.3) charge and magnetic frustration results in the replacement $q_i \to q_i - q_x$ and $v_i \to q_i - f$. (Here $f$ is the magnetic flux piercing an elementary plaquette expressed in units of the quantum of flux $\Phi_0$.)

The presence of a uniform frustration induces a background of elementary excitations (charges of vortices). The influence of a magnetic frustration on the BKT transition, separating the superconducting and resistive phase, has been studied widely. It leads to a richly structured phase diagram. Similarly, a uniform applied gate voltage, leading to charge frustration, gives rise to a lobe structure in the phase diagram with Mott insulating phases characterized by a Wigner crystal-like arrangement of the induced charges.[26] Furthermore new phases may appear. This includes a supersolid phase where diagonal (solid) and off-diagonal (superfluid) long-range order coexist. Predicted in the sixties, supersolid phases have been searched for a long time in superfluid Helium.[29] Josephson arrays[26,30] as well as optical lattices,[31] are systems where a supersolid state could be found. However, in the phase diagram the supersolid occupies only a narrow region between the Mott and the superconducting phases. Hence, it is very fragile in the presence of charge disorder, which is difficult to avoid in Josephson arrays.

### 7.4. Vortex Dynamics

The coupled-Coulomb-gas representation, Eq. (7.3) is also a starting point to study the effective dynamics of charges and vortices. In the extreme limits, $E_J = 0$ or $E_C = 0$, the fluctuation of the charges or the vortices are frozen, and one remains with a classical gas of charges or vortices in the insulating and superconducting states, respectively. Quantum fluctuation gives a non-trivial dynamics to the topological excitations, either as the charges or the vortices. In the following sections, we consider the case in which the array is deep in one or the other phase.

### 7.4.1. *Ballistic vortex motion*

In the limit $E_J \gg E_C$, charges are fluctuating strongly and can be integrated out from the partition function (7.2). In this case, we obtain an effective action for the vortices only, but they are influenced by the charging effects. In the limit where the junction capacitance dominates, $C \gg C_0$, the result is

$$S[v] = \int_0^\beta d\tau \sum_{i,j} \left[ \frac{\pi}{8E_C} \dot{v}_i(\tau) G_{ij} \dot{v}_j(\tau) + \pi E_J v_i(\tau) G_{ij} v_j(\tau) \right]. \quad (7.7)$$

In some cases, it is sufficient to assume that each vortex moves in a continuous way and is described completely by the evolution of its center coordinate **r**. Then the resulting effective action for a single vortex reads

$$S_{\text{eff}} = \frac{1}{2} \sum_{a,b=x,y} \int d\tau d\tau' \dot{r}^a(\tau) M_{ab}[\mathbf{r}(\tau) - \mathbf{r}(\tau'), \tau - \tau'] \dot{r}^b(\tau'), \quad (7.8)$$

where the kernel $M$ depends on the charge–charge correlation function. The retarded character of the action reflects the fact that a moving vortex excites phase waves, which contribute to the damping. In general, the action is also nonlinear in the vortex coordinates, this effect stems from the discrete character of charge excitations.

*Vortex mass*: In the limit of low velocities $\dot{r}(\tau)$ the action (7.8) reduces to that of a free particle. The corresponding adiabatic vortex mass, $M_v = \int_0^\beta d\tau M_{xx}(0, \tau)$, first derived by Eckern and Schmid,[32] becomes in the classical limit

$$M_v = \frac{\pi^2}{4E_C}. \quad (7.9)$$

For typical array parameters, the mass can be smaller than the electron mass. On approaching the superconductor–insulator transition the vortex mass may change considerably. Very close to the transition, however, quantum fluctuations of the superconducting phases are so strong that the picture of a rigid vortex moving through the array is no longer valid.

The existence of a vortex mass can be understood by means of a simple argument.[32] A moving vortex leads to phase changes across the junctions and therefore contributes to the electric energy. In principle, all capacitances contribute, however the main effect comes from the junction capacitance, so that $E_{\rm ch} = 1/2 \sum_{\langle ij \rangle} C V_{ij}^2$ where $V_{ij}$ is the voltage drop across the junction between islands $i$ and $j$. In a quasi-static approach, this sum is calculated by comparing the phase configuration at times before and after the vortex has crossed a junction. The electric energy then acts like a kinetic energy term and the proportionality factor defines the vortex mass. In two-dimensional JJAs, the phase distribution due to a vortex is well approximated by the form (7.4), from which one obtains the result for the mass given above.

Viewed with higher resolution, one comes to the conclusion that the energy of the vortex in a JJA depends on the precise position of its center. The energy has a minimum value when the vortex is in the middle of a plaquette, while the maximum is reached when the vortex position is between two plaquettes. This effect can be taken into account in the equation of motion by including a periodic potential. A simple form which captures the main aspects of the vortex pinning by the lattice potential is[33]

$$U_v(x) = \frac{1}{2}\gamma E_{\rm J} \sin(2\pi x),$$

with $\gamma \sim 10^{-1} - 10^{-2}$.

*Lorentz and Magnus forces*: In the presence of an external current a vortex experiences a Lorentz force, moreover the moving vortex is subject to damping due to the excitation of phase waves. The equation of motion for the vortex may be obtained from Eq. (7.8). After including the external current $I$ directed along the $y$-direction, we obtain from Eq. (7.8)

$$2\pi I/I_c = \partial_\tau \int d\tau' M_{xb}(r(\tau) - r(\tau'), \tau - \tau') \dot{r}^b(\tau'), \qquad (7.10)$$

where $I_c$ is the critical current. In the semiclassical regime, the mass tensor is diagonal and the equation of motion assumes the familiar form where the vortex moves perpendicular to the applied current. Its constant-velocity solutions in the presence of an external current determine the nonlinear relation between driving current and vortex velocity and, hence, the current–voltage characteristics. The ballistic vortex motion, as described by the equation of motion (7.10), has been observed both in one-dimensional[34–36] and two-dimensional arrays,[37] and the mass term has been verified in experiments.[38,39]

In addition to the Lorentz force due to the external current as described by Eq. (7.10), a vortex may be subject also to a Magnus force which acts transverse to the vortex velocity. The study of the Magnus force in superfluids has a long and controversial history (see for example, Ref. 40). In Josephson arrays, in the presence of a gate voltage applied to the ground plane, leading to a gate charge $q_x$, particle-hole symmetry is broken, and a vortex feels a Magnus force[21,41] given by

$$\mathbf{F} = 2eq_x\Phi_0 \hat{\mathbf{z}} \times \dot{\mathbf{r}}. \tag{7.11}$$

As a result of the combined effect of the Magnus force and the Lorentz force, the vortices move at a certain angle, the Hall angle, with respect to the current. Its measurement yields information on the various sources of dissipation. In experiments on classical Josephson arrays, the Hall angle is found to be very small (see e.g. Ref. 42). In *quantum* Josephson arrays, Hall measurements have been performed by the Chalmers group.[43]

### 7.4.2. *Quantum vortices*

The dynamics of vortices in Josephson arrays also provides a playground to understand the quantum dynamics of macroscopic objects and their interaction with the environment. As we saw in the previous section, vortices may behave as particles with a mass moving in a periodic potential and subject to dissipation. There have been a number of important experiments testing their quantum properties,[38,39,44–49] including the observation of macroscopic quantum tunneling, of the Aharonov–Casher effect, and of Bloch oscillations of vortices in superlattices. In the following, we will touch upon these examples.

*Macroscopic quantum tunneling of vortices*: Quantum fluctuations in the vortex position become important when the zero-point energy in the minimum of the potential, $\omega_v = \sqrt{8\gamma E_J E_C}$, becomes comparable to the energy barrier $\sim \gamma E_J$. In this regime, quantum tunneling of vortices leads to a resistance of the array. The resistance has been measured as a function of the temperature in arrays for very small values of the frustration in Ref. 44. In this regime the resistance is proportional to the frustration. One therefore can expect that the vortex–vortex interaction is negligible and the dissipation is due to single-vortex motion in the lattice potential. The experiments of the Delft and Chalmers groups[44,45] have shown two distinct regimes. A classical one where the resistance drops exponentially with a thermally activated behavior, and a quantum regime for arrays with a larger charging energy, where the resistance levels off below a temperature $T_{\text{cr}}$. A simple estimate of the crossover temperature to the quantum tunneling regime gives

$$T_{\text{cr}} \sim 0.4\sqrt{\gamma E_C E_J}. \tag{7.12}$$

In order to estimate the behavior of the resistance in the quantum tunneling dominated regime, one has to evaluate the tunneling rate for a vor-

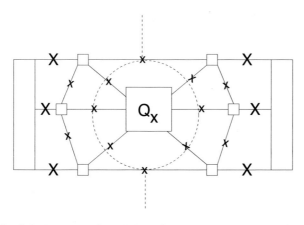

Fig. 7.5. Sketch of the hexagon-shaped Josephson junction array used to measure vortex interference.

tex to jump between two neighboring plaquettes. The exponent in the tunneling rate is given by the instanton action $S_{\text{inst}}$ associated to vortex tunneling. This problem was studied by Korshunov.[50] The instanton action can also be obtained in the language of the coupled Coulomb gas approach (7.7), by evaluating the action associated with the trajectory $\dot{v}_{i,t} = v_{i,\tau+\tau_\epsilon} - v_{i,\tau} = \delta_{\tau,t}[\delta_{i,x+1} - \delta_{i,x}]$ for a hop from $x,\tau$ to $x+1,\tau+\tau_\epsilon$. The result is $S_{\text{inst}} = \frac{1}{2}M_{xx}(0,0)$. In the limit of large Josephson coupling one recovers all known results, i.e. for general capacitance matrix

$$S_{\text{inst}} = \frac{\pi}{4}\sqrt{\frac{E_J}{8E_C}}\left[\sqrt{\pi}\sqrt{\frac{C}{C_0}+4\pi}+\frac{C}{2C_0}\ln\left(2\sqrt{\frac{\pi C_0}{C}}+\sqrt{1+\frac{4\pi C}{C_0}}\right)\right]. \tag{7.13}$$

*Aharonov–Casher effect*: The Aharonov–Casher effect[51] describes the interference of particles with a magnetic moment moving around a line charge. The observation on its vortices in Josephson arrays has been reported by Elion et al.,[49] shortly after the first theoretical proposal.[52,53] The sample in this experiment consists of a hexagon-shaped array with six triangular cells. Large-area junctions couple the hexagon to superconducting banks so that only two paths for vortex motion are possible. The vortices follow the trajectories indicated as dotted lines in Fig. 7.5. A gate controls the charge on the superconducting island in the middle of the hexagon. In this setup, the differential resistance displays clear oscillations as a function of the charge in the middle island.[49]

The Aharonov–Casher effect can be derived from the coupled Coulomb gas description. An external charge yields an additional term in the effective

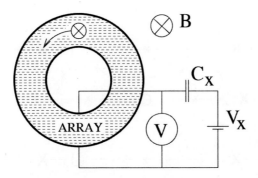

Fig. 7.6. Corbino disk to measure persistent voltages in ring-shaped Josephson arrays.

vortex action of the form

$$S_{\text{AC}} = -i \sum_i \frac{q_{x,i}}{2e} \int_0^\beta d\tau\, \dot{\mathbf{r}} \cdot \nabla\Theta[\mathbf{r}_i - \mathbf{r}(\tau)], \qquad (7.14)$$

where $\Theta$ is defined by (7.4). Inspection of Eq. (7.14) reveals that the effect of the external charge is equivalent to that of a pseudo-vector potential

$$2e\mathbf{A}_Q(\mathbf{r}) = q_x \hat{\mathbf{z}} \times \mathbf{r}/r^2,$$

acting on the vortex motion, similar as a magnetic vector potential influences the charge dynamics.

*Persistent vortex currents*: The charge-vortex duality suggests that in Josephson junction ring arrays a persistent current of vortices should be induced by the presence of charges on the inner island, in complete analogy to the persistent charge currents induced in ring structures by an applied magnetic flux. A persistent vortex current leads to a persistent voltage across the ring, as shown in Fig. 7.6, which is another manifestation of the Aharonov–Casher effect. For a ring defined in a square array the maximum persistent voltage should be of the order of

$$V_{\max} = \frac{2e}{CN^2},$$

where $N$ is the number of junctions along the ring. These effects should be observable in Josephson Corbino-type circuits, but no experimental confirmation has been reported yet.

*Bloch oscillations*: As a final example for the quantum properties of vortices in Josephson arrays, we discuss Bloch oscillations. We consider a particle moving in a periodic potential (leading to a gap in the band structure) which is subject to a constant force. In the absence of damping, with a weak applied force, the wave vector changes linearly in time until it reaches the Brillouin zone edge where it will be Bragg reflected. This periodic motion in momentum space leads to an oscillatory motion also in real space. In the presence of a scattering mechanism, dc transport occurs as well. The combination of both effects leads to a nose-shaped form of the dc current–voltage characteristic.

Quantum vortices in a periodic potential also form energy bands, which opens the possibility to study Bloch oscillation for vortices. While in the charge case, an electric field is needed to accelerate electrons and their motion produces a current, in the case of vortices, an imposed current accelerates the vortices whose motion produces a voltage drop. Experiments have been performed with a quasi one-dimensional JJA.[54] On average the vortex velocity is zero, but the Bloch oscillation frequency is[55]

$$\nu_B = I/(2e),  \qquad (7.15)$$

and the amplitude of the oscillation is

$$x = \int \frac{1}{\phi_0 I} = \frac{E_C}{E_J} \frac{I_c}{\pi I}. \qquad (7.16)$$

A characteristic feature of Bloch oscillating vortices is again the nose-shaped form of the dc current–voltage characteristic, but in the case of vortices as compared to that of charges the role of current and voltage are interchanged.

We only described here quantum properties of isolated vortices. The interplay of lattice pinning and vortex–vortex interaction may lead to a rich phenomenology including the existence of commensurate–incommensurate transitions in a one-dimensional vortex lattice. For a detailed theoretical analysis of this regime, we refer to the work of Bruder *et al.*[56]

## 7.5. Charge Dynamics

Up to now we discussed the dynamic properties of vortices. Due to the high degree of symmetry of the coupled Coulomb gas, similar considerations apply also to charge excitations. It is worth stressing that when referring to charge excitations one has to think of collective excitations of the whole array, in which one island has an extra charge but all the other islands (up to a distance of the order of the screening length) have a non-trivial redistribution of the polarization charges at the junctions.

Starting from Eq. (7.3), in the limit $E_C \gg E_J$, it is possible to integrate out vortex fluctuations leading us to the action of the two-dimensional

Coulomb gas of charges,

$$S[q] = \int_0^\beta d\tau \sum_{i,j} \left[ \frac{1}{2\pi E_{\text{J}}} \dot{q}_i(\tau) G_{ij} \dot{q}_j(\tau) + \frac{1}{2} q_i(\tau) U_{ij} q_j(\tau) \right]. \quad (7.17)$$

From the previous equation, one immediately sees that the charge mass is $M_{\text{q}} = E_{\text{J}}^{-1}$. When the interaction $U_{ij}$ is logarithmic, a charge BKT transition will occur at finite temperatures. The critical temperature will be suppressed by quantum fluctuations induced by the first term in the effective charge-action.

## 7.6. Conclusions

Josephson junction arrays are an ideal playground to study the properties of the Berezinskii–Kosterlitz–Thouless transition. Originally, the interest was focused on classical arrays which are realization of the XY-model. Submicron fabrication progress allowed exploring the BKT transition also in the presence of quantum fluctuations, as well as the properties of charges and vortices in quantum Josephson junction arrays. Here, we have described some aspects of these activities, which were very much stimulated by the intriguing properties of the BKT transition.

## Acknowledgment

When working on topics related to the issues covered in this review we had the pleasure to collaborate with a number of colleagues and we profited from stimulating discussions with many further colleagues. Especially we want to acknowledge C. Bruder, J. E. Mooij, H. W. J. van der Zant, U. Eckern, L. J. Geerligs, J. V. José, A. Schmid, L. L. Sohn, M. Tinkham and B. J. van Wees.

## References

1. R. F. Voss and R. A. Webb, *Phys. Rev. B* **25**, 3446 (1982); R. A. Webb, R. F. Voss, G. Grinstein and P. M. Horn, *Phys. Rev. Lett.* **51**, 690 (1983).
2. V. L. Berezinskii, *Zh. Eksp. Teor. Fiz.* **59**, 907 (1970) [*Sov. Phys. JETP* **32**, 493 (1971)].
3. J. M. Kosterlitz and D. J. Thouless, *J. Phys. C* **6**, 1181 (1973).
4. J. Resnick, J. Garland, J. Boyd, S. Shoemaker and R. Newrock, *Phys. Rev. Lett.* **47**, 1542 (1981); P. Martinoli, P. Lerch, C. Leemann and H. Beck, *J. Appl. Phys.* **26**, 1999 (1987).
5. M. Goldman and S. A. Wolf (eds.), *Percolation, Localization, and Superconductivity*, NATO ASI **108** (1986).

6. J. E. Mooij and G. Schön (eds.), Coherence in superconducting networks, *Physica B* **152**, 1–302 (1988).
7. H. A. Cerdeira and S. R. Shenoy (eds.), Josephson junction arrays, *Physica B* **222**, 253–406 (1996).
8. S. Sachdev, *Quantum Phase Transition* (Cambridge University Press, Cambridge, 1999).
9. L. J. Geerligs, M. Peters, L. E. M. de Groot, A. Verbruggen and J. E. Mooij, *Phys. Rev. Lett.* **63**, 326 (1989).
10. J. E. Mooij and G. Schön, in *Single Charge Tunneling*, eds. H. Grabert and M. Devoret (Plenum Press, 1992), Chapter 8.
11. R. Fazio and H. W. J. van der Zant, *Phys. Rep.* **355**, 235–354 (2001).
12. K. B. Efetov, *Sov. Phys. JETP* **51**, 1015 (1980).
13. S. Doniach, *Phys. Rev. B* **24**, 5063 (1981).
14. E. Simanek, *Phys. Rev. B* **23**, 5762 (1982).
15. L. Jacobs, J. V. José, M. A. Novotny and A. M. Goldman, *Phys. Rev. B* **38**, 4562 (1988).
16. S. V. Syzranov, I. L. Aleiner, B. L. Altshuler and K. B. Efetov, *Phys. Rev. Lett.* **105**, 137001 (2010).
17. J. Villain, *J. Physique* **36**, 581 (1975).
18. J. V. José, L. P. Kadanoff, S. Kirkpatrick and D. R. Nelson, *Phys. Rev. B* **16**, 1217 (1977).
19. R. Savit, *Rev. Mod. Phys.* **52**, 453 (1980).
20. R. Fazio and G. Schön, *Phys. Rev. B* **43**, 5307 (1991).
21. R. Fazio, A. van Otterlo, G. Schön, H. S. J. van der Zant and J. E. Mooij, *Helv. Phys. Acta* **65**, 228 (1992).
22. For clarity we have used a continuous time representation, but the path integral is defined on a discretized time lattice.
23. J. E. Mooij, B. J. van Wees, L. J. Geerligs, M. Peters, R. Fazio and G. Schön, *Phys. Rev. Lett.* **65**, 645 (1990).
24. H. S. J. van der Zant, L. J. Geerligs and J. E. Mooij, *Europhys. Lett.* **19**, 541 (1992).
25. R. Yagi, T. Tamaguchi, H. Kazawa and S. Kobayashi, *J. Phys. Soc. Jpn.* **66**, 2429 (1997)
26. C. Bruder, R. Fazio and G. Schön, *Phys. Rev. B* **47**, 342 (1993); C. Bruder, R. Fazio, A. Kampf, A. van Otterlo and G. Schön, *Physica Scripta* **T42**, 159 (1992).
27. M. V. Feigel'man, S. E. Korshunov and A. B. Pugachev, *JETP Lett.* **65**, 566 (1997).
28. V. M. Vinokur, T. I. Baturina, M. V. Fistul, A. Yu. Mironov, M. R. Baklanov and C. Strunk, *Nature* **452**, 613 (2008).
29. N. Prokofiev, *Adv. Phys.* **56**, 381 (2007).
30. E. Roddick and D. H. Stroud, *Phys. Rev. B* **48**, 16600 (1993); A. van Otterlo and K.-H. Wagenblast, *Phys. Rev. Lett.* **72**, 3598 (1994); G. G. Batrouni, R. T. Scalettar, G. T. Zimanyi and A. P. Kampf, *Phys. Rev. Lett.* **74**, 2527 (1995).
31. T. Lahaye, C. Menotti, L. Santos, M. Lewenstein and T. Pfau, *Rep. Prog. Phys.* **72**, 126401, (2009); L. Pollet, J. D. Picon, H. P. Büchler and M. Troyer, *Phys.*

Rev. Lett. **104**, 125302 (2010); F. Cinti, P. Jain, M. Boninsegni, A. Micheli, P. Zoller and G. Pupillo, *Phys. Rev. Lett.* **105**, 135301 (2010).
32. U. Eckern and A. Schmid, *Phys. Rev. B* **39**, 6441 (1989).
33. C. J. Lobb, D. W. Abraham and M. Tinkham, *Phys. Rev. B* **27**, 150 (1983).
34. A. Fujimaki, K. Nakajima and Y. Sawada, *Phys. Rev. Lett.* **59**, 2985 (1987).
35. H. S. J. van der Zant, T. P. Orlando, S. Watanabe and S. H. Strogatz, *Phys. Rev. Lett.* **74**, 174 (1995).
36. S. Watanabe, S. H. Strogatz, H. S. J. van der Zant and T. P. Orlando, *Phys. Rev. Lett.* **74**, 379 (1995).
37. H. S. J. van der Zant, F. C. Fritschy, T. P. Orlando and J. E. Mooij, *Europhys. Lett.* **18**, 343 (1992).
38. H. S. J. van der Zant, F. C. Fritschy, T. P. Orlando and J. E. Mooij, *Phys. Rev. Lett.* **66**, 2531 (1991).
39. T. S. Tighe, A. T. Jonson and M. Tinkham, *Phys. Rev. B* **44**, 10286 (1991).
40. E. B. Sonin, *Phys. Rev. B* **55**, 485 (1997).
41. M. P. A. Fisher, *Physica A* **177**, 553 (1991).
42. B. J. van Wees, H. S. J. van der Zant and J. E. Mooij, *Phys. Rev. B* **35**, 7291 (1987).
43. P. Delsing, C. D. Chen, D. B. Haviland, T. Bergsten and T. Claeson, in *Superconductivity in Networks and Mesoscopic Structures*, eds. C. Giovannella and C. J. Lambert (American Institute of Physics, 1997).
44. H. S. J. van der Zant, W. J. Elion, L. J. Geerligs and J. E. Mooij, *Phys. Rev. B* **54**, 10081 (1996).
45. C. D. Chen, P. Delsing, D. B. Haviland, Y. Harada and T. Claeson, *Phys. Rev. B* **54**, 9449 (1996).
46. A. van Oudenaarden and J. E. Mooij, *Phys. Rev. Lett.* **76**, 4947 (1996).
47. A. van Oudenaarden, S. J. K. Várdy and J. E. Mooij, *Phys. Rev. Lett.* **77**, 4257 (1996).
48. A. van Oudenaarden, B. van Leeuwen, M. P. P. Robbens and J. E. Mooij, *Phys. Rev. B* **57**, 11684 (1998).
49. W. J. Elion, I. I. Wachters, L. L. Sohn and J. E. Mooij, *Phys. Rev. Lett.* **71**, 2311 (1993).
50. S. E. Korshunov, *JETP Lett.* **46**, 484 (1987); S. E. Korshunov, *Physica B* **152**, 261 (1988).
51. Y. Aharonov and A. Casher, *Phys. Rev. Lett.* **53**, 319 (1984).
52. B. J. van Wees, *Phys. Rev. Lett.* **65**, 255 (1990).
53. T. P. Orlando and K. A. Delin, *Phys. Rev. B* **43**, 8717 (1991).
54. A. van Oudenaarden, S. J. K. Várdy and J. E. Mooij, *Czhec. J. Phys.* **46**, 707 (1996).
55. K. K. Likharev and A. B. Zorin, *J. Low Temp. Phys.* **59**, 347 (1985).
56. C. Bruder, L. I. Glazman, A. I. Larkin, J. E. Mooij and A. van Oudenaarden, *Phys. Rev. B* **59**, 1383 (1999).

## Chapter 8

# Superinsulator–Superconductor Duality in Two Dimensions and Berezinskii–Kosterlitz–Thouless Transition

Valerii M. Vinokur* and Tatyana I. Baturina[†]

*Materials Science Division, Argonne National Laboratory,
9700 S. Cass Ave, Argonne, Illinois 60439, USA*
*vinokour@anl.gov*

For nearly a half century, the dominant orthodoxy has been that the only effect of the Cooper pairing is the state with zero resistivity at finite temperatures, *superconductivity*. In this chapter, we demonstrate that by the symmetry of Heisenberg uncertainty principle relating the amplitude and phase of superconducting order parameter, Cooper pairing can generate the dual state with zero conductivity in the finite temperature range, *superinsulation*. We show that this duality is realized in the planar Josephson junction arrays (JJA) via the duality between the Berezinskii–Kosterlitz–Thouless (BKT) transition in the vortex–antivortex plasma, resulting in phase-coherent superconductivity below the transition temperature, and the charge-BKT transition occurring in the insulating state of JJA, marking the formation of the low-temperature charge-BKT state, *superinsulation*. We find that in disordered superconducting films that are on the brink of superconductor–insulator transition, the Coulomb forces between the charges acquire two-dimensional character, i.e. the corresponding interaction energy depends logarithmically upon charge separation, bringing the same vortex-charge-BKT transition duality, and the realization of superinsulation in disordered films as the low-temperature charge-BKT state. Finally, we discuss possible applications and utilizations of superconductivity–superinsulation duality.

---

[†]A. V. Rzhanov Institute of Semiconductor Physics SB RAS, 13 Lavrentjev Avenue, Novosibirsk, 630090 Russia. E-mail: tatbat@isp.nsc.ru

## 8.1. Introduction

Hundred years ago the discovery of superconductivity, the state possessing a zero resistance at low but finite temperatures, opened a new era in physics. The discovery crowned a systematic study of electronic transport at lowest temperatures available at the time, undertaken at Leiden University by Professor Kamerlingh Onnes, who strived to uncover microscopic mechanisms of conductivity in metals. It was recognized since 1905 that conductivity can be described by the Drude formula $\sigma = ne^2\tau/m$, where $n$ is the charge carrier density, $-e$ is the electron charge, $\tau$ is the scattering time, and $m$ is the mass of the charge carrier. A natural hypothesis was that in pure metals, the scattering time will start to grow indefinitely upon approaching the absolute zero temperature, resulting in infinite conductivity. However, Lord Kelvin argued that mobile charges can freeze out near the zero temperature so that the charge density $n$ in the Drude formula goes to zero and that this effect may win over the increase of $\tau$. Kamerlingh Onnes' finding ruled out Lord Kelvin's hypothesis, and, one can say — adopting extremely loose and non-rigorous interpretation of known microscopic mechanisms of superconductivity — that superconductivity indeed complies with the $\tau$-divergence notion. It is essential for the whole concept of superconductivity that $\tau$ becomes infinite at *finite* temperatures. The idea on the mobile charges gradually freezing out as temperature tends to zero was found to be true in the resistivity of semiconductors (i.e. band insulators). The latter exhibit the metallic type of behavior at high temperatures but at low temperatures their charge density $n$ exponentially decreases with decreasing temperature, and wins over the gradually growing $\tau$. The next step towards what might sound as a realization of the Lord Kelvin idea, was the discovery that there exist two-dimensional superconducting systems, namely, Josephson junction arrays, granular and homogeneously disordered films, which unite both types of behavior, superconducting and insulating. The idea that the material with the Cooper pairing can be transformed into an insulator traces back to Anderson,[1] who considered small superconductors coupled by Josephson link, which was further discussed by Abeles[2] in the context of granular systems. The analytical proof for the existence of superconductor–insulator transition (SIT) in granular superconductors was given by Efetov,[3] and the first theory of the disorder-induced superconductor–insulator transition in the 3D disordered Bose condensate was developed by Gold.[4,5] Since then the study of the SIT, which is an exemplary quantum phase transition, has been enjoying intense experimental and theoretical attention and has become one of the mainstreams of contemporary condensed matter physics. In this work, we will describe further advances in understanding the SIT, the realization that at the transition point, superconductivity transforms into a mirror image of superconductivity, the *superinsulating state*, possessing *infinite* resistivity at *finite* temperatures.

This chapter is organized as follows. Section 8.2 establishes the possibility of superinsulation from the uncertainty principle while Sec. 8.3 discusses the general features of two-dimensional systems enabling experimental realization of the superinsulating state; in Sec. 8.4, we briefly remind the properties of Berezinskii–Kosterlitz–Thouless (BKT) transition and the relation between this transition and the formation of the superinsulating state, presenting the phase diagram of the system in the close vicinity of superconductor–superinsulator transition; Sec. 8.5 is devoted to the realization of 2D electrostatics in disordered superconducting films that brings to existence the charge-BKT transition; in Sec. 8.6 we discuss a possible microscopic mechanism behind the superconducting state; we present in Sec. 8.7 a qualitative consideration of the conductivity in the superinsulating state; Sec. 8.8 discusses possible applications and utilizations of the superconductor–superinsulator duality; and finally, we summarize the content of our review in Sec. 8.9.

## 8.2. Superconductivity and Superinsulation from the Uncertainty Principle

The possibility of the existence of the superinsulating state can be established from the most fundamental quantum-mechanical standpoint. Superconducting state is characterized by an order parameter, $\Psi = |\Psi| \exp(i\varphi)$, where the phase, $\varphi$, is well defined across the whole system. The uncertainties in the phase $\Delta\varphi$ and the number of Cooper pairs in the condensate, $N_s = 2|\Psi|^2$, compete according to the Heisenberg uncertainty principle $\Delta\varphi \Delta N \geq 1$.[1,6–8] This brings about the basic property of a superconductor, its ability of carrying a loss-free current. Indeed, the absence of an uncertainty in the phase, $\Delta\varphi = 0$, i.e. the definite phase, implies that the number of condensate particles is undefined. It means that the particles that comprise condensate cannot scatter and dissipate energy, since every scattering event can be viewed as the measurement, and, in principle, could be used to determine the number of particles. In two-dimensional systems, the global phase coherence leading to true superconductivity establishes itself at some finite temperature, $T_{SC}$, which is below the temperature $T_c$, where the finite modulus of the order parameter appears.[9,10] The phase coherence then remains stable against moderate perturbations: the loss-free current is maintained as long as it does not exceed the certain value $I_c$ determined by the material parameters.

Now let us assume that we have managed to tune the parameters of a 2D superconductor in such a way that the Cooper pairs were pinned within the system. This would eliminate the uncertainty of the total charge setting $\Delta N = 0$. The Heisenberg principle then requires the finite uncertainty of the global phase, i.e. pinning the Cooper pairs results in $\Delta \varphi \neq 0$. It implies that due to the Josephson relation, $d\varphi/dt = (2e/\hbar)V$, where $\hbar$ is the Planck constant, there may be a finite voltage, $V$, across the whole system while the flow of current is blocked. This establishes the possibility for the existence of a distinct superinsulating state,[11,12] which is dual to the superconducting one and has zero conductivity over the finite temperature range from the critical temperature $T_{SI}$ down to $T = 0$. Of course, one has to bear in mind that the term "zero conductivity" is used in the same idealized sense as the "zero resistance" of a two-dimensional superconductor. We would like to emphasize here once again the conceptual difference between the notions of the "superinsulator" and merely the "insulator," which is that the former has *zero conductivity* in the finite temperatures range, while the latter would have possessed the finite, although exponentially small, linear conductivity $\sigma \propto \exp(-E_g/k_B T)$ ($E_g$ is the energy gap) at all finite temperatures, except for $T = 0$.

## 8.3. Two-Dimensional Superconducting Systems

To gain an insight into the origin of a superinsulating state, we will focus on two-dimensional tunable superconducting systems, Josephson junction arrays (JJA), the systems comprised of small superconducting islands connected by Josephson junctions. A theoretical and experimental study of large regular JJAs has been one of the main directions of condensed matter physics for decades,[13-26] see Refs. 27–29 for a review. The interest was motivated by the appeal of dealing with the experimentally accessible systems whose properties can be easily controlled by tuning two major parameters quantifying the behavior of the array, the Josephson coupling energy between the two adjacent superconducting islands comprising a JJA, $E_J$, and the charging energy of a single junction, i.e. the Coulomb energy cost for transferring the Cooper pair between the neighbouring islands, $E_C = (2e)^2/(2C)$, where $C$ is the junction capacitance.[1] Moreover, JJA offers a generic model that captures the essential features of two-dimensional disordered superconductors including homogeneously disordered and granular films. The remarkable feature of JJAs is that they experience a phase transition, which historically was described in terms of the zero-temperature transition between the superconducting and *insulating* states, superconductor-to-insulator transition (SIT), which occurs as $E_\text{C}$ compares to $E_\text{J}$. In the arrays that are already near the critical region where $E_\text{C} \approx E_\text{J}$, the SIT can also be induced by the magnetic field. The SIT studies in JJAs were paralleled by the investigations on the superconductor–insulator transition in thin granular films, which revealed similar behavior.[2,3,30–41] What more, even homogeneously disordered superconducting films[42–75] and layered high-$T_c$ superconductors[76–82] exhibited all the wealth of the SIT-related phenomena which were viewed as characteristic to granular superconducting systems. This led to the idea that in the vicinity of SIT, disorder may cause the electronic phase separation (often referred to as the "self-induced granularity") such that the strong disorder induces an inhomogeneous spatial structure of isolated superconducting islands in thin homogeneously disordered films.[47,48,83–87] Numerical simulations of the homogeneously disordered superconducting films confirmed that indeed in the presence of sufficiently strong disorder, the system breaks up into superconducting islands separated by an insulating sea.[88–91] Recent scanning tunneling spectroscopy measurements of the local density of states in TiN and InO films[92,93] and in high-$T_c$ superconductors[94,95] have offered strong support to this hypothesis. We will discuss these issues in more detail in Sec. 8.4.

All the above studies have showed that the Josephson junction arrays indeed offer a generic model that captures most essential features of the superconductor–insulator transition in a wide class of systems ranging from artificially manufactured Josephson junction arrays to superconducting granular systems and even the homogeneously disordered superconducting films allowing to be considered all of them on the common ground. The properties of the films were discussed in terms of the dimensionless [measured in the quantum units $e^2/(2\pi\hbar)$] conductance $g$, which played the role of the tuning parameter for films replacing the ratio $E_\text{J}/E_\text{C}$ characterizing JJAs. Large $g$ correspond to superconducting domain, as small $g$ films become insulating. And at some critical value $g_\text{c} \simeq 1$ disordered films are viewed to undergo superconductor–insulator transition.

Using JJA as an exemplary system allows to understand better the microscopic mechanism of the realization of the charge-phase duality of the uncertainty principle giving rise to superinsulation. As already indicated in Ref. 96, the charge-vortex duality reflects the duality between the Aharonov–Bohm[97] and Aharonov–Casher[98] effects. In the Aharonov–Bohm effect, the charges moving in a field-free region surrounding a magnetic flux acquires the phase proportional to the number of flux quanta piercing the area. Accordingly, the Aharonov–Casher effect is the reverse[99]: magnetic vortices moving in a 2D system around a charge acquires a phase proportional to that charge. Now, consider an insulating side, $E_C \gg E_J$, where all the Cooper pairs are pinned at the granules by Coulomb blockade, we can conjecture, following the line of reasoning of Refs. 100 and 101, that in the fluxons tunnel *coherently* through the Josephson links between the granules that corresponds to *constructive* interference of their tunnelling paths i.e. to the process having the largest quantum mechanical amplitude. Thus the phase synchronizes across the entire system and remains undefined. In other words, one can possibly view the superinsulating state where the current is completely blocked as a state mediated by the coherent quantum phase slips discussed recently in the context of Josephson junction chains[102,103] and disordered wire ring.[104] The uncertainty of the phase implies the superfluid state of fluxons at the insulating side dual to superfluid state of Cooper pair condensate, by the same token as uncertainty in the charge implies superconductivity as discussed above. Indeed, the possibility of measuring the relative phase at different junctions that lifts the phase uncertainty, implies the destruction of their coherent motion, which in its turn would mean that Cooper pairs are not blocked at their respective granules any more.

We would like to emphasize here the crucial role played by charge pinning in the formation of the superinsulating state. It has been already recognized earlier[105] that at very low temperatures, the system of vortices transforms into a vortex-superfluid state in which an infenitizimal current induces an infinite voltage, i.e. the system acquires an infinite resistance and should be called a superinsulator. However, the fact that in the absence of the charge pinning the charge fluctuations would destroy the superinsulating state was not appreciated in Ref. 105.

### 8.4. Superinsulation and Berezinskii–Kosterlitz–Thouless Transition

As early as in 1963, Salzberg and Prager,[106] in the course of their study on thermodynamics of the 2D electrolyte, derived the equation of state and noticed that due to logarithmic interaction between the "plus" and "minus" ions, there appears a singular temperature where the pressure of ions turns zero. They ruled that at this temperature, the ion pair is formed. In 1970, Berezinskii[107,108] and later Kosterlitz and Thouless[109,110] offered a more refined consideration to what had become known ever since as the binding–unbinding Berezinskii–Kosterlitz–Thouless (BKT) phase transition. The physical idea behind the transition is as follows.[109] The particle–antiparticle attraction contributes the term $\mathcal{E}_0 \ln(r/r_0)$ into the free energy of the system, where $\mathcal{E}_0$ is the energy parameter characterizing the interaction and $r_0$ is the microscopic spatial cut off parameter. At the same time, the entropy contribution to the free energy is $-k_\mathrm{B} T \ln[(r/r_0)^2]$, as there are $\simeq (r/r_0)^2$ ways of placing two particles apart within distance $r$. At low temperatures, the attraction between the particles and antiparticles wins and they remain bounded. At $T_\mathrm{BKT} \simeq \mathcal{E}_0/(2k_\mathrm{B})$ the entropy term balances the attraction, and at $T > T_\mathrm{BKT}$ the particles and antiparticles get unbound. We thus see that the BKT transition is the consequence of the logarithmic interaction between the constitutive entities in two-dimensional systems (2D vortex–antivortex system, 2D charge–anticharge plasma, dislocation–antidislocation array), irrespectively to their particular nature. Below, we will demonstrate that it is the BKT transition that provides the mechanism by which the 2D JJA falls into either superconducting or superinsulating states at low temperatures. The choice of the particular ground state is dictated by the relation between $E_\mathrm{J}$ and $E_\mathrm{C}$.[2,3,14,16,17,28,29]

The absence of phase coherence at high temperatures and its appearance at low temperatures in a strong coupling regime, $E_\mathrm{J} > E_\mathrm{C}$, can be described in terms of vortex–antivortex plasma dynamics and the corresponding BKT transition. In two-dimensional superconductors and Josephson junction arrays, finite temperatures $T$ generate fluctuational vortices and antivortices, the number of former and the latter remaining equal in order to conserve "flux neutrality." At high temperatures, $T > T_\mathrm{BKT}$ vortices and antivortices diffuse freely, and since each vortex (antivortex) carries the phase $2\pi$, their motion leads to the breaking down of the global phase coherence. Below the BKT transition temperature, vortices and antivortices get bound into the neutral dipole pairs and cannot diffuse independently any more. As the phase gain when encircling a vortex–antivortex dipole is zero, the diffusion of the dipoles as a whole does not cause phase fluctuations on large spatial scales, and the global phase coherence sets. The crucial feature that ensures binding of vortex–antivortex pairs and setting down the phase coherence at low temperatures is the so-called vortex–antivortex *confinement*: the energy of vortex–antivortex interaction *grows* (logarithmically) with the separation between them, as long as the vortex–antivortex spacing does not exceed London penetration depth $\lambda = \hbar^2/[(2e)^2 \mu_0 E_\mathrm{J}]$. This condition reflects the restriction of the notion of 2D superconductivity: in the system with the lateral dimensions exceeding $\lambda$, unbound vortices would be present at *all* finite temperatures (except for $T = 0$) and true superconductivity, i.e. the state which possesses the zero resistance below some finite transition temperature is absent.

Now going to the insulating side, where $E_J < E_C$, we note that the charges of Cooper pairs in the planar Josephson junction array also interact according to the logarithmic law[16]: in two dimensions, the Coulomb interaction between the charges grows logarithmically with distance, $r$, separating them, $\mathcal{E}_{\text{Coulomb}} \propto \ln(r/r_0)$. Thermal fluctuations generate excessive Cooper pairs with charge $-2e$ each and equal number of "anti-Cooper-pairs" (i.e. local deficit of Cooper pairs) with charges $+2e$, to conserve the electroneutrality. As long as these charge separations, $r$, do not exceed the electrostatic screening length, $\Lambda = a(C/C_0)^{1/2}$, where $C_0$ is the self-capacitance of a superconducting island (capacitance to the ground) with $a$ being the characteristic size of a single junction, the interaction energy of charges is proportional to $\ln(r/r_0)$, where in the case of JJA $r_0 = a$. Therefore, planar JJAs possess a remarkable duality between the logarithmically confined magnetic vortices at the superconducting side and the logarithmically confined Cooper pairs in the insulating state. By the same "logarithmic token" as above, the Cooper pairs–anti-Cooper pairs 2D plasma undergoes the charge-BKT (CBKT) transition binding, at $T < T_{\text{CBKT}}$, Cooper pairs and "anti-pairs" into the neutral Cooper pairs dipoles (CPD) that do not carry any charge. The charge confinement at $E_J < E_C$ can be expressed in terms of the *macroscopic* Coulomb blockade that impedes the free motion of the Cooper pairs and breaks loose the global phase coherence of the condensate, according to the uncertainty principle, and, thus, gives rise to the superinsulating state of the JJA.[12] In particular, at $T = 0$ the magnetic vortex–Cooper pairs charge duality in JJA leads to a quantum phase transition between superconductivity and superinsulation at the self-dual point where $E_J \approx E_C$ was first noticed in Ref. 111.

The authors of Ref. 111 considered a two-dimensional Josephson junction array and showed that the zero-temperature behavior of planar JJAs in the self-dual approximation is governed by an Abelian gauge theory with the periodic mixed Chern–Simmons term describing the charge-vortex coupling. The periodicity requires the existence of (Euclidean) topological excitations, which determine the quantum phase structure of the model. Symmetry between the logarithmically interacting vortices and logarithmically interacting charges give rise to a quantum phase transition at the self-dual point. This is the transition between the superconducting phase and the phase with the confined charges, dual to superconductivity and having zero linear conductivity in the thermodynamic limit. This phase, by analogy with the

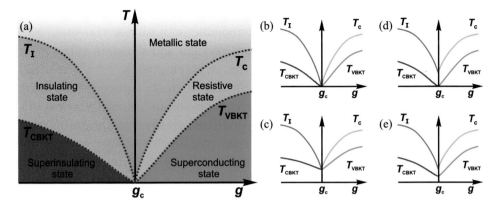

Fig. 8.1. (a) The sketch of the phase diagram for the superinsulator–superconductor transition in two dimensions in the close proximity to the critical point. A generic diagram is plotted in coordinates temperature ($T$)–critical parameter ($g$). In disordered films, $g$ is a conductance in the metallic phase, whereas in the Josephson junction arrays (JJA), it represents the ratio of coupling- and charging energies, i.e. $g = E_J/E_C$, where $E_J$ is the Josephson coupling energy of the two adjacent superconducting islands, and the charging energy $E_C$, the energy cost to transfer a Cooper pair charge $2e$ between the neighboring islands. In the JJA, the condition $g = g_c$ means $E_J \approx E_C$. At $g > g_c$ the finite value of the modulus of the order parameter $|\Psi|$ appears at the superconducting transition temperature $T_c$. The temperature $T_{\text{VBKT}}$ is the Berezinskii–Kosterlitz–Thouless (BKT) transition temperature in the 2D vortex–antivortex plasma. At $T_{\text{VBKT}} < T < T_c$ the system is in a resistive state since in this temperature range vortex–antivortex pairs are unbound and can diffuse freely breaking down the global phase coherence. At $T < T_{\text{VBKT}}$ vortex–antivortex pairs are bound, the global phase coherence establishes and the superconducting state forms. At $g < g_c$ and $T < T_I$, the system becomes an insulator and in the range $T_{\text{CBKT}} < T < T_I$, it exhibits the exponentially low conductivity due to tunneling transfer of electric charges, $T_{\text{CBKT}}$ being the temperature of the charge-BKT transition. At $T < T_{\text{CBKT}}$, where the negative and positive charges are bound into dipoles, the system becomes superinsulating. The blue sector of $T > T_c$ and $T > T_I$ corresponds to a metallic state. Both low-temperature states, superinsulating and superconducting, are non-dissipative. (b)–(e) Schematic possible realizations of this phase diagram: (b) the same as (a), for the case where at the critical point, all the transition temperatures turn zero; (c) all the characteristic temperatures coincide at $g = g_c$, but remain finite. This situation corresponds to the *first order* transition between the superconductor and superinsulator; (d) $T_{\text{CBKT}}(g_c) = T_{\text{VBKT}}(g_c) = 0$, but $T_I(g_c) = T_c(g_c) \neq 0$; and (e) $T_{\text{CBKT}}(g_c) = T_{\text{VBKT}}(g_c) \neq 0$ and $T_I(g_c) = T_c(g_c) \neq 0$.

superconductors, where linear resistivity is zero, was termed "superinsulator".[111] Therefore, the quantum phase transition extensively discussed in the literature and referred to as SIT is in fact *superconductor–superinsulator transition*.

Turning now to the description of the phase diagram of the JJA in the vicinity of the superconductor–superinsulator transition (Fig. 8.1), we note first that in the 2D superconductors the symmetry of the phase-charge uncertainty relation can be recast into a picture of the vortex-Cooper pair duality (the concept of quantum superconductor–insulator transition as a transition between the state with pinned vortices and the state with pinned Cooper pairs was first introduced in the pioneering works [112,113] as a SIT in disordered superconducting film). In this context, superconductivity is viewed as an ensemble of Cooper pair condensate and pinned vortices. Inversely, the insulator consists of Bose condensed vortices and pinned Cooper pairs. The distinctive feature of the superconductor–insulator transition is that this is the transition at which, by its very definition, one expects a singularity in transport properties. Thus the symmetry in the ground states is paralleled by the symmetry of the electronic transport[96,114] which can be mediated either by the tunneling of the fluxons or by the tunneling of the Cooper pairs. Now both sides of the phase diagram of JJA near the superconductor–superinsulator transition (see Fig. 8.1) can be described from the unified viewpoint: If the linear size of the JJA is less than both characteristic screening lengths, the system would undergo both vortex- and charge-BKT transitions (VBKT and CBKT) at temperatures $T_{\text{VBKT}} \approx E_J$ and $T_{\text{CBKT}} \approx E_C$, respectively (the effects of the finite screening lengths on VBKT and CBKT transitions were discussed in Refs. 28 and 115). The VBKT transition marks the transition between the resistive (at $T > T_{\text{VBKT}}$) and global phase coherent superconducting (at $T < T_{\text{VBKT}}$) states. Accordingly, transition at $T = T_{\text{CBKT}}$ at the insulating side separates the "high-temperature state," insulator, where Cooper pair dipoles (CPD) get unbound, and the "low-temperature phase," superinsulator, where dipole excitations are bound.

The insulating phase at $T_{\text{CBKT}} < T < T_{\text{I}}$ exhibits thermally activated conductivity $\sigma \propto \exp(-\Delta_C/k_B T)$. The charge transfer is mediated by the tunnelling of Cooper pairs, so this state is referred to as a Cooper pair insulator. Not too far above $T_{\text{CBKT}}$, where the density of the unbound CPD is still low, $\Delta_C = E_C \ln(L/a)$ (where $L$ is the linear size of the array, and $L < \Lambda$).[11] That the activation energy $\Delta_C$ characterizing the insulating behavior exhibits such an unusual and peculiar dependence on the size of the system reflects the fact that the test excessive charge placed locally at any superconducting island polarizes the whole JJA.[16] In other words, the characteristic Coulomb energy cost associated with the excitation of an excess Cooper pair is determined by the total system capacitance $C_{\text{tot}} \approx C/\ln(L/a)$.[12] Thus, since it is the whole system that participates in building the electrostatic barrier impeding the free charge propagation, we refer to this peculiar Coulomb blockade effect as a *macroscopic* Coulomb blockade. On the experimental side, the size-scaling of the activation energy of the insulator has not been demonstrated on JJAs yet and is waiting to be revealed. At the same time, the scaling of the activation energy as the logarithm of the sample size was recently observed in disordered InO films[68] and in TiN films.[71] We will discuss these observations in more detail in Sec. 8.4 below.

Upon further temperature growth, screening of an excess charge by the free unbound charges becomes more efficient and the characteristic activation energy reduces to the charging energy of a single island, $E_C$. The sketch of the phase diagram showing the possible states in the planar JJA in $g$-$T$ coordinates, where $g = E_J/E_C$ is a critical parameter, is presented in Fig. 8.1 and generalizes the phase diagram proposed first by Fazio and Schön,[16] see also Refs. 15 and 29. We expect the similar diagram describes the behavior of disordered superconducting films. In this case, the role of the critical parameter is taken by the dimensionless conductance $g$, and the superinsulator–superconductor transition occurs at the critical conductance $g = g_c$. The metal that forms near $g_c$ in the $T \to 0$ limit has the conductance close to the quantum value $e^2/(2\pi\hbar) = (25.8\,\mathrm{k\Omega})^{-1}$. The corresponding distinct state has been detected by applying strong magnetic field completely destroying superconductivity in disordered films of Be, TiN, and InO, and is often referred to as "quantum metallicity".[54,59,62,65,69]

An important comment is in order. Two-dimensional systems display perfect superconductivity–superinsulation duality because of the logarithmic interaction of charges in 2D. One can ask whether this duality consideration can be extended to three-dimensional systems and whether one can expect superinsulating behavior in 3D. One can see straightforwardly that while 3D Abrikosov vortices still interact according to the logarithmic law, the charges follow the conventional 3D Coulomb law and therefore charge–anticharge confinement is absent in 3D. Hence, absent is the 3D superinsulating state.

## 8.5. Two-Dimensional Universe in Superconducting Films

As we have already discussed, Josephson junction arrays with the junction capacitances well exceeding their respective capacitances to the ground are two-dimensional with respect to their electrostatic behavior since most of the electric force lines remain trapped within the junctions themselves. When considering the behavior of homogeneously disordered superconducting films, the question can arise, to what extent can they be described by the physics derived for JJAs.

The analogy between the planar JJA and the critically disordered thin superconducting films can be perceived from the fact that the experimentally observed in TiN films magnetic-field dependences of the activation energy, $\Delta_{\rm C}(B)$, and the threshold voltage $V_{\rm T}(B)$[66] are remarkably well described by formulas obtained in the framework of JJA model.[11] This hints that in the critical vicinity of the SIT, a film can be viewed as an array of superconducting islands coupled by the weak links. This incipient conjecture of the early work,[47] discussed further in Refs. 12, 51, 66–69, 74, 75, 88–91 that near the disorder-driven SIT the electronic phase separation occurs leading to the formation of the droplet-like (or island-like) texture, superconducting islands coupled via the weak links, was supported by recent scanning tunnelling spectroscopy (STS) findings.[92,93] In general, formation of regular textures and, in particular, the spontaneous self-organization of electronic nanometer-scale structures, due to the existence of competing states is ubiquitous in nature and is found in a wealth of complex systems and physical phenomena ranging from magnetism,[116–120] superconductivity and superfluidity to liquid crystals, see Ref. 121 for a review. The arguments that the long-range fields — for example, elastic or Coulomb — can promote phase separation were given in Ref. 122 where the island texture due to elastic interaction between the film and the substrate was found. More refined calculations for the JJA model[123] showed that in the presence of the long-range Coulomb forces, the SIT could become a first order transition, supporting the idea of possible electronic phase separation at the SIT.

Furthermore, while at the superconducting side of the transition. vortices interact logarithmically, thus guaranteeing all BKT physics to occur, provided the size of the system remains less than the magnetic screening length, the question arises whether the charges in the same film but at the insulating side would also interact according to the logarithmic law. In simple words, in order for the film to demonstrate the 2D behavior, the electric force lines are to be trapped within the film over the appreciable distance. One can derive from the fundamentals of electrostatics that if the test charge is placed within the dielectric film of thickness $d$ and with dielectric constant $\varepsilon$, the characteristic length over which the electric field remains trapped within the film is of about $\varepsilon d$. The quantitative description of the Coulomb interaction in a thin film was first given by Rytova,[124] the logarithmic asymptotic of the solution and its implications were discussed by Chaplik and Entin,[125] and a refined calculation for a film with the large dielectric constant $\varepsilon$ sandwiched between the two dielectric media with dielectric constants $\varepsilon_1$ and $\varepsilon_2$ (see the left inset in Fig. 8.2) was done by Keldysh.[126] The dependence of the electrostatic potential of the charge $-e^\star$ on the distance $r \gg d$, is given by

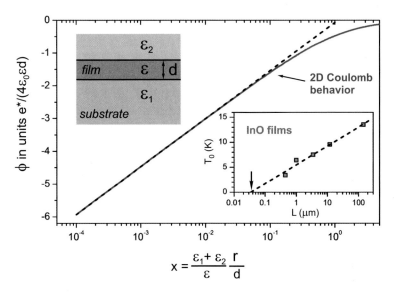

Fig. 8.2. The electrostatic potential (shown as the solid line) induced by the charge $-e^\star$ in the film with the dielectric constant $\varepsilon$ placed in between of two media with the dielectric constants $\varepsilon_1$ and $\varepsilon_2$, $\varepsilon_1 + \varepsilon_2 \ll \varepsilon$ (see the left inset) as function of the logarithm of the reduced charge separation $x = [(\varepsilon_1 + \varepsilon_2)/\varepsilon](r/d)$ according to Eq. (8.1) in the units of $e^\star/(4\varepsilon_0\varepsilon d)$. The dashed line shows the logarithmic asymptote of the potential offering a perfect approximation in the region $d \ll r \ll \varepsilon d/(\varepsilon_1 + \varepsilon_2)$. The right inset displays the fit of the data on the dependence of the activation energy of the InO films on the size of the sample from Ref. [68] by Eq. (8.3) (the dashed line) according to.[11] The intersection of the line with the x-axis gives the low-distance cutoff of 35 nm, which agrees fairly well with the InO film thickness 25 nm quoted in.[68] The slope of the fit allows then for determination of $\varepsilon$.

(see Fig. 8.2):

$$\phi(\vec{r}) = -\frac{e^\star}{4\varepsilon_0\varepsilon d}\left[\mathcal{H}_0\left(\frac{\varepsilon_1 + \varepsilon_2}{\varepsilon}\frac{r}{d}\right) - \mathcal{N}_0\left(\frac{\varepsilon_1 + \varepsilon_2}{\varepsilon}\frac{r}{d}\right)\right], \qquad (8.1)$$

where $\mathcal{H}_0$ and $\mathcal{N}_0$ are the Struve and Neumann (or the Bessel function of the second kind) functions, respectively. This formula corresponds to the choice $\phi(r) \to 0$ as $r \to \infty$. In the region $d \ll r \ll \Lambda$, where

$$\Lambda = (\varepsilon d)/(\varepsilon_1 + \varepsilon_2) \qquad (8.2)$$

is the electrostatic screening length, the electric force lines are trapped by the film, and the interaction energy between the charges $e^\star$ and $-e^\star$ obtained from Eq. (8.1) acquires the form

$$V(\vec{r}) = \frac{(e^\star)^2}{2\pi\varepsilon_0\varepsilon d}\ln\left(\frac{r}{d}\right), \quad d \ll r \ll \Lambda. \qquad (8.3)$$

The universal behavior of the electrostatic potential as a function of the reduced coordinate $x = [(\varepsilon_1 + \varepsilon_2)/\varepsilon](r/d) \equiv r/\Lambda$ is shown in Fig.8.2 by the solid line. The dashed line shows the logarithmic approximation valid in the spatial region $d \ll r \ll \Lambda$. The right inset displays the fit of the experimental data to the dependence of the activation energy of the InO films on the size of the sample from Ref. 68 by Eq. (8.3) (the dashed line) according to Ref. 11. The intersection of the line with the x-axis gives the low-distance cutoff of 35 nm which agrees fairly well with the InO film thickness 20÷25 nm quoted in Ref. 68. The slope of the fit allows then for the determination of $\varepsilon$ and, taking $e^\star = 2e$, yields $\varepsilon = 2.3 \cdot 10^3$. One can then estimate the charging energy of an "effective single junction" $E_C = (2e)^2/(4\pi\varepsilon_0\varepsilon d)$ and find the value for $T_{\text{CBKT}} \simeq E_C/k_B \approx 0.8\,\text{K}$ for InO films. Using the data for the size-dependence of the activation energy from Ref. 71 we can carry out similar estimates for TiN films as well. The intersection of the linear fit is similar to that shown in the right inset of Fig.8.2, one finds $d = 3\,\text{nm}$ which is fairly close to their nominal thickness of 5 nm. Then the slope of the fit gives $\varepsilon \simeq 4 \cdot 10^5$, and accordingly, $T_{\text{CBKT}} \simeq E_C/k_B \approx 0.06\,\text{K}$. The latter estimate is in pretty good agreement with the experimental findings of Refs. 71 and 75 that gave $T_{\text{CBKT}}^{\text{IV}} \lesssim 60\,\text{mK}$ for the TiN film. One can further estimate the electrostatic screening length $\Lambda$. Using the parameters determined from the fit and $\varepsilon_1 = 4$ for $SiO_2$ and $\varepsilon_2 = 1$, one finds $\Lambda \simeq 240\,\mu\text{m}$. This macroscopic value of $\Lambda$ implies that all the TiN samples used in experiments[12,66,67,69] having 50 $\mu$m width fall well into a domain of validity of two-dimensional electrostatics. One of the conclusions that follows from our estimates is that the condition $\varepsilon_1 + \varepsilon_2 \ll \varepsilon$ is an important requirement for the observation of the superinsulating behavior; so one has to take special care in taking the substrate with a reasonably low $\varepsilon_1$. Another comment is that while our results allow for the fairly reliable determination of $\varepsilon$ and are remarkably self-consistent, yet the direct measurements of the film dielectric constant in the critical vicinity of the SIT are highly desirable.

An important feature of the evaluated parameters are the pretty high values of the dielectric constant, especially if talking about the "conventional" bulk materials. However, it ceases to be surprising once we recollect that the experiments were carried out in close proximity to the SIT. Indeed, in a seminal paper of 1976, Dubrov et al.[127] presented two complimentary considerations based on the percolation theory and on the theory of the effective media showing that the static dielectric constant should become infinite at the metal–insulator transition (see also Ref. 128). The authors were motivated by the experimental findings of Castner and collaborators[129]

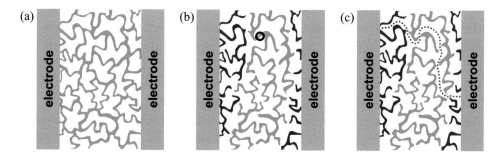

Fig. 8.3. A sketch of the electronic phase separation in a disordered film. (a) An insulating state, where superconducting regions (shown as white areas) are separated by the insulating spacers. (b) An insulating state on the very verge of the percolation transition: the green stripe highlights the "last" or "critical" insulating interface dividing two large superconducting clusters. The circle marks the critical ("last") insulating "bond." (c) The same system at the extreme proximity to the SIT on the superconducting side. As soon as the "critical" insulating bond is broken, the path connecting electrodes through the superconducting clusters shown by the dotted line appears and the film turns superconducting. The system capacitance $C$ on the insulating side is proportional to the length of the insulating interface separating two adjacent superconducting clusters, which grows near the transition as $b^{1+\psi}$, $\psi > 0$, i.e. faster than $b$. Thus, the dielectric constant $\varepsilon \sim C/b$ diverges upon approach to superconductor–superinsulator transition.

who revealed the onset of a divergence in the static dielectric constant of $n$-type silicon at the insulator–metal transition (this phenomenon known as *polarization catastrophe* was predicted as early as in 1927,[130]). Furthermore, Dubrov and collaborators investigated numerically and experimentally the model system, the square conducting network, where each bond was a parallel resistor–capacitor circuit. Breaking randomly the connections between the elements, they demonstrated that near the percolation transition, the dielectric constant diverges in a power-law fashion. In what follows, we present the arguments, following the considerations of Ref. 127 by which the two-phase systems that undergo the transition between the conducting and non-conducting states should exhibit the diverging dielectric constant when approaching this transition. A qualitative picture of this phenomenon is illustrated in Fig. 8.3. Figure 8.3(a) shows the spatial electronic structure of an insulating film in a close proximity to the percolation-like superconductor–insulator transition[131] approaching it from the insulating side. The white areas denote the superconducting regions which are separated from each other by the dark insulating areas. Figure 8.3(b) displays the same film but "one step before" the transition: upon breaking the "last" insulating separation (the circled segment in the panel) between the superconducting clusters, the path connecting electrodes through superconducting clusters emerges and the film becomes superconducting as shown in Fig. 8.3(c). The dielectric constant is proportional to the capacitance of the system, $C$, which in its turn, is proportional to the total length of the line separating the adjacent critical superconducting clusters. If we let the linear size of the superconducting cluster be $b$, this length would grow as $b^{1+\psi}$ with the exponent $\psi > 0$ (see Ref. 128). Then the dielectric constant $\varepsilon \propto C/b \propto b^{\psi}$. As the size of the critical cluster diverges on approaching the transition, so does the dielectric constant $\varepsilon$ (for more refined derivation of this divergence, see Ref. 127). Such a behavior, which is called the "dielectric catastrophe" and is very well known in semiconductor physics, has been observed experimentally (see, for example, Ref. 132, where the $\varepsilon$ divergence was observed near the metal–insulator transition). Therefore, one would expect that in the critical vicinity of the superconductor–insulator transition, the dielectric constant grows large enough so that the Coulomb interaction between the charges became logarithmic over the appreciable scale comparable to the size of the system, and the 2D plasma of CPD excitations experiences the charge-BKT transition leading to the formation of a low-temperature superinsulating state.

We now turn to the structure of the phase diagram of Fig. 8.1. At the very transition, the dielectric constant $\varepsilon \to \infty$, and, accordingly, the charging energy $E_C \to 0$. Therefore, at the self-dual point, $g = g_c$, both characteristic temperatures vanish, $T_{\text{CBKT}} \equiv T_{\text{VBKT}} = 0$. Upon decreasing $g$ and departing from the transition to the insulating side, $\varepsilon$ decreases and $T_{\text{CBKT}} \simeq E_C$ grows while going deeper into the insulating side as shown in Fig. 8.1. Note that the activation energy $\Delta_C = E_C \ln(L/d)$ also grows upon decreasing $g$ (or increasing the resistance of the film) as $(g_c - g)^{\nu}$, with the exponent $\nu > 0$ and so does the crossover line $T_I \simeq \Delta_C$. The width of the insulating domain in the phase diagram confined between $T_{\text{CBKT}}$ and $T_I$ increases as well. Upon further decrease $g$ and the corresponding decrease in $\varepsilon$, the electrostatic screening length $\Lambda \propto \varepsilon d$ shrinks, and as it drops down appreciably below the sample size, the superinsulating domain at the phase diagram ceases to exist.

Having discussed the behavior of the films on the insulating side, we now review, for completeness, the superconducting side. Suppression of the superconducting transition temperature, $T_c$, by disorder in thin superconducting films can be traced back to pioneering works by Shal'nikov,[133,134] where it was noticed for the first time, that $T_c$ decreases with the decrease of the film thickness. The next important step was made by Strongin and collaborators,[42] who found that $T_c$ correlates with the sheet resistance

$R_\Box = 3\pi^2\hbar/(e^2 k_F^2 \ell d) = 1/g$, much better than with the thickness of the film or its resistivity, and thus has to serve as a measure of disorder in the film.

Since then, numerous experimental works have revealed a drastic suppression of $T_c$ in various superconducting films with growing sheet resistance, such as Pb,[42,43,46,135,136] Al,[46] Bi,[42,43,46,49,51] W-Re alloys,[137] MoGe,[138,139] InO,[45,48,140] Be,[55] MoSi,[141] Ta,[142] NbSi,[143] TiN,[59,60,66,67,92,144,145] and PtSi,[146] to name a few. The observed behavior appeared to be in good quantitative accord with the theoretical predictions by Maekawa and Fukuyama,[147] who first connected quantitatively the reduction in $T_c$ as compared to its bulk value $T_{c0}$ with $R_\Box$, and with those of the subsequent work by Finkel'stein,[148] who came up with the comprehensive formula for $T_c(R_\Box)$:

$$\ln\left(\frac{T_c}{T_{c0}}\right) = \gamma + \frac{1}{\sqrt{2r}} \ln\left(\frac{1/\gamma + r/4 - \sqrt{r/2}}{1/\gamma + r/4 + \sqrt{r/2}}\right), \qquad (8.4)$$

where $r = R_\Box e^2/(2\pi^2 \hbar)$ and $\gamma = \ln[\hbar/(k_B T_{c0} \tau)]$. The reduction of $T_c$ with the increasing sheet resistance according to this formula is shown in Fig. 8.4 for several values of the product $T_{c0}\tau$.

The physical picture behind suppressing superconductivity by disorder is that in quasi-two-dimensional systems disorder inhibits electron mobility and thus impairs dynamic screening of the Coulomb interaction. This implies that disorder enhances effects of the Coulomb repulsion between electrons which, if strong enough, breaks down Cooper pairing and destroys superconductivity. Importantly, according to Ref. 148, the degree of disorder at which the Coulomb repulsion would balance the Cooper pair coupling is not sufficient to localize normal carriers. Therefore, at the suppression point, superconductors transform into a metal. The latter can transform into an insulator upon further increase of disorder. Therefore, this mechanism, which is referred to as a fermionic mechanism, results in a sequential superconductor–metal–insulator transition.[148] However, several experiments[60,92] have demonstrated that in the vicinity of the SIT, the dependence $T_c(R_\Box)$ follows perfectly well the Finkel'stein formula. We would like to reiterate here that in spite of the fact that Finkelstein's theory treats superconductor–metal transition, and moreover, the films in the critical region may develop strong mesoscopic fluctuations,[87] the experimentally observed $T_c$ near the SIT where $T_c \to 0$, is still perfectly fitted by formula (8.4).

At $g = g_c$, the superfluid density of the Cooper pair condensate is zero so at the transition point $T_{\text{VBKT}} = 0$. An increase in $g$ suppresses the effects of quantum fluctuations near the transition and, correspondingly, $T_{\text{VBKT}}$ increases. Using the dirty-limit formula which relates the two-dimensional

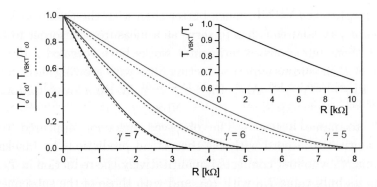

Fig. 8.4. Reduced critical temperature $T_c/T_{c0}$ versus sheet resistance according to Eq. (8.4) (solid lines) for different parameters $\gamma$. The values $\gamma = 5, 6, 7$ correspond to $\tau = 10.3 \cdot 10^{-15}$ sec, $4.8 \cdot 10^{-15}$ sec, and $1.4 \cdot 10^{-15}$ sec, respectfully, at $T_{c0} = 5$ K. The inset shows $T_{\text{VBKT}}/T_c$ versus resistance BMO relation calculated from Eqs. (8.5) and (8.6), which is used to find $T_{\text{VBKT}}/T_{c0}$ shown by the dashed lines.

magnetic screening length to the normal-state resistance $R_N$, Beasley, Mooij, and Orlando (BMO)[9] proposed the universal expression for ratio $T_{\text{VBKT}}/T_c$ (see inset to Fig. 8.4):

$$\frac{T_{\text{VBKT}}}{T_c} f^{-1}\left(\frac{T_{\text{VBKT}}}{T_c}\right) = 0.561 \frac{\pi^3}{8}\left(\frac{\hbar}{e^2}\right)\frac{1}{R_N}, \qquad (8.5)$$

$$f\left(\frac{T}{T_c}\right) = \frac{\Delta(T)}{\Delta(0)} \tanh\left[\frac{\beta\Delta(T)T_c}{2\Delta(0)T}\right], \qquad (8.6)$$

where $\Delta(T)$ is the temperature dependence of the superconducting gap and parameter $\beta = \Delta(0)/(k_B T_c)$. As it was shown in studies on InO[149] and TiN[144] the BMO formula well agrees with the experiment and correctly describes a decrease in the ratio $T_{\text{VBKT}}/T_c$ with increasing disorder, provided the proper choice of $R_N$ and $\beta$ based on the experimental data is made. Two comments are in order. Note that the merging tails of $T_{\text{VBKT}}$ and $T_c$, where they both tend to zero in Fig. 8.4) is, in fact, the image of the same transition point of the phase diagram of Fig. 8.1 where the curves were plotted as functions of the conductance. Second, we would like to emphasize here that the behavior of the superconducting films near the SIT is fully controlled by the unique parameter, the sheet resistance.

To conclude here, we have demonstrated that close to the superconductor–superinsulator transition the superconducting and insulating sides of the phase diagram are the mirror images of each other with the correspondence between the vortex- and charge-BKT transition. As we have mentioned above, this duality is paralleled by the symmetry of transport properties: the VBKT current–voltage ($I$-$V$) characteristics, $V \propto I^\alpha$, is mirrored by the CBKT $I \propto V^\alpha$ power-law behavior with the interchanging current and voltage.[12] Accordingly, the critical current of a superconducting state, setting the upper bound for the loss-free currents which the superconductor can still support, maps onto the threshold voltage which marks the dielectric breakdown of the superinsulating state and the appearance of finite conductivity. This duality is illustrated by recent studies of the VBKT transition in TiN films [144] which revealed the classical BKT transport behavior: the power-law $V \propto I^\alpha$ current–voltage characteristics with the exponent $\alpha$ jumping from $\alpha = 1$ to $\alpha = 3$ at $T = T_{\text{VBKT}}$, in accordance with the Halperin–Nelson prediction,[10] and then rapidly growing with further decrease in temperature. The measurements of the current–voltage ($I$–$V$) dependences on the same material but at the insulating side of the transition,[71,75] revealed the dual $I$–$V$ characteristics which evolve upon lowering the temperature in precisely BKT-like manner, going from ohmic to power-law, $I \propto V^\alpha$, behavior with the exponent $\alpha$ switching from $\alpha = 1$ to $\alpha = 3$ at $T = T_{\text{CBKT}}$ and then rapidly growing from $\alpha \approx 3$ to well above unity with decreasing temperature which is again an inherent feature of the BKT transition.

The dual similarity between the two states manifests itself further in that both states are the loss-free ones as the condition for the Joule loss $P = IV = 0$ holds both in superconductors and superinsulators. This and the mirror correspondence between the $I$–$V$ curves hold high potential for applications, which we will discuss below in Sec. 8.7.

## 8.6. Microscopic Mechanism of Conductivity in the Insulating State

Extensive studies on the role of strong correlations and effects of disorder in insulating behavior can be traced back to the classical works,[150–154] see also Refs. 155 and 156. Recent years have seen an explosive growth in investigations on the interplay between many-body and disorder effects leading to the formation of non-conventional disorder-induced insulators, the properties of which are intimately connected with the localization phenomena.[12,157–162] Yet, at present we are able to describe the tunneling transport in insulators only with exponential accuracy — using euristic considerations — as an Arrhenius or Mott and/or Efros-Shklovskii behavior. The challenge posed by experimental observations, the transition from an activated behavior in the insulating state to the practically complete suppression of the tunneling current in the superinsulator remains unfulfilled.

such a sequential, *cascade* relaxation in 1D granular arrays was developed in Refs. 164–166. One of the important conclusions of this theory is that the interaction of the environment with the infinite number of degrees of freedom with disorder gives rise to broadening of the levels of the tunneling carriers. Another important result is that as soon as the spectrum of the environment excitations shows gaps, the relaxation vanishes and the tunneling current becomes completely suppressed. One can thus conjecture that it is the localization transition of the environmental excitation spectrum that marks the appearance of the gap in the local spectrum of environmental excitations and thus the suppression of the tunnelling charge transfer.

In a two-dimensional array the dipole excitations form the two-dimensional Coulomb plasma, such a transition, and, respectively, the suppression of the relaxation rate take place at the temperature of the charge of BKT transition, $T_{\mathrm{CBKT}} \simeq \bar{E}_{\mathrm{C}}$, where $\bar{E}_{\mathrm{C}}$ is the average charging energy of a single granule. Below this temperature, charges and anti-charges get bound into the neutral CPD, the glassy state forms, the gap in the *local* density of states of the environmental excitation spectrum appears and the tunneling current vanishes.[11,12,164–166] Although the detailed analytical calculations in two dimensions are not available at this point, the conjecture that the microscopic mechanisms of suppression of conductivity in one and two dimensions is of similar nature and are due to the appearance of the gap in the local density in the CPD excitation spectrum (i.e. localization in the energy space) promises to offer a route towards a quantitative microscopic mechanism behind the formation of the superinsulating state.

Recently, the transition from the insulating to superinsulating phase at low but finite temperature due to the suppression of the energy relaxation was also found for the model of superconductor–insulator transition on the Bethe lattice.[167,168]

Remarkably, the employed concept of the relaxation mediated by the emission/absorption of dipole excitation explains also the simultaneous (i.e. occurring at about the same temperature) suppression of both, Cooper pairs and the normal excitations tunneling currents. However, the detailed study of this appealing topic still remains a challenging task.

## 8.7. Conductivity of the Cooper-Pair Insulator and Superinsulator: Qualitative Consideration

Based on the ideas of the cascade relaxation discussed in the previous section we can offer a qualitative description of the temperature evolution of conductivity on the insulating side. A crude estimate can be obtained following heuristic considerations of Ref. 12. The tunneling current in an array of superconducting granules can be written in the following form[164,166,169]

$$I \propto \exp(-E/W), \qquad (8.7)$$

where $E$ is the characteristic energy barrier controlling the charge transfer between the granules, $W = \hbar/\tau_{\rm W}$, and $\tau_{\rm W}$ is the relaxation time, i.e. the time characterizing the rate of the energy exchange between the tunneling charges and the environment. One would expect that the rate of relaxation is proportional to the density of the CPD excitations, which, in its turn, can be taken proportional to the Bose distribution function (CPD excitations are bosons). The width of the local energy gap is $\bar{E}_{\rm C}$, therefore one can take

$$\frac{\hbar}{\tau_{\rm W}} \simeq \frac{\bar{E}_{\rm C}}{\exp(E_{\rm C}/T) - 1}. \qquad (8.8)$$

In other words, the relevant energy scale characterizing the tunneling rate is the energy gap that enters with the weight equal to the Bose filling factor implying the probability of exciting the unbound charges. Above the charge BKT transition Eq. (8.8) gives $W \simeq T$; this is nothing but the equipartition theorem telling us that the number of unbound charges is merely proportional to $T/E_{\rm C}$. To complete the estimate, we have to evaluate the characteristic energy $E$. In a 2D JJA or in a 2D disordered film in the vicinity of SIT, where the electrostatic screening length $\Lambda$ is large and exceeds the linear sample size, $L$, the characteristic energy $E = E_{\rm C} \ln(L/a)$, where $a$ is the size of a single Josephson junction. One realizes that well above $T_{\rm CBKT}$, the logarithmic charge interaction is screened so that $\Lambda \approx a$ and $E$ is reduced to $E_{\rm C}$. In this temperature region, the system exhibits "bad metal" behavior. However, if one is not too far from the charge-BKT transition, $T \gtrsim T_{\rm CBKT}$, one still has $\Lambda > L$ and conductivity acquires Arrhenius thermally activated form with the activation energy that scales as $\ln(L/a)$:

$$\sigma \propto \exp[-E_{\rm C} \ln(L/a)/T], \quad T \gtrsim T_{\rm CBKT}. \qquad (8.9)$$

Notably, Eq. (8.9) looks like a formula for thermally activated conductivity. Yet, one has to remember that the physical mechanism behind the considered charge transfer is quantum mechanical tunneling process which can take place only if the mechanisms for the energy relaxation are switched on.

Turning now to very low temperatures, $T \ll T_{\rm CBKT} \simeq E_{\rm C}$, one sees from Eq. (8.8) that the characteristic energy $W \simeq E_{\rm C} \exp(-\bar{E}_{\rm C}/T)$, i.e. the relaxation rate becomes exponentially low. The unbound charges that have to mediate the energy relaxation from the tunneling carriers are in exponentially short supply, and the estimate for conductivity yields:

$$\sigma \sim \exp[-\ln(L/a)\exp(E_C/T)], \quad T \ll T_{\text{CBKT}}, \qquad (8.10)$$

in accordance with the earlier estimate.[11] One has to bear in mind though that the latter formula has indeed a character of a very crude estimate showing that conductivity at $T < T_{\text{CBKT}}$ is practically zero since the spectrum of the environment excitations acquires the gap $\simeq E_C$. In reality, this "zero-conductivity" regime will be shunted at very low temperatures by phonons and quantum fluctuations (the discussion of which we leave for the forthcoming publication) which will yield finite conductivity.

Let us note that interpolation formula (8.8) becomes very inaccurate in the close vicinity of the transition, since it does not take into account the fact that the typical distance between the free unbound charges (called the correlation length, $\xi_{\text{CBKT}}$) diverges upon approaching the transition as:[13,28]

$$\xi_{\text{CBKT}} \propto \exp\sqrt{\frac{b}{(T/T_{\text{CBKT}})-1}}, \qquad (8.11)$$

where $b$ is a constant of the order of unity. It implies that the density of the environment excitations starts to drop exponentially $\propto \xi_{\text{CBKT}}^{-2}$ and that one has to observe an appreciable increase and deviation from its Arrhenius form resistance as a precursor of the charge-BKT when approaching this transition from above. This should be seen as an upturn in the $\ln R$ versus $1/T$ curves and one can attribute the hyperactivated behaviour found[69] in the 5 nm thin disordered titanium nitride (TiN) film, which was, by its degree of disorder, in the extreme proximity to the disorder-driven SIT, approaching it from the insulating side. To test this we took $\ln R_\square(T)$ versus $1/T$ plots from Ref. 69 for zero field and $B = 0.3\,\text{T}$ and fitted it with

$$R = R_0 \exp\left(A\exp\sqrt{\frac{b}{(T/T_{\text{CBKT}})-1}}\right), \qquad (8.12)$$

following from the fact that at $T \approx T_{\text{CBKT}}$, $W \propto \xi_{\text{CBKT}}^{-2}$. One sees an excellent agreement of this formula (solid curves) with the experimental data. The dashed line indicates the thermally activated behavior, and one sees that the upturn smoothly transforms into the Arrhenius slope, this justifies our assumption that indeed the activated behavior is observed in the temperature region where the density of free dipole excitations is low and the logarithmic interaction remains unscreened.

As early as in '90-s, an upturn in the $\log R(T)$ versus $1/T$ dependence indicating the "stronger than activation" behavior was observed in JJAs.[21,22,25]

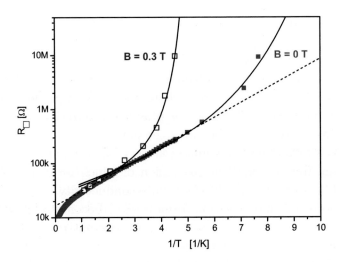

Fig. 8.6. Plots of the logarithm of the sheet resistance $R_\square$ versus $1/T$ taken at two values of the magnetic field, $B = 0\,\text{T}$ and $B = 0.3\,\text{T}$ (data from Ref. 69) (red filled and blue open symbols, respectfully). The solid lines represent the fit given by Eq. (8.12) with parameters $T_{\text{CBKT}} = 0.062\,\text{mK}$, $b = 1$, at $B = 0\,\text{T}$, and $T_{\text{CBKT}} = 0.175\,\text{mK}$, $b = 0.5$ at $B = 0.3\,\text{T}$. The pre-exponential factors are the same for both curves and are taken as $R_0 = 8\,\text{k}\Omega$ and $A = 1$. The dashed straight line corresponds to $T_I = 0.63\,\text{K}$. Equation (8.12) is valid till the square root in the exponent exceeds unity. In the presented data, this condition is satisfied at $T \leq 0.31\,\text{K}$ for $B = 0$ and at $T \leq 0.35\,\text{K}$ for $B = 0.3\,\text{T}$. The overlap between the ranges of applicability of Eq. (8.12) and activated behavior means that indeed the screening length remains large enough down to $T \approx 1\,\text{K}$. Note that $T_{\text{CBKT}} = 0.062\,\text{mK}$, at zero field well coincides with the above estimate (60 mK) for a similar sample derived from the data on the size-dependent activation energy.

They also viewed it as a precursor of the charge-BKT behavior and fitted it with $R \propto A \exp \sqrt{b/[(T/T_{\text{CBKT}}) - 1]}$ analogously to the procedure adopted for describing the vortex-BKT data.[28] We have found that such a procedure also gives a reasonable fit to the hyperactivated behavior data of Ref. 69, with slightly different fitting parameters. This is not surprising since we work in the temperature range which does not include the extreme proximity to $T_{\text{CBKT}}$, i.e. in the range where $\xi_{\text{CBKT}}$ is not excessively large.

## 8.8. Applications

The existence of the superinsulating state opens a route for a new class of devices for cryogenic electronics, utilizing the duality between the superinsulation (SI) and superconductivity (SC). The left panel of Fig. 8.7 sketches the dual-diode current–voltage ($I$–$V$) characteristics of a film or Josephson junction array corresponding to superconducting and superinsulating states, respectively. Application of a moderate voltage $V$ below the threshold voltage $V_T$ realizes the "logical unit" operational mode of the device. When in a superconducting state ("on" regime), the device carries a loss-free supercurrent, in the superinsulating state ("off" regime) the current is blocked since $V < V_T$. Building the superswitch into an electric circuit, one designs a binary logical unit, where (0) corresponds to, say, the SI zero-current and (1) corresponds to the current-carrying SC state. Note that in both, "on" and "off" regimes the superswitch does not lose any power, remaining non-dissipative in both operating states.

Superinsulators can serve as a working body of a detector or bolometer (see Fig. 8.7). To realize this mode, one applies the bias $V$ less but close to $V_T$ and takes $T \lesssim T_{SI}$ as a working temperature. Heating by an irradiation shifts $V_T$ below the voltage of the working point and the current jump occurs over several orders of magnitude. The choice of $T \lesssim T_{SI}$ as a working point ensures that the current jump is controlled by the heating instability solely, eliminating effects of disorder and imperfections, which would randomly shift the voltage at which the switching to the "hot" branch takes place if the working point were chosen deep in the superinsulating state. The important feature of the prospective superinsulator-based electronic devices is that one can expect their sensitivity to be very high since the current in the closed mode is completely blocked.

The switching between the regimes, i.e. between the superconducting and superinsulating states, can be implemented with the aid of the magnetic field which modulates the Josephson coupling $E_J$, changing thus the actual $E_J/E_C$ ratio. Figure 8.7(b) displays the energy-magnetic field phase diagram for a particular case of the square JJA array, the $E_J(B)$ dependence for which was calculated in Ref. 170 in the nearest neighbors approximation.

From the point of view of applications, the most appealing device would be the one that could be operated by means of the electric field. The technologically attractive possibility for implementing such a device is to construct a planar Josephson junction array in which the electric field tunes the ratio $E_J/E_C$, and, therefore, switches the device between the SI and SC states. A practical design of such a superswitch which is realized by utilizing the field effect transistor principle, a superconductor–superinsulator field effect transistor (SSFET) is shown in Fig. 8.8. In this device, in the absence of voltage at the gate, the superconducting islands are decoupled due to high tunneling resistance, $R_T$, of the dielectric separating them, and the array is in a superinsulating state. Applying the sufficient gate voltage, the insulating channels can become conducting (or, at least, their $R_T$ becomes reduced). This will give rise to increase of ratio $E_J/E_C$ and will drive the

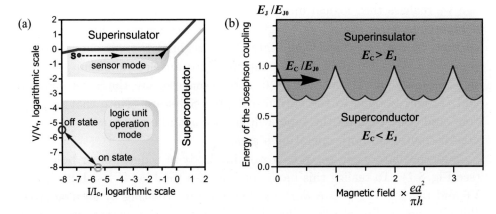

Fig. 8.7. Switching between superconducting and superinsulating states. (a) Exemplary dual threshold current–voltage ($I$–$V$) characteristics in the superconducting and superinsulating states of a 5 nm thin TiN film in the double-log coordinates. The current and voltage are measured in units of the critical current, $I_c$, and threshold voltage, $V_T$, respectively. The lower inset illustrates switching between the "off" (superinsulating) and "on" (superconducting) states in the logic unit operation mode. The upper inset illustrates the sensor (bolometer) mode utilizing the threshold character of the $I$–$V$ characteristics in the superinsulating state: when some pre-threshold voltage is applied to the film, it remains a superinsulating state with zero current. As soon as the film is heated by some irradiation, the threshold voltage decreases below the applied value and the current jumps over six orders of magnitude. An advantage of using a superinsulator as the working body for a sensor over the superconductor-based sensor is that due to huge resistance, the current in the superinsulating state is extremely low, and, therefore, superinsulator is less sensitive to background noise. (b) A sketch of the magnetic field controlled switching between two non-dissipative states in a square Josephson junction array (JJA). The line separating the superinsulating and superconducting domains depicts the dependence of Josephson coupling energy, $E_J$ (measured in the units of the Josephson coupling energy, $E_{J0}$, in the zero magnetic field), on the applied magnetic field, $B$. The arrow shows the evolution of the system, with the ratio $E_C/E_{J0}$ taken slightly less then unity, where $E_C$ is the charging energy of a single junction. If $E_C/E_{J0} < 1$, the system is a superconductor, and if $E_C/E_{J0} > 1$, the system is superinsulator in the zero magnetic field. Upon increasing field, the system evolves along the arrow and, starting from the superconducting state transits into the superinsulator as soon as it crosses the line $E_J(B)$, i.e. as soon as $E_J(B)$ becomes less than the charging energy $E_C$.

array in a superconducting state. Thus, the proposed switching technique is built on changing the conductivity of the medium confined between the adjacent superconducting islands by tuning the gate voltage.

The first steps on the realization of the electrostatically-driven SIT have already been undertaken. The system of superconducting islands placed on the graphene layer was used in a recent work[171] in order to tune the $E_J/E_C$ ratio as proposed above and the pronounced SIT with the resistance as high as $10^7\,\Omega$ at 40 mK was observed in the insulating state. The change

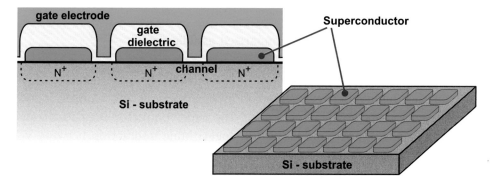

Fig. 8.8. Superswitch utilizing the field effect transistor technology. A sketch of the cross-section (the left panel) and the view of the working body (the right panel) of the superconductor–superinsulator field effect transistor (SSFET). The SSFET is comprised of the array of the superconducting islands placed on the Si-substrate. Highly doped regions of the substrate marked as $N^+$. The islands are covered by a dielectric separating them from the gate on top.

of the resistance as a function of the gate voltage was about seven orders of magnitude.

The story, however, would have been incomplete, without mentioning the approach based on the electric field-induced change of the properties of the superconducting films themselves.[172] Such an approach, a direct varying of the charge carrier density in the superconducting film by applying a gate voltage in a field effect transistor configuration was considered in many other works.[82,173–180] At present two kinds of materials are used as gate insulators: $SrTiO_3$ (STO) crystal and/or various electrolytes (or ionic liquids). We would like to note that STO is an insulator with very high dielectric constant $\varepsilon \sim 20000$. Such a dielectric sandwiching or covering the superconducting film is detrimental to the formation of the superinsulating state since it catastrophically reduces the electrostatic screening length, see Eq. (8.2). Indeed, in none of the quoted works, the sheet resistance in the non-superconducting state does not exceed $40\,K\Omega$, thus the very possibility of the transition into an insulating state remains to be investigated. Turning to the electrolytes as gate insulators, their disadvantage, from the viewpoint of practical applications, is that the operating mode requires thermocycling up to room temperatures when changing the gate voltage and influencing the properties of superconducting films.[82] Yet, both novel techniques lend the way for controlling properties of superconducting films and open a route to a new generation of switching devices with outstanding performance characteristics.

## 8.9. Conclusions

The above discussion can be summarized as follows:

(1) On the fundamental level, the phenomenon of superinsulation rests on the duality between superconducting vortices and charges in two-dimensional systems experiencing the superconductor–insulator transition. At the superconducting side logarithmic interaction between vortices (i.e. vortex confinement) leads to superconductivity, the state with the zero linear resistance below the vortex-BKT transition where global phase coherence establishes. Correspondingly, the logarithmic interaction between charges (charge confinement) at the insulating side leads to a superinsulating state with zero linear conductivity. This duality follows from the Heisenberg uncertainty principle: at the superconducting side, the phase uncertainty is zero implying that charges that comprise Cooper pair condensate move without scattering. At the superinsulating side, all the charges are pinned below the charge-BKT transition implying, according to the Aharonov–Casher effect, the coherent quantum tunneling of all the fluxons (the synchronized quantum slips of globally uncertain phase) i.e. fluxons are in a superfluid state.
(2) The condition for a charge confinement i.e. logarithmic growth of the interaction between the charges as a function of the distance between them in the insulating state of a planar Josephson junction array is realized if the electrostatic screening length exceeds the linear size of the array. This requires that a single junction capacitance well exceeds the capacitance to the ground. This logarithmic interaction leads to the logarithmic scaling of the activation energy of the Cooper pair insulator in Josephson junctions array with the linear size of the array. We refer to this phenomenon of the emergence of the large energy scale controlling the activated behavior and growing with the size of the system as a *macroscopic Coulomb blockade*.
(3) Disordered thin films in the critical vicinity of the superconductor–superinsulator transition have huge, diverging on approach to the transition to the superconducting state, dielectric constant. This leads to the two-dimensional logarithmic Coulomb interaction between the unscreened charges in the film, which, in its turn results in the charge Berezinskii–Kosterlitz–Thouless transition. The logarithmic interaction holds as long as the distance separating charges does not exceed the electrostatic screening length. Accordingly, the CBKT transition is most pronounced in the systems whose linear size does not exceed this length.

(4) One can expect that in the critical vicinity of the superconductor–superinsulator transition where long-range Coulomb interactions become relevant, electronic phase separation can occur and the texture of weakly coupled superconducting islands may form. This indicates that near the transition the transport behavior of disordered films is analogous to that of planar Josephson junctions arrays.

(5) Logarithmic interaction between charges in the Cooper-pair insulating phase of the two-dimensional Josephson junction array is dual to the logarithmic interactions between the vortices at the superconducting side. This implies the dual phase diagram in the vicinity of the superconductor–superinsulator transition: Transformation of the resistive state into superconductivity possessing the global phase coherence occurring at $T_{\text{VBKT}}$ is mirrored by the transition from the insulator to superinsulator at the charge BKT transition temperature $T_{\text{CBKT}}$. We identify a superinsulator as a low-temperature charge-BKT phase, possessing infinite resistivity in the finite temperature range in the same sense as the low-temperature vortex BKT phase is a superconductor having infinite conductivity in a finite temperature range as well.

(6) In the Cooper-pair insulator, current occurs via tunneling of the Cooper pairs across the Josephson links. To ensure tunneling transport in the random system, the energy relaxation mechanism providing the energy exchange between the charge carriers and some bosonic environment is required. At low temperatures, tunneling in Josephson junction array is mediated by the *self-generated* bosonic environment of the Cooper-pair dipole excitations, comprised of the same Cooper pairs that tunnel and mediate the charge transfer. Importantly, being a system with infinite number degrees of freedom, the dipole environment plays the role of the thermostat. Opening the gap in the *local* dipole excitations spectrum suppresses energy relaxation and impedes the tunneling current. In the two-dimensional Josephson junction array this gap appears below $T_{\text{CBKT}}$. We expect that the gap in the local dipole spectrum is associated with formation of a low-temperature glassy state. This constitutes the microscopic mechanism of superinsulation.

**Summary**

In this chapter, we have shown that the existence of the superinsulating state is a fundamental phenomenon intimately connected with the very existence of superconductivity and that it follows from the symmetry of the uncertainty principle, setting a competition between the conjugated quantities, the charge and the phase of the wave function of the Cooper pair condensate. This symmetry can be expressed in terms of the charge-vortex duality in two-dimensional superconducting systems and manifests itself via the reversal between the Aharonov–Bohm and the Aharonov–Casher effects. We have described the conditions for superinsulation emerging in real experimentally accessible two-dimensional systems, lateral Josephson junction arrays and highly disordered thin superconducting films and have demonstrated that the critical component of the phenomenon is the high dielectric constant of the insulating phase of the system in question which develops in the close vicinity of superconductor–insulator transition. The latter ensures the logarithmic interaction between the charges and brings with it the effect of the macroscopic Coulomb blockade, which, in its turn manifests as a size-dependent characteristic energy that controls the activation processes in the insulating phase. We have examined these concepts, testing them on the original experimental findings, and have demonstrated that they work perfectly well offering self-consistent physical picture describing collective behavior of two-dimensional superconducting systems in the vicinity of superconductor–insulator transition. We have constructed and analyzed the phase diagram for a planar Josephson junction array and/or disordered superconducting film in the vicinity of the superconductor insulator transition and identified superinsulating state as a low temperature charge-BKT phase. And, finally, the perspectives for practical applications of the superinsulating systems as a working body for new generation of electronic devices utilizing the duality between superinsulation and superconductivity have been discussed.

## Acknowledgments

We are delighted to thank D. Averin, N. Chtchelkatchev, E. Chudnovsky, R. Fazio, A. Glatz, L. Glazman, K. Matveev, A. Melnikov, V. Mineev, Yu. Nazarov, and Ch. Strunk for enlightening discussions. The work was supported by the U.S. Department of Energy Office of Science through the contract DE-AC02-06CH11357. The work of T. Baturina was partly supported by the Program "Quantum Mesoscopic and Disordered Systems" of the Russian Academy of Sciences and by the Russian Foundation for Basic Research (Grant No. 12-02-00152).

## References

1. P. W. Anderson, *Lectures on the Many-Body Problem, Vol. 2*, ed. by E. R. Caianiello (Academic, New York 1964), p. 113.

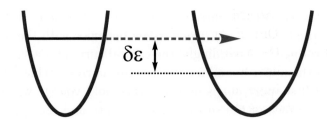

Fig. 8.5. A sketch of tunneling between two mesoscopic potential wells. Since the energy, levels in these wells are, in general, vary due to structural disorder, quantum mechanical tunneling is possible if and only if the tunneling particle would emit (or absorb) some bosonic excitation which would accommodate the energy difference $\delta\varepsilon$ between the respective energy levels.

In what follows, we will discuss one of the aspects of tunneling transport in disorder-induced insulators which may be the key to understanding dynamic insulator–superinsulator transition. Namely, in the case of disorder-induced Cooper pair insulators, the charge transfer can be viewed as a tunneling between the localized Cooper pair sites possessing essentially different energy levels, see Fig. 8.5. Tunneling is possible only in the presence of some energy relaxation mechanism which is able to accommodate the energy differences between the Cooper pair states at the neighboring localized sites hosting the tunneling process, see Fig. 8.5.

One can propose that an appropriate theory of transport in the insulating state should be constructed via the incorporation of ideas of relaxation physics into a general model of transport in granular materials. As a starting point, one can take a model that had already been successfully used by Efetov [3] to demonstrate the very existence of the disorder-driven SIT. The customary relaxation mechanism in "conventional" conductors is the energy exchange between the tunneling charge carriers and phonons, comprising a thermal bath. In granular materials, at low temperatures however, the role of the major relaxation processes, ensuring the tunneling charge transfer, is taken by emission (and/or absorption) of dipole excitations. Importantly, these excitations are the same particles that mediate the charge transfer. In the Cooper-pair insulators, the dipoles are thus made up of the local excess (-2e) and local deficit (+2e) in the Cooper-pairs number and form the bosonic environment.[11,12,160,163–166] The important features of this dipole excitations environment is that the dipoles are generated in the process of tunneling and that the dipole environment possesses an infinite number of degrees of freedom, and as such it efficiently takes away the energy from the tunneling particles and plays the role of the thermostat itself. Eventually, dipole excitations relax energy to the phonon bath. A theory of

2. B. Abeles, Effect of charging energy on superconductivity in granular metal films, *Phys. Rev. B* **15** (5), 2828–2829 (1977).
3. K. B. Efetov, Phase transitions in granulated superconductors, *Zh. Eksp. Teor. Fiz.* **78** (5), 2017–2032 (1980) [*Sov. Phys. JETP.* **51** (5), 1015–1022 (1980)].
4. A. Gold, Impurity-induced phase transition in the interacting Bose gas: 1. Analytical results, *Z. Phys. B — Condensed Matter.* **52**, 1–8 (1983).
5. A. Gold, Dielectric properties of a disordered Bose condensate, *Phys. Rev. A* **33**, 652–659 (1986).
6. R. D. Feynman, *Statistical Mechanics: A Set of Lectures* (Addison Wesley, 1981).
7. M. Sugahara, Superconductive granular thin film and phase quantum tunnel device, *Jpn. J. Appl. Phys.* **24**, 674–678 (1985).
8. J. E. Mooij and Yu. V. Nazarov, Superconducting nanowires as quantum phase-slip junctions, *Nature Phys.* **2**, 169–172 (2006).
9. M. R. Beasley, J. E. Mooij and T. P. Orlando, Possibility of vortex-antivortex pair dissociation in two-dimensional superconductors, *Phys. Rev. Lett.* **42**, 1165–1168 (1979).
10. B. I. Halperin and D. R. Nelson, Resistive transitions in superconducting films, *J. Low. Temp. Phys.* **36**, 599–616 (1979).
11. M. V. Fistul, V. M. Vinokur and T. I. Baturina, Collective Cooper-pair transport in the insulating state of Josephson-junction arrays, *Phys. Rev. Lett.* **100**, 086805 (2008).
12. V. M. Vinokur, T. I. Baturina, M. V. Fistul, A. Yu. Mironov, M. R. Baklanov and C. Strunk, Superinsulator and quantum synchronization, *Nature* **452**, 613–615 (2008).
13. P. Minnhagen, The two-dimensional Coulomb gas, vortex unbinding, and superfluid-superconducting films, *Rev. Mod. Phys.* **59**, 1001–1066 (1987).
14. L. J. Geerligs, M. Peters, L. E. M. de Groot, A. Verbruggen and J. E. Mooij, Charging effects and quantum coherence in regular Josephson junction arrays, *Phys. Rev. Lett.* **63**, 326–329 (1989).
15. J. E. Mooij, J. van Wees, L. J. Geerligs, M. Peters, R. Fazio and G. Schön, Unbinding of charge-anticharge pairs in two-dimensional arrays of small tunnel junctions, *Phys. Rev. Lett.* **65**, 645–648 (1990).
16. R. Fazio and G. Schön, Charge and vortex dynamics in arrays of tunnel junctions, *Phys. Rev. B* **43**, 5307–5320 (1991).
17. B. J. van Wees, Duality between Cooper-pair and vortex dynamics in two-dimensional Josephson-junction arrays, *Phys. Rev. B* **44**, 2264–2267 (1991).
18. H. S. J. van der Zant, F. C. Fritschy, W. J. Elion, L. J. Geerligs and J. E. Mooij, Field-induced superconductor-to-insulator transitions in Josephson-junction arrays, *Phys. Rev. Lett.* **69**, 2971–2974 (1992).
19. T. S. Tighe, M. T. Tuominen, J. M. Hergenrother and M. Tinknam, Measurements of charge soliton motion in two-dimensional arrays of ultrasmall Josephson junctions, *Phys. Rev. B* **47**, 1145–1148 (1993).
20. P. Delsing, C. D. Chen, D. B. Haviland, Y. Harada and T. Claeson, Charge solitons and quantum fluctuations in two-dimensional arrays of small Josephson junctions, *Phys. Rev. B* **50**, 3959–3971 (1994).

21. A. Kanda, S. Katsumoto and S. Kobayashi, Charge-soliton transport properties in two-dimensional array of small Josephson junctions, *J. Phys. Soc. Jpn.* **63**, 4306–4309 (1994).
22. A. Kanda and S. Kobayashi, Precursor of charge KTB transition in normal and superconducting tunnel junction array, *J. Phys. Soc. Jpn.* **64**, 19–21 (1995).
23. A. Kanda and S. Kobayashi, Effect of self-capacitance on charge Kosterlitz-Thouless transition in small tunnel-junction arrays, *J. Phys. Soc. Jpn.* **64**, 3172–3174 (1995).
24. H. S. J. van der Zant, W. J. Elion, L. J. Geerligs and J. E. Mooij, Quantum phase transitions in two dimensions: Experiments in Josephson-junction arrays, *Phys. Rev. B* **54**, 10081–10093 (1996).
25. T. Yamaguchi, R. Yagi, S. Kobayashi and Y. Ootuka, Two-dimensional arrays of small Josephson junctions with regular and random defects, *J. Phys. Soc. Jpn.* **67**, 729–731 (1998).
26. Y. Takahide, R. Yagi, A. Kanda, Y. Ootuka and S. Kobayashi, Superconductor-insulator transition in a two-dimensional array of resistively shunted small Josephson junctions, *Phys. Rev. Lett.* **85**, 1974–1977 (2000).
27. C. Giovanella and M. Tinkham, Macroscopic quantum phenomena and coherence in superconducting networks (World Scientific, Singapore, 1995).
28. R. S. Newrock, C. J. Lobb, U. Geigenmüller and M. Octavio, The two-dimensional physics of Josephson junction arrays, *Solid State Phys.* **54**, 263–512 (2000).
29. R. Fazio and H. van der Zant, Quantum phase transitions and vortex dynamics in superconducting networks, *Phys. Rep.* **355**, 235–334 (2001).
30. R. C. Dynes, J. P. Garno and J. M. Rowell, Two-dimensional electrical conductivity in quench-condensed metal films, *Phys. Rev. Lett.* **40**, 479–482 (1978).
31. A. E. White, R. C. Dynes and J. P. Garno, Destruction of superconductivity in quench-condensed two-dimensional films, *Phys. Rev. B* **33**, 3549–3552 (1986).
32. B. G. Orr, H. M. Jaeger, A. M. Goldman and C. G. Kuper, Global phase coherence in two-dimensional granular superconductors, *Phys. Rev. Lett.* **56**, 378–381 (1986).
33. N. Yoshikawa, T. Akeyoshi, M. Kojima and M. Sugahara, Dual conduction characteristics observed in highly resistive NbN granular thin films, *Proc. LT-18, Kyoto 1987, Jpn. J. Appl. Phys.* **26** (Suppl. 26-3), 949–950 (1987).
34. N. Yoshikawa, T. Akeyoshi and M. Sugahara, Field-effect induced sinusoidal conductivity variation of NbN granular thin films, *Jpn. J. Appl. Phys.* **26**, L1701–L1702 (1987).
35. A. Widom and S. Badjou, Quantum displacement-charge transitions in two-dimensional granular superconductors, *Phys. Rev. B* **37**, 7915–7916 (1988).
36. H. M. Jaeger, D. B. Haviland, B. G. Orr and A. M. Goldman, Onset of superconductivity in ultrathin granular metal films, *Phys. Rev. B* **40**, 182–196 (1989).
37. R. P. Barber, Jr. and R. E. Glover III, Hyper-resistivity to global-superconductivity transition by annealing in quench-condensed Pb film, *Phys. Rev. B* **42**, 6754–6757 (1990).

38. W. Wu and P. W. Adams, Electric-field tuning of the superconductor-insulator transition in granular Al films, *Phys. Rev. B* **50**, 13065–13068 (1994).
39. C. Christiansen, L. M. Hernandez and A. M. Goldman, Evidence of collective charge behavior in the insulating state of ultrathin films of superconducting metals, *Phys. Rev. Lett.* **88**, 037004 (2002).
40. A. Frydman, The superconductor insulator transition in systems of ultrasmall grains, *Physica C* **391**, 189–195 (2003).
41. R. P. Barber, Jr., Shih-Ying Hsu, J. M. Valles, Jr., R. C. Dynes and R. E. Glover III, Negative magnetoresistance, negative electroresistance, and metallic behavior on the insulating side of the two-dimensional superconductor-insulator transition in granular Pb films, *Phys. Rev. B* **73**, 134516 (2006).
42. M. Strongin, R. S. Thompson, O. F. Kammerer and J. E. Crow, Destruction of superconductivity in disordered near-monolayer films, *Phys. Rev. B* **1**, 1078–1091 (1970).
43. D. B. Haviland, Y. Liu and A. M. Goldman, Onset of superconductivity in the two-dimensional limit, *Phys. Rev. Lett.* **62**, 2180–2183 (1989).
44. A. F. Hebard and M. A. Paalanen, Magnetic-field-tuned superconductor-insulator transition in two-dimensional films, *Phys. Rev. Lett.* **65**, 927–930 (1990).
45. D. Shahar and Z. Ovadyahu, Superconductivity near the mobility edge, *Phys. Rev. B* **46**, 10917–10922 (1992).
46. Y. Liu, D. B. Haviland, B. Nease and A. M. Goldman, Insulator-to-superconductor transition in ultrathin films, *Phys. Rev. B* **47**, 5931–5946 (1993).
47. D. Kowal and Z. Ovadyahu, Disorder induced granularity in an amorphous superconductor, *Solid State Comm.* **90**, 783–786 (1994).
48. V. F. Gantmakher, M. V. Golubkov, J. G. S. Lok and A. K. Geim, Giant negative magnetoresistance of semi-insulating amorphous indium oxide films in strong magnetic fields, *Zh. Eksp. Teor. Fiz.* **109**, 1765–1778 (1996) [*JETP* **82**, 951–958 (1996)].
49. N. Marković, C. Christiansen and A. M. Goldman, Thickness-magnetic field phase diagram at the superconductor-insulator transition in 2D, *Phys. Rev. Lett.* **81**, 5217–5220 (1998).
50. V. F. Gantmakher, M. V. Golubkov, V. T. Dolgopolov, G. E. Tsydynzhapov and A. A. Shashkin, Destruction of localized electron pairs above the magnetic-field-driven superconductor-insulator transition in amorphous In-O films, *JETP Lett.* **68**, 363–369 (1998).
51. J. A. Chervenak and J. M. Valles, Jr., Observation of critical amplitude fluctuations near the two-dimensional superconductor-insulator transition, *Phys. Rev. B* **59**, 11209–11212 (1999).
52. V. F. Gantmakher, M. V. Golubkov, V. T. Dolgopolov, G. E. Tsydynzhapov and A. A. Shashkin, Scaling analysis of the magnetic field-tuned quantum transition in superconducting amorphous In-O films, *JETP Lett.* **71**, 160–164 (2000).
53. V. F. Gantmakher, M. V. Golubkov, V. T. Dolgopolov, A. A. Shashkin

and G. E. Tsydynzhapov, Observation of the parallel-magnetic-field-induced superconductor-insulator transition in thin amorphous In-O films, *JETP Lett.* **71**, 473–478 (2000).
54. V. Yu. Butko and P. W. Adams, Quantum metallicity in a two-dimensional insulator, *Nature* **409**, 161–164 (2001).
55. E. Bielejec, J. Ruan and Wenhao Wu, Hard correlation gap observed in quench-condensed ultrathin beryllium, *Phys. Rev. Lett.* **87**, 036801 (2001).
56. E. Bielejec, J. Ruan and W. Wu, Anisotropic magnetoconductance in quench-condensed ultrathin beryllium films, *Phys. Rev. B* **63**, 100502(R) (2001).
57. E. Bielejec and W. Wu, Field-tuned superconductor-insulator transition with and without current bias, *Phys. Rev. Lett.* **88**, 206802 (2002).
58. G. Sambandamurthy, L. W. Engel, A. Johansson and D. Shahar, Superconductivity-related insulating behavior, *Phys. Rev. Lett.* **92**, 107005 (2004).
59. T. I. Baturina, D. R. Islamov, J. Bentner, C. Strunk, M. R. Baklanov and A. Satta, Superconductivity on the localization threshold and magnetic-field-tuned superconductor-insulator transition in TiN films, *Pis'ma Zh. Eksp. Teor. Fiz.* **79**, 416–420 (2004) [*JETP Lett.* **79**, 337–341 (2004)].
60. N. Hadacek, M. Sanquer and J-C. Villégier, Double reentrant superconductor-insulator transition in thin TiN films, *Phys. Rev. B* **69**, 024505 (2004).
61. T. I. Baturina, J. Bentner, C. Strunk, M. R. Baklanov and A. Satta, From quantum corrections to magnetic-field-tuned superconductor-insulator quantum phase transition in TiN films, *Physica B* **359-361**, 500–502 (2005).
62. M. Steiner and A. Kapitulnik, Superconductivity in the insulating phase above the field-tuned superconductor-insulator transition in disordered indium oxide films, *Physica C* **422**, 16–26 (2005).
63. G. Sambandamurthy, L. W. Engel, A. Johansson, E. Peled and D. Shahar, Experimental evidence for a collective insulating state in two-dimensional superconductors, *Phys. Rev. Lett.* **94**, 017003 (2005).
64. J. S. Parker, D. E. Read, A. Kumar and P. Xiong, Superconducting quantum phase transitions tuned by magnetic impurity and magnetic field in ultrathin a-Pb films, *Europhys. Lett.* **75**, 950–956 (2006).
65. T. I. Baturina, C. Strunk, M. R. Baklanov and A. Satta, Quantum metallicity on the high-field side of the superconductor-insulator transition, *Phys. Rev. Lett.* **98**, 127003 (2007).
66. T. I. Baturina, A. Yu. Mironov, V. M. Vinokur, M. R. Baklanov and C. Strunk, Localized superconductivity in the quantum-critical region of the disorder-driven superconductor-insulator transition in TiN thin films, *Phys. Rev. Lett.* **99**, 257003 (2007).
67. T. I. Baturina, A. Bilušić, A. Yu. Mironov, V. M. Vinokur, M. R. Baklanov and C. Strunk, Quantum-critical region of the disorder-driven superconductor-insulator transition, *Physica C* **468**, 316–321 (2008).
68. D. Kowal and Z. Ovadyahu, Scale dependent superconductor-insulator transition, *Physica C* **468**, 322–325 (2008).
69. T. I. Baturina, A. Yu. Mironov, V. M. Vinokur, M. R. Baklanov and

C. Strunk, Hyperactivated resistance in TiN films on the insulating side of the disorder-driven superconductor-insulator transition, *JETP Lett.* **88**, 752–757 (2008).
70. Y.-H. Lin and A. M. Goldman, Hard energy gap in the insulating regime of nominally granular films near the superconductor-insulator transition, *Phys. Rev. B* **82**, 214511 (2010).
71. D. Kalok, A. Bilušić, T. I. Baturina, V. M. Vinokur and C. Strunk, Intrinsic non-linear conduction in the super-insulating state of thin TiN films, arXiv:1004.5153v2 (2010).
72. Y.-H. Lin and A. M. Goldman, Magnetic-field-tuned quantum phase transition in the insulating regime of ultrathin amorphous Bi films, *Phys. Rev. Lett.* **106**, 127003 (2011).
73. A. Johansson, I. Shammassa, N. Stander, E. Peled, G. Sambandamurthy and D. Shahar, Angular dependence of the magnetic-field driven superconductor-insulator transition in thin films of amorphous indium-oxide, *Solid State Comm.* **151**, 743–746 (2011).
74. O. Cohen, M. Ovadia and D. Shahar, Electric breakdown effect in the current-voltage characteristics of amorphous indium oxide thin films near the superconductor-insulator transition, *Phys. Rev. B* **84**, 100507(R) (2011).
75. D. Kalok, A. Bilušić, T. I. Baturina, A. Yu. Mironov, S. V. Postolova, A. K. Gutakovskii, A. V. Latyshev, V. M. Vinokur and C. Strunk, Non-linear conduction in the critical region of the superconductor-insulator transition in TiN thin films, *J. Phys.: Conference Series (JPCS)*, accepted for publication.
76. D. Mandrus, L. Forro, C. Kendziora and L. Mihaly, Two-dimensional electron localization in bulk single crystals of $Bi_2Sr_2Y_xCa_{1-x}Cu_2O_8$, *Phys. Rev. B* **44**, 2418–2421 (1991).
77. G. T. Seidler, T. F. Rosenbaum and B. W. Veal, Two-dimensional superconductor-insulator transition in bulk single-crystal $YBa_2Cu_3O_{6.38}$, *Phys. Rev. B* **45**, 10162–10164 (1992).
78. S. Tanda, S. Ohzeki and T. Nakayama, Bose glass-vortexglass phase transition and dynamics scaling for high-$T_c$ $Nd_{2-x}Ce_xCuO_4$ thin films, *Phys. Rev. Lett.* **69**, 530–533 (1992).
79. B. Beschoten, S. Sadewasser, G. Güntherodt and C. Quitmann, Coexistence of superconductivity and localization in $Bi_2Sr_2(Ca_z, Pr_{1-z})Cu_2O_{8+y}$, *Phys. Rev. Lett.* **77**, 1837–1840 (1996).
80. Y. Ando, A. N. Lavrov, S. Komiya, K. Segawa and X. F. Sun, Mobility of the doped holes and the antiferromagnetic correlations in underdoped high-$T_c$ cuprates, *Phys. Rev. Lett.* **87**, 017001 (2001).
81. V. F. Gantmakher, S. N. Ermolov, G. E. Tsydynzhapov, A. A. Zhukov and T. I. Baturina, Suppression of 2D superconductivity by the magnetic field: Quantum corrections vs. the superconductor-insulator transition, *JETP Lett.* **77**, 424–428 (2003).
82. A. T. Bollinger, G. Dubuis, J. Yoon, D. Pavuna, J. Misewich and I. Božović, Superconductor-insulator transition in $La_{2-x}Sr_xCuO_4$ at the pair quantum resistance, *Nature* **472**, 458–460 (2011).
83. A. I. Larkin and Yu. N. Ovchinnikov, Effect of inhomogeneities on the proper-

ties of superconductors, *Zh. Eksp. Teor. Fiz.* **61**, 1221–1230 (1971) [*Sov. Phys. JETP* **34**, 651 (1972)].

84. L. B. Ioffe and A. I. Larkin, Properties of superconductors with a smeared transition temperature, *Zh. Eksp. Teor. Fiz.* **81**, 707–718 (1981) [*Sov. Phys. JETP* **54**, 378–384 (1981)].

85. M. Ma and P. A. Lee, Localized superconductors, *Phys. Rev. B* **32**, 5658–5667 (1985).

86. Y. Imry, M. Strongin and C. C. Homes, An inhomogeneous Josephson phase in thin film and high-$T_c$ superconductors, *Physica C* **468**, 288–293 (2008).

87. M. A. Skvortsov and M. V. Feigelman, Superconductivity in disordered thin films: Giant mesoscopic fluctuations, *Phys. Rev. Lett.* **95**, 057002 (2005).

88. A. Ghosal, M. Randeria, and N. Trivedi, Role of spatial amplitude fluctuations in highly disordered s-wave superconductors, *Phys. Rev. Lett.* **81**, 3940–3943 (1998).

89. A. Ghosal, M. Randeria and N. Trivedi, Inhomogeneous pairing in highly disordered s-wave superconductors, *Phys. Rev. B* **65**, 014501 (2001).

90. Y. Dubi, Y. Meir and Y. Avishai, Nature of the superconductor-insulator transition in disordered superconductors, *Nature* **449**, 876–880 (2007).

91. K. Bouadim, Y. L. Loh, M. Randeria and N. Trivedi, Single- and two-particle energy gaps across the disorder-driven superconductor-insulator transition, *Nature Phys.* **7**, 884–889 (2011).

92. B. Sacépé, C. Chapelier, T. I. Baturina, V. M. Vinokur, M. R. Baklanov and M. Sanquer, Disorder-induced inhomogeneities of the superconducting state close to the superconductor-insulator transition, *Phys. Rev. Lett.* **101**, 157006 (2008).

93. B. Sacépé, T. Dubouchet, C. Chapelier, M. Sanquer, M. Ovadia, D. Shahar, M. Feigel'man and L. Ioffe, Localization of preformed Cooper pairs in disordered superconductors, *Nature Phys.* **7**, 239–244 (2011).

94. A. C. Fang, L. Capriotti, D. J. Scalapino, S. A. Kivelson, N. Kaneko, M. Greven and A. Kapitulnik, Gap-inhomogeneity-induced electronic states in superconducting $Bi_2Sr_2CaCu_2O_{8-\delta}$, *Phys. Rev. Lett.* **96**, 017007 (2006).

95. K. K. Gomes, A. N. Pasupathy, A. Pushp, Sh. Ono, Y. Ando and A. Yazdani, Visualizing pair formation on the atomic scale in the high-$T_c$ superconductor $Bi_2Sr_2CaCu_2O_{8+\delta}$, *Nature* **447**, 569–572 (2007).

96. R. Fazio, Condensed-matter physics — Opposite of a superconductor, *Nature* **452**, 542–543 (2008).

97. Y. Aharonov and D. Bohm, Significance of electromagnetic potentials in the quantum theory, *Phys. Rev.* **115**, 485–491 (1959).

98. Y. Aharonov and A. Casher, Topological quantum effects for neutral particles, *Phys. Rev. Lett.* **53**, 319–321 (1984).

99. B. Reznik and Y. Aharonov, Question of the nonlocality of the Aharonov–Casher effect, *Phys. Rev. D* **40**, 4178–4183 (1989).

100. D. A. Ivanov, L. B. Ioffe, V. B. Geshkenbein and G. Blatter, Interference effects in isolated Josephson junction arrays with geometric symmetries, *Phys. Rev. B* **65**, 024509 (2001).

101. J. R. Friedman and D. V. Averin, Aharonov–Casher-Effect of macroscopic tunneling of magnetic flux, *Phys. Rev. Lett.* **88**, 050403 (2002).
102. K. A. Matveev, A. I. Larkin and L. I. Glazman, Persistent current in superconducting nanorings, *Phys. Rev. Lett.* **89**, 096802 (2002).
103. I. M. Pop, I. Protopopov, F. Lecocq, Z. Peng, B. Pannetier, O. Buisson and W. Guichard, Measurement of the effect of quantum phase slips in a Josephson junction chain, *Nature Phys.* **6**, 589–592 (2010).
104. O. V. Astafiev, L. B. Ioffe, S. Kafanov, Yu. A. Pashkin, K. Yu. Arutyunov, D. Shahar, O. Cohen and J. S. Tsai, Coherent quantum phase slip, *Nature Phys.* **484**, 355–358 (2012).
105. A. Krämer and S. Doniach, Superinsulator phase of two-dimensional superconductors, *Phys. Rev. Lett.* **81**, 3523–3526 (1998).
106. A. Salzberg and S. Prager, Equation of state for a two-dimensional electrolyte, *J. Chem. Phys.* **38**, 2587 (1963).
107. V. L. Berezinskii, Violation of long range order in one-dimensional and two-dimensional systems with a continuous symmetry group. I. Classical systems, *Zh. Eksp. Teor. Fiz.* **59**, 907–920 (1970) [*Sov. Phys. JETP* **32**, 493–500 (1971)].
108. V. L. Berezinskii, Destruction of long-range order in one-dimensional and two-dimensional systems possessing a continuous symmetry group. II. Quantum systems, *Zh. Eksp. Teor. Fiz.* **61**, 1144–1155 (1971) [*Sov. Phys. JETP* **34**, 610–616 (1972)].
109. J. M. Kosterlitz and D. Thouless, Long range order and metastability in two dimensional solids and superfluids, *J. Phys. C* **5**, L124–L126 (1972).
110. J. M. Kosterlitz and D. Thouless, Ordering, metastability and phase transitions in two-dimensional systems, *J. Phys. C* **6**, 1181–1203 (1973).
111. M. C. Diamantini, P. Sodano and C. A. Trugenberger, Gauge theories of Josephson junction arrays, *Nucl. Phys. B* **474**, 641–677 (1996).
112. M. P. A. Fisher, G. Grinstein and S. M. Girvin, Presence of quantum diffusion in two dimensions: Universal resistance at the superconductor-insulator transition, *Phys. Rev. Lett.* **64**, 587–590 (1990).
113. M. P. A. Fisher, Quantum phase transitions in disordered two-dimensional superconductors, *Phys. Rev. Lett.* **65**, 923–926 (1990).
114. B. I. Halperin, G. Refael and E. Demler, Resistance in superconductors, *Int. J. Mod. Phys. B* **24**, 4039–4080 (2010).
115. Y. Miyachi and S. Kurihara, Effect of finite screening length on charge Kosterlitz–Thouless–Berezinskii transition, *J. Phys. Soc. Jpn.* **69**, 2356–2357 (2000).
116. L. N. Bulaevskii, A. A. Sobyanin and D. I. Khomskii, Superconducting properties of systems with local pairs, *Zh. Eksp. Teor. Fiz.* **87**, 1490–1500 (1984) [*Sov. Phys. JETP* **60**, 856–862 (1984)].
117. M. Yu. Kagan, D. I. Khomskii and M. V. Mostovoy, Double-exchange model: phase separation versus canted spins, *Eur. Phys. J. B* **12**, 217–223 (1999).
118. M. Yu. Kagan, K. I. Kugel and D. I. Khomskii, Phase separation in systems with charge ordering, *JETP* **93**, 415–423 (2001).
119. K. I. Kugel, A. L. Rakhmanov, A. O. Sboychakov and D. I. Khomskii, Doped

orbitally ordered systems: Another case of phase separation, *Phys. Rev. B* **78**, 155113 (2008).
120. A. O. Sboychakov, K. I. Kugel, A. L. Rakhmanov and D. I. Khomskii, Phase separation in doped systems with spin-state transitions, *Phys. Rev. B* **80**, 024423 (2009).
121. E. Dagotto, Complexity in strongly correlated electronic systems, *Science* **309**, 257–262 (2005).
122. A. Glatz, I. S. Aranzon, T. I. Baturina, N. M. Chtchelkatchev and V. M. Vinokur, Self-organized superconducting textures in thin films, *Phys. Rev. B* **84**, 024508 (2011).
123. S. V. Syzranov, I. L. Aleiner, B. L. Altshuler and K. B. Efetov, Coulomb interaction and first-order superconductor-insulator transition, *Phys. Rev. Lett.* **105**, 137001 (2010).
124. N. S. Rytova, Screening potential of the point charge in a thin film, *Vestnik MSU.* **3**, 30–37 (1967) (in Russian).
125. A. V. Chaplik and M. V. Entin, Charged impurities in very thin layers, *Zh. Eksp. Teor. Fiz.* **61**, 2496–2503 (1971).
126. L. V. Keldysh, Coulomb interaction in thin semiconductor and semimetal films, *JETP Lett.* **29**, 658–661 (1979).
127. V. E. Dubrov, M. E. Levinstein and M. S. Shur, Permittivity anomaly in metal-dielectric transitions. Theory and simulation, *Zh. Eksp. Teor. Fiz.* **70**, 2014–2024 (1976) [*Sov. Phys. JETP* **43**, 1050–1056 (1976)].
128. D. Stauffer and A. Aharony, *Introduction to Percolation Theory*, 2nd edn. (Taylor and Francis, London 1994) (second printing).
129. T. G. Castner, N. K. Lee, G. S. Cieloszyk and G. L. Salinger, Dielectric anomaly and the metal-insulator transition in n-type silicon, *Phys. Rev. Lett.* **34**, 1627–1630 (1975).
130. K. F. Herzfeld, On atomic properties which make an element a metal, *Phys. Rev.* **29**, 701–705 (1927).
131. E. Shimshoni, A. Auerbach and A. Kapitulnik, Transport through quantum melts, *Phys. Rev. Lett.* **80**, 3352–3355 (1998).
132. H. F. Hess, K. DeConde, T. F. Rosenbaum and G. A. Thomas, Giant dielectric constants at the approach to the insulator-metal transition, *Phys. Rev. B* **25**, 5578–5580 (1982).
133. A. I. Shal'nikov, Superconducting thin films, *Nature (London)* **142**, 74 (1938).
134. A. I. Shal'nikov, Superconducting properties of thin metallic layers, *Zh. Eksp. Teor. Fiz.* **10**, 630–640 (1940).
135. R. C. Dynes, A. E. White, J. M. Graybeal and J. P. Garno, Breakdown of Eliashberg theory for two-dimensional superconductivity in the presence of disorder, *Phys. Rev. Lett.* **57**, 2195–2198 (1986).
136. P. Xiong, A. V. Herzog and R. C. Dynes, Superconductivity in ultrathin quench-condensed Pb/Sb and Pb/Ge multilayers, *Phys. Rev. B* **52**, 3795–3801 (1995).
137. H. Raffy, R. B. Laibowitz, P. Chaudhari and S. Maekawa, Localization and interaction effects in two-dimensional W-Re films, *Phys. Rev. B* **28**, 6607–6609 (1983).

138. J. M. Graybeal and M. R. Beasley, Localization and interaction effects in ultrathin amorphous superconducting films, *Phys. Rev. B* **29**, 4167–4169 (1984).
139. H. Kim, A. Ghimire, S. Jamali, T. K. Djidjou, J. M. Gerton and A. Rogachev, Effect of magnetic Gd impurities on the superconducting state of amorphous Mo-Ge thin films with different thickness and morphology, *Phys. Rev. B* **86**, 024518 (2012).
140. A. F. Hebard and M. A. Paalanen, Diverging characteristic lengths at critical disorder in thin-film superconductors, *Phys. Rev. Lett.* **54**, 2155–2158 (1985).
141. S. Okuma, T. Terashima and N. Kokubo, Anomalous magnetoresistance near the superconductor-insulator transition in ultrathin films of $a$-Mo$_x$Si$_{1-x}$, *Phys. Rev. B* **58**, 2816–2819 (1998).
142. Y. Qin, C. L. Vicente and J. Yoon, Magnetically induced metallic phase in superconducting tantalum films, *Phys. Rev. B* **73**, 100505(R) (1998).
143. C. A. Marrache-Kikuchi, H. Aubin, A. Pourret, K. Behnia, J. Lesueur, L. Bergé and L. Dumoulin, Thickness-tuned superconductor-insulator transitions under magnetic field in $a$-NbSi, *Phys. Rev. B* **78**, 144520 (2008).
144. T. I. Baturina, S. V. Postolova, A. Yu. Mironov, A. Glatz, M. R. Baklanov and V. M. Vinokur, Superconducting phase transitions in ultrathin TiN films, *EPL* **97**, 17012 (2012).
145. T. Suzuki, Y. Seguchi and T. Tsuboi, Fermi liquid effect on tricritical superconducting transitions in thin TiN films under the spin paramagnetic limitation, *J. Phys. Soc. Jpn.* **69**, 1462–1471 (2000).
146. K. Oto, S. Takaoka and K. Murase, Superconductivity in PtSi ultrathin films, *J. Appl. Phys.* **76**, 5339–5342 (1994).
147. S. Maekawa and H. Fukuyama, Localization effects in two-dimensional superconductors, *J. Phys. Soc. Jpn.* **51**, 1380–1385 (1982).
148. A. M. Finkel'stein, Superconducting transition temperature in amorphous films, *Pis'ma Zh. Eksp. Teor. Fiz.* **45**, 37–40 (1987) [*Sov. Phys. JETP Lett.* **45**, 46–49 (1987)]; Suppression of superconductivity in homogeneously disordered systems, *Physica B* **197**, 636–648 (1994).
149. A. F. Hebard and G. Kotliar, Possibility of the vortex-antivortex transition temperature of a thin-film superconductor being renormalized by disorder, *Phys. Rev. B* **39**, 4105–4109 (1989).
150. N. F. Mott and R. Peierls, Discussion of the paper by de Boer and Verwey, *Proc. Phys. Soc. London* **49**, 72 (1937).
151. N. F. Mott, The basis of the electron theory of metals, with special reference to the transition metals, *Proc. Phys. Soc. London, Ser. A.* **62**, 416 (1949).
152. P. W. Anderson, Absence of diffusion in certain random lattices, *Phys. Rev.* **109**, 1492–1505 (1958).
153. W. Kohn, Theory of insulating state, *Phys. Rev.* **133**, A171–A181 (1964).
154. E. Abrahams, P. W. Anderson, D. C. Licciardello and T. V. Ramakrishnan, Scaling theory of localization: Absence of quantum diffusion in two dimensions, *Phys. Rev. Lett.* **42**, 673–676 (1979).
155. M. Imada, A. Fujimori and Y. Tokura, Metal-insulator transitions, *Rev. Mod. Phys.* **70**, 1039–1263 (1998).
156. T. Brandes and S. Kettemann, *The Anderson Transition and Its Ramifications*

— *Localisation, Quantum Interference, and Interactions* (Berlin: Springer Verlag, 2003).
157. A. I. Larkin and V. M. Vinokur, Bose and vortex glasses in high temperature superconductors, *Phys. Rev. Lett.* **75**, 4666–4669 (1995).
158. D. M. Basko, I. L. Aleiner and B. L. Altshuler, Metal-insulator transition in a weakly interacting many-electron system with localized single-particle states, *Ann. Phys.* **321**, 1126–1205 (2006).
159. I. V. Gornyi, A. D. Mirlin and D. G. Polyakov, Interacting electrons in disordered wires: Anderson localization and low-$T$ transport, *Phys. Rev. Lett.* **95**, 206603 (2005).
160. I. S. Beloborodov, A. V. Lopatin and V. M. Vinokur, Coulomb effects and hopping transport in granular metals, *Phys. Rev. B* **72**, 125121 (2005).
161. G. M. Falco, T. Nattermann and V. L. Pokrovsky, Weakly interacting Bose gas in a random environment, *Phys. Rev. B* **80**, 104515 (2009).
162. I. L. Aleiner, B. L. Altshuler and G. V. Shlyapnikov, A finite-temperature phase transition for disordered weakly interacting bosons in one dimension, *Nature Phys.* **6**, 900–904 (2010).
163. A. V. Lopatin and V. M. Vinokur, Hopping transport in granular superconductors, *Phys. Rev. B* **75**, 092201 (2007).
164. N. M. Chtchelkatchev, V. M. Vinokur and T. I. Baturina, Hierarchical energy relaxation in mesoscopic tunnel junctions: Effect of a nonequilibrium environment on low-temperature transport, *Phys. Rev. Lett.* **103**, 247003 (2009).
165. N. M. Chtchelkatchev, V. M. Vinokur and T. I. Baturina, Nonequilibrium transport in superconducting tunneling structures, *Physica C* **470**, S935–S936 (2010).
166. N. M. Chtchelkatchev, V. M. Vinokur and T. I. Baturina, Low temperature transport in tunnel junction arrays: Cascade energy relaxation, in: *Physical Properties of Nanosystems*, eds. J. Bonca and S. Kruchinin, NATO Science for Peace and Security Series B: Physics and Biophysics (Springer Science+Business Media B.V., Dordrecht, 2011), Chapter 3, pp. 25–44.
167. L. B. Ioffe and M. Mézard, Disorder-driven quantum phase transitions in superconductors and magnets, *Phys. Rev. Lett.* **105**, 037001 (2010).
168. M. V. Feigel'man, L. B. Ioffe and M. Mézard, Superconductor-insulator transition and energy localization, *Phys. Rev. B* **82**, 184534 (2010).
169. G.-L. Ingold and Yu. V. Nazarov, Charge tunneling rates in ultrasmall junctions, in single charge tunneling, eds. H. Grabert and M. H. Devoret, NATO ASI, Ser. B, Vol. 294 (Plenum, New York, 1991), Chapter 2, pp. 21–107.
170. M. Tinkham, D. W. Abraham and C. J. Lobb, Periodic flux dependence of the resistive transition in two-dimensional superconducting arrays, *Phys. Rev. B* **28**, 6578–6581 (1983).
171. A. Allain, Z. Han and V. Bouchiat, Electrical control of the superconducting-to-insulating transition in graphene-metal hybrids, *Nature Materials* **11**, 590–594 (2012).
172. C. H. Ahn, A. Bhattacharya, M. Di Ventra, J. N. Eckstein, C. D. Frisbie, M. E. Gershenson, A. M. Goldman, I. H. Inoue, J. Mannhart, A. J. Millis, A. F.

Morpurgo, D. Natelson and J.-M. Triscone, Electrostatic modification of novel materials, *Rev. Mod. Phys.* **78**, 1185–1212 (2006).
173. K. A. Parendo, K. H. Sarwa, B. Tan and A. M. Goldman, Electrostatic and parallel-magnetic-field tuned two-dimensional superconductor-insulator transitions, *Phys. Rev. B* **73**, 174527 (2006).
174. D. Matthey, N. Reyren, J.-M. Triscone and T. Schneider, Electric-field-effect modulation of the transition temperature, mobile carrier density, and in-plane penetration depth of $NdBa_2Cu_3O_{7-\delta}$ thin films, *Phys. Rev. Lett.* **98**, 057002 (2007).
175. A. D. Caviglia, S. Gariglio, N. Reyren, D. Jaccard, T. Schneider, M. Gabay, S. Thiel, G. Hammerl, J. Mannhart and J.-M. Triscone, Electric field control of the $LaAlO_3/SrTiO_3$ interface ground state, *Nature* **456**, 624–627 (2008).
176. K. Ueno, S. Nakamura, H. Shimotani, A. Ohtomo, N. Kimura, T. Nojima, H. Aoki, Y. Iwasa and M. Kawasaki, Electric-field-induced superconductivity in an insulator, *Nature Materials* **7**, 855–858 (2008).
177. C. Bell, S. Harashima, Y. Kozuka, M. Kim, B. G. Kim, Y. Hikita and H. Y. Hwang, Dominant mobility modulation by the electric field effect at the $LaAlO_3/SrTiO_3$ interface, *Phys. Rev. Lett.* **103**, 226802 (2009).
178. J. Biscaras, N. Bergeal, A. Kushwaha, T. Wolf, A. Rastogi, R. C. Budhani and J. Lesueur, Two-dimensional superconductivity at a Mott insulator/band insulator interface $LaTiO_3/SrTiO_3$, *Nature Commun.* **1**, 89 (2010).
179. J. T. Ye, S. Inoue, K. Kobayashi, Y. Kasahara, H. T. Yuan, H. Shimotani and Y. Iwasa, Liquid-gated interface superconductivity on an atomically flat film, *Nature Materials* **9**, 125–128 (2010).
180. A. S. Dhoot, S. C. Wimbush, T. Benseman, J. L. MacManus-Driscoll, J. R. Cooper and R. H. Friend, Increased $T_c$ in electrolyte-gated cuprates. *Adv. Mater.* **22**, 2529–2533 (2010).

## Chapter 9

# BKT Physics with Two-Dimensional Atomic Gases

Zoran Hadzibabic

*Cavendish Laboratory, University of Cambridge, JJ Thomson Avenue, Cambridge CB3 0HE, United Kingdom*

Jean Dalibard

*Laboratoire Kastler Brossel, CNRS, UPMC, École normale supérieure, 24 rue Lhomond, 75005 Paris, France*

> In this chapter we review the recent advances in experimental investigation and theoretical understanding of Berezinskii–Kosterlitz–Thouless (BKT) physics in ultracold trapped atomic gases. We explain how to produce quasi two-dimensional (2D) samples by freezing out one degree of freedom with laser beams, and we detail the modeling of atomic interactions in this constrained geometry. We discuss some remarkable properties of the equation of state of dilute 2D Bose gases, such as its scale invariance. We also comment on the "competition" between the superfluid, BKT-driven phase transition in a homogenous 2D fluid of bosons, and the more conventional Bose–Einstein condensation that is expected for an ideal 2D Bose gas confined in a harmonic potential. We present some experimental techniques that allow one to measure the equation of state of the gas, monitor the occurrence of a BKT-driven phase transition, and observe thermally activated quantized vortices, which constitute the microscopic building block of the BKT mechanism.

## 9.1. Introduction

In recent years, ultracold atomic gases have emerged as a versatile playground for studies of fundamental many-body physics.[1–5] The appeal of these systems stems from a high degree of control available to the experimentalists in designing their properties such as the strength of inter-atomic interactions or the geometry of the external potential the atoms live in. In this respect, some of the most exciting developments have been associated with the possibility to trap and study quantum degenerate atomic samples in reduced dimensionality.[3]

The physics explored with atomic gases is believed to be universal in the sense that fundamentally analogous phenomena occur in a range of more "conventional" condensed-matter systems, such as liquid helium and solid state materials. At the same time, the experimental methods in atomic physics, and in particular, the probes of many-body behavior, are quite different from those used in the traditional condensed matter research. This often offers a complementary view on some long-studied problems.

Here, we will give an overview of the progress in theoretical understanding and experimental investigation of the Berezinskii–Kosterlitz–Thouless[6,7] (BKT) physics occurring in atomic Bose gases confined to two spatial dimensions (2D). In this context, these systems are conceptually closest to $^4$He films, in which BKT superfluid phenomena have first been observed more than 30 years ago.[8] In order to highlight the new insights into the physics of 2D Bose fluids, whether already experimentally demonstrated or still just theoretically anticipated, we will, in particular, focus on the differences between atomic gases and liquid helium films. These are primarily associated with three experimental aspects of ultracold atom research:

(1) In contrast to the spatially uniform helium films, ultracold atomic gases are produced in harmonic (in-plane) trapping potentials.
(2) The inter-particle interactions in a dilute atomic gas are usually significantly weaker than in liquid helium. Moreover the strength of interactions varies between different experimental setups, and can, in principle, even be continuously tuned in a single experiment, using so-called Fano–Feshbach resonances.[9]
(3) The experimental probes of atomic physics, in particular, matter-wave interferometry, can give direct access to the coherence (phase) properties of a 2D gas.

Before discussing the physics of 2D Bose gases in greater detail in the following sections, we briefly anticipate some salient points:

The importance of the in-plane harmonic trapping potential is clear from the very general point of view of the properties of phase transitions in 2D systems. Historically, the BKT theory was necessary to explain how a phase transition to a superfluid state can occur in a 2D Bose fluid at a non-zero temperature $T$, despite the fact that the Mermin–Wagner theorem[10,11] forbids the emergence of a true long range order (LRO) associated with Bose–

The description of interactions in terms of the single constant parameter $\tilde{g}$ is usually sufficient for addressing the physics of cold atomic gases. However, this description cannot be made fully consistent, as it is based on the notion of a contact interaction potential. Strictly speaking this potential is ill-defined in two dimensions because it leads to an ultra-violet divergence for the quantum two-body problem.[29] A more accurate way to evaluate the interaction energy consists of starting from the description of a binary collision between two atoms, with initial and final states whose $z$-dependence coincides with the ground state of the potential $U(z)$. The scattering amplitude $f(k)$ can then be written[28,30]

$$\frac{1}{f(k)} \approx \frac{1}{\tilde{g}} - \frac{1}{2\pi}\ln(ka_z) + \frac{i}{4}, \qquad (9.4)$$

where $\tilde{g}$ is still given by (9.3). The expression (9.1) for the interaction energy assumes that $f(k)$ is energy-independent and $\approx \tilde{g}$, which is legitimate when $\tilde{g} \ll 1$.

### 9.2.3.2. The healing length

The length scale associated with the interaction energy $gn$ is the *healing length* $\xi = 1/\sqrt{gn}$. It gives the characteristic size over which the density is significantly depleted around a point where $n$ vanishes. In particular, $\xi$ can be viewed as the typical size for a vortex core. The number of "missing atoms" due to the presence of a vortex is thus given by $\delta N \approx \pi \xi^2 n = \pi/\tilde{g}$. In the weakly interacting regime, $\tilde{g} \ll 1$ hence $\delta N \gg 1$, and a vortex corresponds to a significant density dip in the spatial distribution. In a strongly interacting gas $\delta N \sim 1$, which means that a vortex is an atomic-size defect in this case, with no detectable dip in the density profile of the fluid.

### 9.2.3.3. Fano–Feshbach resonances

A remarkable feature of cold atom physics is the possibility to control the interaction strength (i.e. the scattering length $a_s$) using a scattering resonance, also called a Fano–Feshbach resonance.[9] To reach the resonance, one tunes the ambient magnetic field $B$ in order to vary the energy of the ingoing channel of two colliding atoms with respect to another (closed) channel. The resonance occurs for a particular value $B_0$ of the ambient field, such that the asymptotic energy of the ingoing channel coincides with a bound state of the closed channel. Close to the resonance, the phase shift accumulated by the atom pair during the collision varies rapidly with $B$. The scattering length diverges at the resonance point $B = B_0$ and scales as $(B - B_0)^{-1}$ in its vicinity. In the 2D context a Fano–Feshbach resonance has been used[26] to tune the coupling strength $\tilde{g}$ of a cesium gas between 0.05 and 0.26.

### 9.2.4. *Direct implementation of the XY model*

Most of this chapter is devoted to the description and the analysis of experiments performed with continuous 2D atomic gases. Here, we briefly describe another cold-atom setup[31] that can be viewed as an atomic realization of a 2D array of Josephson junctions, and thus constitutes a direct implementation of the XY model. The experiment starts with a regular hexagonal array of parallel tubes created in a 2D optical lattice. Each tube can be labeled by its coordinates $i, j$ in the plane perpendicular to the tube axis. It contains a large number of atoms forming a condensate, which can be described by a classical complex field $\psi_{ij}$. In this limit, the amplitude $|\psi_{ij}|$ can be considered as constant and the only relevant dynamical variable is the phase of the field $\theta_{ij}$. The tunneling between adjacent tubes, described by the matrix element $J$, is equivalent to the coupling across a Josephson junction. The phase pattern $\theta_{ij}$ is probed by turning off the lattice, which allows neighboring condensates to merge and interfere. In particular, vortices in the initial phase distribution of the tubes are converted into density holes of the recombined condensate. The measurements of the surface density of these vortices reveal that it is a function of the ratio $J/T$ only, and this function is in good agreement with the BKT prediction.

## 9.3. Theoretical Understanding of Interacting Atomic Gases

We present in this section the main features of an atomic 2D Bose gas, restricting ourselves to the regime where interactions can be described by the dimensionless, energy-independent parameter $\tilde{g}$ given in Eq. (9.3). We first discuss the equation of state of a uniform gas and we comment on its various regimes. We then turn to the case of a harmonically trapped gas and show how this situation can be addressed using local density approximation. Finally we contrast the prediction of a superfluid BKT transition in the central region of the trap with that of a standard Bose–Einstein condensation, as expected for an ideal gas.

### 9.3.1. *The equation of state of a homogeneous 2D Bose gas*

The equation of state of a fluid consists in the expression of relevant state variables, such as the phase space density, in terms of other variables such as the chemical potential $\mu$ and the temperature $T$. Taking the ideal Bose gas as an example,[13] we can write its phase space density $\mathcal{D}$ as

$$\mathcal{D} \equiv n\lambda^2 = -\ln[1 - e^{-\mu/k_B T}], \tag{9.5}$$

where $\lambda = (2\pi\hbar^2/mk_B T)^{1/2}$ is the thermal wavelength.

The approximation that consists in describing the interactions in terms of the dimensionless parameter $\tilde{g}$ entails that the equation of state of the 2D gas $\mathcal{D}(\mu, T, \tilde{g})$ must present a scale invariance[32]: since $\mathcal{D}$ is dimensionless, it cannot be an arbitrary function of the two variables $\mu$ and $k_B T$ and it can depend only on the ratio of these two quantities. This (approximate) scale invariance occurs for all interaction strengths $\tilde{g} \ll 1$ and is specific to the two-dimensional case.

The determination of the function $\mathcal{D}(\mu/k_B T, \tilde{g})$ for arbitrary values of $\mu/k_B T$ requires a numerical calculation. Prokof'ev and Svistunov[32] computed it for $\tilde{g} \ll 1$ using a classical field Monte Carlo method. Two limiting cases can be recovered analytically. In the regime $\mathcal{D} \lesssim 1$, one can use the Hartree–Fock approximation[33]: interactions are treated at the mean-field level and each particle acquires the energy $2gn$ from its interaction with the neighboring atoms. The factor 2 here is due to the exchange term in the Hartree–Fock approach[a]; it reflects the fact that in the non-degenerate limit, bosons are bunched and $\langle n^2 \rangle = 2n^2$. We then replace $\mu$ by $\mu + 2gn$ in the ideal gas result (9.5) and obtain:

$$\mathcal{D} \lesssim 1: \quad \mathcal{D} \approx -\ln[1 - e^{-\mu/k_B T} e^{-\tilde{g}\mathcal{D}/\pi}]. \tag{9.6}$$

In practice, this approximate form of the equation of state is used to determine the value of $T$ and $\mu$ of the trapped gas, exploiting, via the local density approximation, the spatial variation of the atomic distribution at the edges of the cloud.

The opposite regime of a strongly degenerate gas can also be handled analytically. This is the so-called *Thomas–Fermi regime*, in which density fluctuations are suppressed ($\langle n^2 \rangle = n^2$) and kinetic energy is negligible. The only relevant energy is due to interactions, so that $\mu = gn$ or equivalently

$$\mathcal{D} \gg 1: \quad \mathcal{D} = \frac{2\pi}{\tilde{g}} \frac{\mu}{k_B T}. \tag{9.7}$$

### 9.3.2. *The superfluid transition*

#### 9.3.2.1. *Vortex pairing*

For a large enough phase space density, the interacting Bose gas is expected to become superfluid with a transition that is of the BKT type. We briefly

---

[a]The same factor of 2 occurs in the description of the Hanbury–Brown and Twiss effect.

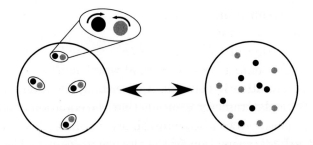

Fig. 9.1. Vortices exist in a two-dimensional Bose gas as possible thermal excitations. Left: In the thermodynamic limit and for a temperature below the critical point, vortices can only exist in the form of bound pairs with a total zero circulation. The gas has a non-zero superfluid component. Right: For a temperature larger than the critical point, the proliferation of free vortices brings the superfluid fraction to zero.

summarize the physical mechanism[6,7] at the heart of the transition, i.e. vortex pairing. Let us first assume that density fluctuations are essentially frozen in the gas, so that the dynamics is dominated by thermal phase fluctuations. We shall return to this assumption later in this section. Among possible sources for phase fluctuations, vortices play an essential role. For simplicity, we consider only vortices with unit topological charge, for which the local phase of the fluid has a winding of $\pm 2\pi$ around the vortex core. For a temperature below the transition point, the free energy of an isolated vortex is positive and infinite in the thermodynamic limit. In this regime, vortices can only exist in the form of bound pairs, containing two vortices of opposite circulation (Fig. 9.1). The one-body correlation function $g_1(r)$ decays algebraically at long distance, $g_1(r) \propto r^{-\alpha}$, and a non-zero superfluid density $n_s$ can appear in the fluid. The superfluid density can be related to the exponent characterizing the decay of $g_1$: $\alpha = 1/\mathcal{D}_s$, where $\mathcal{D}_s = n_s \lambda^2$. For a given total density $n$, the average distance between the two members of a pair is an increasing function of temperature. There exists a critical temperature at which this distance diverges, signaling a proliferation of free vortices. This plasma of free vortices, with positive and negative circulations, destroys superfluidity so that $n_s$ is zero above the critical point.

### 9.3.2.2. *The critical point for the superfluid transition*

One can show that the superfluid phase space density $\mathcal{D}_s = n_s \lambda^2$ undergoes a universal jump at the transition point[34]

$$\mathcal{D}_s = 4 \longrightarrow \mathcal{D}_s = 0. \tag{9.8}$$

This remarkable result is independent of the strength of interactions $\tilde{g}$. Note that this is an implicit result that does not allow one to determine unambiguously the position of the critical point. Indeed the total phase space density $\mathcal{D}$ and the temperature $T$ in (9.8) are at this point still unknown, and the only information is the obvious inequality $\mathcal{D} > 4$.

Numerical calculations[32] indicate that the *total* phase space density $\mathcal{D}$ is a smooth function of $\mu/k_\mathrm{B}T$ around the critical point. It shows no direct signature of the BKT phase transition, which is in line with the fact that this transition is of infinite order. Using a calculation of the superfluid density via a classical Monte Carlo simulation, Prokof'ev et al.[35] have determined the position of this critical point for $\tilde{g} \ll 1$

$$\mathcal{D}^\mathrm{crit} \approx \ln(380/\tilde{g}), \quad \frac{\mu^\mathrm{crit}}{k_\mathrm{B}T} \approx \frac{\tilde{g}}{\pi} \ln(13.2/\tilde{g}). \tag{9.9}$$

This result complements analytical predictions[36] obtained earlier for asymptotically low values of $\tilde{g}$. In the vicinity of the critical point, the behavior of the system is universal[32] in the sense that $\mathcal{D} - \mathcal{D}^\mathrm{crit}$ is a function of the single parameter $(\mu - \mu^\mathrm{crit})/(\tilde{g}\, k_\mathrm{B}T)$.

### 9.3.2.3. *Density fluctuations*

An important ingredient for the BKT mechanism to operate in an atomic gas is the reduction of density fluctuations for phase space densities $\mathcal{D}$ around $\mathcal{D}^\mathrm{crit}$. For $\mathcal{D} < \mathcal{D}^\mathrm{crit}$, the gas with reduced density fluctuations constitutes a presuperfluid medium, in which the dynamics is essentially governed by the phase fluctuations dynamics. We quote again the results of the classical field Monte Carlo calculation[35]:

$$\frac{n_\mathrm{qc}}{n} = \frac{7.16}{\ln(380/\tilde{g})} \tag{9.10}$$

at the critical point. Here the "quasi-condensate" density $n_\mathrm{qc}$, defined by $n_\mathrm{qc}^2 = 2n^2 - \langle n^2 \rangle$, captures the reduction of density fluctuations: (i) $n_\mathrm{qc}$ vanishes in the ideal gas limit where atom bunching leads to $\langle n^2 \rangle = 2n^2$. (ii) $n_\mathrm{qc} = n$ when density fluctuations are negligible ($\langle n^2 \rangle = n^2$). The result (9.10) shows that for practical values of $\tilde{g}$ in atomic physics (say larger than 0.01), the quasi-condensate density is close to the total density. One therefore expects that density fluctuations are strongly suppressed in the vicinity of the transition point. Note that with this definition, the notion of a quasi-condensate makes no direct reference to the coherence properties of the gas.

### 9.3.3. *The case of trapped gases*

9.3.3.1. *The local density approximation*

Experiments with atomic gases are mostly performed in the presence of a harmonic confinement $V(\boldsymbol{r})$ in the $xy$ plane. The local density approximation (LDA) amounts to assuming that the thermodynamic variables of the gas at any point $\boldsymbol{r}$ in the trap, such as the phase space density $\mathcal{D}(\boldsymbol{r})$, take the same value as in a homogenous system with the same temperature $T$ and a chemical potential $\mu - V(\boldsymbol{r})$. Therefore, instead of being a nuisance, the non-homogeneity of the trapping potential is often an asset; it provides in a single-shot measurement the equation of state of the fluid for chemical potentials ranging from $-\infty$ (at the edges of the cloud) to $\mu$ (in the centre).

The LDA is valid if the various thermodynamic quantities do not change significantly over the length scales set by the healing length $\xi$, the thermal wavelength $\lambda$ and the characteristic size of the ground state of the single particle motion in the $xy$ plane $a_{xy} = [\hbar/(m\omega_{xy})]^{1/2}$. It can be tested experimentally by measuring at given $T$ the radial density distributions of the gas for several values of the particle number, hence several values of the chemical potential ($\mu_j$), and by checking that all distributions get superimposed when plotted as function of $\mu_j - V(r)$. LDA was also validated using Quantum Monte Carlo simulations.[37] These simulations were performed for a number of bosons and a trap geometry comparable to the experimental parameters, taking into account, in particular, the residual excitation of the $z$ degree of freedom. The superfluid fraction was obtained from the value of the non-classical moment of inertia. The results were in good agreement with the predictions[32] for a homogenous system combined with LDA, and the small residual deviations have been analyzed by various authors.[38-41]

9.3.3.2. *The competition between BKT physics and Bose–Einstein condensation*

The simple fact of placing the atomic gas in a trap immediately raises the question of possible finite-size effects. This question also emerges for 3D trapped gases, but it is even more relevant in 2D, since it is well known that approaching the thermodynamic limit is an extremely difficult task in this case. In the context of 2D magnetism, Bramwell and Holdsworth[42] wrote the famous statement: "With a magnetization at the BKT critical point smaller than 0.01 as a reasonable estimate for the thermodynamic limit, the sample would need to be bigger than the state of Texas for the Mermin–Wagner theorem to be relevant!" Needless to say, cold atomic gases are very far from this limit. The ratio between the characteristic size of the gas and the microscopic lengths $\xi$ and $\lambda$ does not exceed a few hundreds, and finite size effects cannot be ignored.

The situation is made even more intricate by the presence of the harmonic confinement $V(r) = m\omega^2 r^2/2$ in the $xy$ plane (in this section, we set $\omega \equiv \omega_{xy}$ for simplicity). In the ideal gas case where no BKT transition is expected, an assembly of bosons can undergo a standard Bose–Einstein condensation (BEC) in the thermodynamic limit,[43] where $N \to \infty$, $\omega \to 0$, with $N\omega^2$ kept constant. The BEC transition takes place when the temperature is below the critical value

$$k_B T_{\text{ideal}}^{\text{crit}} = (\sqrt{6}/\pi)\,\hbar\omega\,N^{1/2}, \qquad (9.11)$$

or equivalently when $N$ exceeds the critical atom number

$$N_{\text{ideal}}^{\text{crit}} = (\pi^2/6)(k_B T/\hbar\omega)^2 \qquad (9.12)$$

for a gas maintained at fixed temperature $T$. This result can be inferred from (9.5) by (i) replacing $\mu$ by the local chemical potential $\mu - V(r)$ in the expression for the density:

$$n(r)\lambda^2 = -\ln[1 - e^{-[\mu - V(r)]/k_B T}], \qquad (9.13)$$

(ii) requiring that the chemical potential is equal to $V(0) = 0$ at the condensation point, and (iii) calculating $N = \int n\, d^2r$.

We now switch to the case of an interacting Bose gas in a harmonic potential, and meet the natural question: when increasing the atom number in a gas at fixed temperature, what is (are) the phase transition(s) that one may encounter: a Bose–Einstein transition, with a possible shift of the critical atom number with respect to (9.12), or (and) a BKT-driven transition? To address this question, we first recall the simpler 3D case. In a uniform ideal gas, BEC occurs at the density $n_{3,\text{ideal}}^{\text{crit}} = 2.612\,\lambda^{-3}$. In a harmonically confined ideal gas, BEC occurs when the density at the center reaches the same threshold, which requires the presence of $N_{3,\text{ideal}}^{\text{crit}} = 1.202\,(k_B T/\hbar\omega)^3$ atoms in the trap. In the presence of weak repulsive interactions, BEC occurs for (approximately) the same central density[1] $n_{3,\text{ideal}}^{\text{crit}}$. Achieving this density requires to put more atoms in the trap than in the ideal case since repulsion between atoms tends to broaden the spatial distribution. Therefore in the 3D case, the main effect of interactions is[1] an upwards shift of the critical atom number with respect to $N_{\text{ideal}}^{\text{crit}}$ at fixed $T$, or equivalently a downwards shift of the critical temperature with respect to $T_{\text{ideal}}^{\text{crit}}$ at fixed $N$.

In the 2D case, the ideal gas BEC in a harmonic trap is a marginal and fragile effect, in the sense that it is associated with an infinite spatial density $n$ (hence an infinite phase space density $\mathcal{D}$) in the center. This result is a direct consequence of Eq. (9.13) when one lets $\mu \to 0$. Now, in the presence of repulsive interactions between atoms, the divergence of the spatial density in the center cannot occur anymore, and one cannot hope to recover a situation similar to the ideal case by simply increasing the atom number. Therefore, our first conclusion is that for a 2D trapped gas, the standard BEC does not survive in the presence of interactions, at least when the discrete nature of the single-particle eigenstates can be omitted (i.e. when $k_B T \gg \hbar\omega$).

On the other hand, the BKT superfluid transition requires a non-infinite phase space density to occur, and we thus expect that it can still take place in a trapped Bose gas with repulsive interactions. To determine in which conditions this transition is observable, we can rely on LDA. We use Eq. (9.9) to provide the density threshold $n^{\mathrm{crit}} = \ln(380/\tilde{g})\,\lambda^{-2}$ at which the gas in the center of the trap becomes superfluid. Using the Hartree–Fock result (9.6), this criterion can be transposed[44] into the critical atom number[b]

$$N_{\mathrm{BKT}}^{\mathrm{crit}} = N_{\mathrm{ideal}}^{\mathrm{crit}} \left[1 + \frac{3\tilde{g}}{\pi^3}(\mathcal{D}^{\mathrm{crit}})^2\right]. \qquad (9.14)$$

It is noteworthy that the critical atom number for observing a BKT transition in a harmonic trap has the same scaling with respect to $T$ and $\omega$ as the critical number for BEC of an ideal gas. In addition Eq. (9.14) implies that the BKT threshold for an interacting gas requires more particles than the BEC threshold for an ideal gas at the same temperature.

To summarize we expect that a single, BKT-driven, phase transition takes place in a trapped interacting 2D gas.[13] When taking the limit $\tilde{g} \to 0$, the critical phase space density $\mathcal{D}^{\mathrm{crit}}$ given in Eq. (9.9) for a uniform system tends to infinity, and the critical atom number $N_{\mathrm{BKT}}^{\mathrm{crit}}$ for a trapped gas tends to the ideal gas threshold (9.12) for BEC. The ideal gas BEC in a 2D harmonic potential can thus be viewed as a special, non-interacting limit of the BKT transition. We also note that in the presence of interactions, there exists in addition to the BKT transition, a cross-over in which the density fluctuations gradually decrease. As for the homogeneous case this cross-over mostly takes place for phase space densities lower than $\mathcal{D}^{\mathrm{crit}}$, and it produces a presuperfluid medium in which the BKT mechanism can operate.

---

[b]Note that Eq. (9.14) is an upper bound on $N_{\mathrm{BKT}}^{\mathrm{crit}}$ because it assumes that the interaction energy per particle is $2gn$. It is, in reality, smaller due to the reduction of density fluctuations.

### 9.3.3.3. *The condensed fraction*

Due to finite-size effects, the conclusion that we just reached concerning the non-existence of a standard BEC transition is still compatible with a significant condensed fraction in the trapped gas. The condensed fraction $\Pi_0$ is defined as the largest eigenvalue of the one-body density matrix $\hat\rho$, with $\langle \mathbf{r}|\hat\rho|\mathbf{r}'\rangle = g_1(|\mathbf{r}-\mathbf{r}'|)$ in a uniform system. When $g_1(r)$ is significantly different from zero for values of $r$ comparable to the characteristic radius of the trapped gas $R$, then we expect that this gas exhibits a detectable macroscopic quantum coherence.

Let us first address the case where the superfluid threshold (9.14) has been reached and suppose that $R$ stands for the radius of the disk where the superfluid fraction is non-zero. Since $g_1$ decays algebraically in the superfluid region with an exponent that is smaller than $1/4$, we have $\Pi_0 \approx g_1(R) \gtrsim (\xi/R)^{1/4}$. The radius $R$ can be estimated using the Thomas–Fermi approximation[1]: in the limit where the kinetic energy is negligible, the spatial density $n(r)$ in the superfluid disk is obtained from the simple relation $gn(r) + V(r) = \mu$, hence $R = a_{xy}(4\tilde{g}N/\pi)^{1/4}$, where $N$ is the number of atoms in the disk. Taking as typical values $a_{xy} = 3\,\mu$m, $\tilde{g} = 0.1$, $N = 2\times 10^4$, we find $R \approx 20\,\mu$m. The healing length in the center of the trap $\xi$ is $\approx 0.6\,\mu$m, so that $\Pi_0 \gtrsim 0.4$.

The emergence of a detectable condensed fraction can significantly precede the point where the critical density (9.9) is reached in the center of the trap. This is due to the particular behavior of the correlation length $\ell$ characterizing the exponential decay of $g_1$ in the normal (non-superfluid) region. In a uniform 2D fluid, the length $\ell$ is predicted to diverge exponentially at the critical point ($\ell \sim \exp[b/(T-T_c)^{1/2}]$, where $b$ is a constant), hence $\ell$ may take a value comparable to the size $R$ of the trapped cloud notably before the threshold (9.14) is reached. One can thus predict that the transition from a fully thermal to a significantly coherent gas is actually a crossover, whose relative width $\Delta T/T_c$ ranges from 5% to 20% for realistic trap parameters.[13] An evidence for this crossover is provided by classical field Monte Carlo calculations for a trapped gas.[45,46]

## 9.4. Experimental Investigation of 2D Physics in Atomic Gases

In ultracold atom research, essentially all raw data take the form of images of atomic clouds. The most commonly used technique is absorption imaging, whereby a laser beam tuned close to resonance with an atomic transition is passed through the cloud, and the "shadow" of the cloud is imaged on a camera. From the fractional attenuation of the laser beam at any point in the image plane, one infers the column atomic density along the "line of sight" of the laser. While this probe always couples to the atomic density, what the observed density distribution actually represents depends on what one does with the atomic cloud just before imaging it. We can distinguish three typical scenarios which are relevant for our discussion here:

(1) If one takes an image of a trapped atomic cloud then one indeed simply obtains the (column) density distribution.
(2) In the "time of flight" (TOF) technique, the cloud is released from the trap some time before the image is taken, and the atoms are allowed to freely expand. For sufficiently long expansion times (typically tens of milliseconds), the cloud becomes significantly larger than its initial in-trap size. In this limit, the observed density distribution reflects the *momentum* distribution of the atoms.
(3) Finally, in matter-wave interference experiments, two initially separate clouds, or two parts of the same cloud, are made to overlap during the expansion. In this way, the phase properties of the wave function(s) describing the cloud(s) are converted into density information which can be extracted by imaging. This option is particularly appealing for studies of 2D physics since it allows direct visualization of both phonons and vortices, as well as quantitative measurements of correlation functions.

### 9.4.1. *Measurements of density and momentum distributions*

An obvious way to experimentally obtain the equilibrium density distribution of a trapped 2D gas is to take an absorption image along the direction ($z$) perpendicular to the planar gas. In contrast to the case of a 3D gas, in the 2D case, the line-of-sight integration by the imaging beam does not result in any loss of information, since along $z$ the gas is simply in the ground-state of the trapping potential $U(z)$.

Besides the average density distribution, one can in principle, in this way also study the density fluctuations which are suppressed in the degenerate regime. Moreover, taking advantage of the non-uniform density profile in a harmonic trap, from the spatial variation of the (average) density one can deduce the compressibility of the gas, and hence quantify the level of density fluctuations in that way. Qualitatively, the compressibility of the gas is directly related to the interaction energy density, and hence reveals whether $\langle n^2 \rangle$ is closer to $2\langle n \rangle^2$ (corresponding to a fully fluctuating gas) or to $\langle n \rangle^2$ (corresponding to suppressed density fluctuations).

The in-trap density measurements however do not provide information on the phase fluctuations, neither phonons nor vortices. Vortices, in principle, do have a signature in the density distribution, but in practice, the size $\xi$ of a vortex core is smaller than a micrometer, which is usually too small compared to the imaging resolution to be directly observed.

To measure the momentum distribution, one can use the time-of-flight (TOF) technique. In its simplest version, a TOF measurement entails a sudden switch-off of all confining potentials, and a subsequent expansion of the gas along all three spatial dimensions ("3D TOF"). For studies of 2D gases, this technique has two qualitatively different variants:

- One can choose to turn off only the confining potential $V(r)$ in the $xy$ plane, while leaving the tight confinement $U(z)$ on during the expansion. In this "2D TOF" approach the gas lives in two dimensions at all times.
- One can switch off the confinement $U(z)$ along the initially strongly confined direction $z$, while keeping the potential $V(r)$ in the $xy$ plane ("1D TOF").

After long TOF, the observed density distribution reflects the momentum distribution in the cloud. However, this momentum distribution is, in general, not the same as the original momentum distribution in a trapped gas. The reason for this is that the interaction energy of a trapped gas is converted into (additional) kinetic energy during the early stages of the expansion. We can now already anticipate that this conversion of interaction into kinetic energy is different, if the gas expands along one, two or three directions.

### 9.4.1.1. *In situ images*

A series of experiments have investigated the equilibrium state of an atomic 2D Bose gas using in-trap images.[25,26,47] The measured density profiles are in good agreement with the prediction of classical field Monte Carlo calculations, associated with LDA. The scale invariance of the equations of state for the phase space density[26] and the pressure[47] of the gas have been confirmed, which also allows one to deduce the equation of state for the entropy of the gas.[47] In addition, one can extract from these measurements the compressibility of the gas[25,26] and thus directly confirm the suppression of the density fluctuations.

It is important to stress that in these *in situ* measurements, nothing remarkable happens at the BKT transition point when the atom number reaches $N_{\text{BKT}}^{\text{crit}}$. Also for an atom number well above this critical value, the average density distribution varies smoothly in space, including around the radius $r$ where the phase space density equals the critical value (9.9). As already mentioned, this absence of singularity is a consequence of the fact that the BKT transition is of infinite order.

### 9.4.1.2. *2D TOF*

In 2D TOF something remarkably simple happens — the density distribution evolves in a self-similar way[48] and the picture of the cloud simply gets rescaled. Explicitly, if the equilibrium in-trap density profile is $n_{\rm eq}(\boldsymbol{r})$, then in 2D TOF we get:

$$n(\boldsymbol{r},t) = \eta_t^2\, n_{\rm eq}(\eta_t \boldsymbol{r}), \quad \text{where} \quad \eta_t = (1+\omega^2 t^2)^{-1/2}, \qquad (9.15)$$

where $t$ is the expansion time. Such perfect scaling is a direct consequence of the fact that the coupling strength $\tilde{g}$ is a dimensionless constant, and it entails that purely 2D measurements, in principle, provide identical information, whether they are performed in-trap or in 2D TOF.

An evidence for this scaling has been obtained experimentally.[17] Small deviations (not yet observed) are expected[49,50] because of the logarithmic dependence of the coupling strength with respect to the energy of the particles [Eq. (9.4)], which itself results from the needed regularization of the contact interaction potential in 2D.

### 9.4.1.3. *1D and 3D TOFs*

1D and 3D TOFs provide a way to observe directly the occurrence of the BKT transition in a trapped gas. Indeed in this case, the high trap anisotropy results in an expansion of the cloud along $z$ in a time $\sim \omega_z^{-1}$ that is short compared to any time constant associated with the motion in the $xy$ plane. We can then conceptually divide the expansion into two stages: in the initial stage, all the interaction energy is taken out of the system by the fast expansion along $z$, before almost any motion of the atoms in the $xy$ plane occurs, and without affecting the in-plane momentum distribution. The subsequent slower in-plane evolution of the density distribution thus corresponds either to the evolution in the potential $V(r)$ (1D TOF) or to the free expansion of an ideal gas (3D TOF). In both cases, the initial momentum distribution is the same as in the trapped gas.

With both methods, one observes that the momentum distribution becomes sharply peaked for atom numbers above a critical value which is in good agreement with the BKT prediction[15,25,51] [see e.g. Fig. 9.2(a)]. Since the momentum distribution is the Fourier transform of $g_1$, this sharp peak

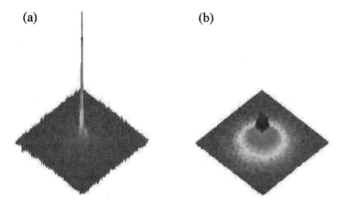

Fig. 9.2. (a) Momentum distribution of a 2D Bose gas, for an atom number above the critical value. (b) *In situ* density distribution for near-identical conditions (image from Ref. 25 courtesy of Eric Cornell).

signals the appearance of a significant condensed fraction, with a phase ordering that extends over a large fraction of the trapped cloud. The signaling of the BKT transition via the appearance of this sharp peak in momentum space is very reminiscent of the emergence of a non-zero magnetization at the BKT critical point for a finite size XY system.[42]

In the context of cold atomic gases, it is particularly interesting to perform both in-trap and 1D/3D TOF density/momentum measurements on identically prepared samples.[25] In this case, the 1D/3D measurements provide the striking qualitative signature of the phase transition [Fig. 9.2(a)], while the in-trap measurements [Fig. 9.2(b)] allow quantitative confirmation that the condensate momentum peak emerges at the critical phase space density predicted by the BKT theory.

### 9.4.2. *Visualizing phase fluctuations*

To directly visualize the phase fluctuations in a 2D Bose gas, both the phonons which (in the thermodynamic limit) destroy the LRO even at very low $T$, and the vortices which drive the BKT transition, in atomic physics, one can employ matter-wave interference experiments. It is in principle possible to interfere two different parts of the same 2D sample,[15] but here, we will illustrate the key ideas by considering the more intuitive case of interference between two parallel, independent 2D clouds.[24] This scenario is illustrated in Fig. 9.3.

We consider two planes of atoms which are separated by a distance $d_z$ along $z$. The potential barrier between the planes is sufficiently large so

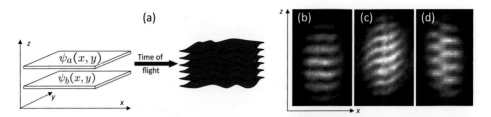

Fig. 9.3. (a) Principle of an experiment in which two independent 2D Bose gases, described by fluctuating macroscopic wave functions $\psi_a$ and $\psi_b$, are released from the potential that was initially confining them along the $z$ direction. The overlap of the expanding matter waves leads to interference patterns that are observed by looking at the absorption of a probe laser beam. (b–d) Examples of observed interference patterns.[24] The contrast and the waviness of the patterns provide information on the phase distributions of the initial states $\psi_{a,b}$.

that there is no quantum tunneling between the two planes and the phase fluctuations in them are independent. At the same time, the planes can be considered to be statistically identical, i.e. they have very similar temperature and atomic density. Such a configuration can readily be realized in a one-dimensional optical lattice, as explained in Sec. 9.2.1. Each plane $(a, b)$ is thus described by the wave function $\psi_{a/b}(x, y) = |\psi_{a/b}| e^{i\theta_{a/b}(x,y)}$, where $|\psi_a| \approx |\psi_b|$, while $\theta_a(x, y)$ and $\theta_b(x, y)$ are independently fluctuating.

If we perform a 3D TOF of sufficiently long duration $t$, the two clouds expand into each other and interfere. Moreover, there exists an intermediate range of times $t$ for which we can still assume that no expansion in the $xy$ plane has taken place. In this way, we obtain an interference pattern in which the local interference phase depends on the local difference $\theta_a(x, y) - \theta_b(x, y)$. More explicitly, neglecting the trivial global envelope function (which describes the expansion along $z$ of each cloud) the total density distribution after TOF is:

$$n \propto |\psi_a|^2 + |\psi_b|^2 + (\psi_a \psi_b^* e^{i2\pi z/D_z} + \text{c.c.}), \quad (9.16)$$

where $D_z = ht/md_z$ is the period of the interference fringes. In practice, this density distribution is then projected onto the $xz$ plane by absorption imaging along the $y$ direction.

### 9.4.2.1. Phonons

At very low $T$, the functions $\theta_a$ and $\theta_b$ are almost constant, and hence so is their difference. Although this difference is not known, or even defined, prior to the actual measurement, the interference of two such planes with quasi-constant phases inevitably leads to straight interference fringes, as shown in Fig. 9.3(b). The absolute difference between $\theta_a$ and $\theta_b$ simply corresponds to the absolute position (along $z$) of the fringes on the camera, and varies randomly between different experimental realizations.

On the other hand, if the phase describing each plane slowly varies in space, this results in the meandering interference pattern, such as shown in Fig. 9.3(c). Experimentally, it is observed that the fringes indeed gradually become more wavy as the temperature is increased.

#### 9.4.2.2. Vortices

An important outcome of the experiments[24] depicted in Fig. 9.3 is the direct visual evidence for the proliferation of free vortices near the BKT transition. If a free vortex is present in one of the two interfering clouds, while the phase of the other cloud varies smoothly across the same region in the $xy$ plane, the vortex appears as a sharp dislocation in the interference pattern. An example of such a dislocation is shown in Fig. 9.3(d). The reason for the dislocation is that as one moves along $x$, the relative phase $\theta_a - \theta_b$ suddenly jumps by $\pi$ at the position of the vortex.

These $\pi$-dislocations are absent at low temperature and the probability of their occurrence rapidly rises as one approaches the BKT critical temperature from below. Once the critical temperature is crossed and the gas becomes strongly (phase-)fluctuating, the interference pattern is no longer observable.

### 9.4.3. Correlation functions

From two-plane interference experiments one can also extract off-diagonal correlation functions, by studying the spatial variation of the phase of the complex interference contrast in Eq. (9.16).

A particularly elegant method for quantitative analysis of correlation functions was proposed by Polkovnikov et al.,[52] and we outline it here. As we mentioned above, the interference pattern can never be measured at a single point in the $xy$ plane, but is always integrated by imaging along $y$. Additionally, one can choose to integrate it over an arbitrary distance along $x$. Qualitatively, we can see that the contrast of the resulting integrated image will decay with the integration area whenever the fringes show some phase fluctuations. The more fluctuating the gases are, the more rapid will the decay of the contrast be. More quantitatively, we can define an average integrated contrast, obtained by averaging over many experimental realizations:

$$C^2(A) = \frac{1}{A^2} \left\langle \left| \int_A \psi_a(\mathbf{r}) \psi_b^*(\mathbf{r}) \, d^2r \right|^2 \right\rangle, \quad (9.17)$$

where $A$ is the integration area, which here we assume to be square. Since $\psi_a$ and $\psi_b$ are uncorrelated and have statistically identical fluctuations, for a uniform system, this factors out to yield:

$$C^2(A) = \frac{1}{A} \int_A |g_1(\mathbf{r})|^2 \, d^2r, \quad (9.18)$$

where $g_1(\mathbf{r})$ is the first order correlation function describing either of the two planes. Similarly, from higher moments of the distribution of $C^2(A)$ one can extract higher order correlation functions.

Focusing on the first order correlation function, we can distinguish two qualitatively different cases. In the normal state $g_1$ is rapidly (exponentially) decaying, so for sufficiently large $A$ we get $C^2(A) \propto A^{-1}$. On the other hand, in the superfluid state $g_1$ decays only slowly (algebraically), $g_1 \propto r^{-\alpha}$, with $\alpha = 1/(n_s \lambda^2)$. We thus get $C^2(A) \propto A^{-\alpha}$. Since $\alpha \leq 1/4$, at the BKT transition, the exponent characterizing the decay of $C^2(A)$ jumps between 1 and 1/4. This jump corresponds to the universal jump in the superfluid density,[52] and is a direct coherence analogue of the jump observed in the transport experiments with liquid helium films.[8]

Although adapting these calculations to a harmonically trapped gas is a difficult task,[53] one can use Eq. (9.18) as a qualitative guide. Experimental measurements of $C^2(A)$ indeed qualitatively agree with the expected change from exponential to algebraic decay of $g_1$ in the vicinity of the BKT transition temperature.[24] Specifically, within experimental precision, this change coincides with the proliferation of interference defects signaling the presence of free vortices in 2D clouds.

## 9.5. Outlook

So far we have advertised the measurements performed with 2D atomic gases as complementary to those performed on other physical systems displaying BKT physics. In particular, at the level of theory, coherence measurements on atomic systems are a direct analogue of the transport measurements on liquid helium films. It is however important to also stress that superfluidity in the traditionally defined transport sense has not yet been observed in these systems. In fact, a quantitative measurement of the superfluid density, is more generally, an open question in the field of ultracold atomic gases. This is an active area of current research,[54,55] and one big hope is that in the near feature it will be possible to perform both types of measurements on identically prepared atomic samples, thus providing a direct experimental comparison of different theoretical definitions of superfluidity.

To conclude, we briefly mention two further experimental possibilities that have recently emerged in the field of ultracold atomic gases, and could in the future be applied to 2D systems:

First, atomic gases are naturally very "clean", but it is possible to controllably introduce various degrees of disorder into these systems.[56] A particularly flexible approach is to use either quasi-periodic or truly random optical potentials, so that the strength of the disordered potential is simply proportional to the laser intensity. In the case of random, laser speckle potentials, the disorder can be truly microscopic in the sense that the correlation length of the random potential can be comparable to the healing length $\xi$. Recently, application of optically induced disorder allowed studies of Anderson localization[57,58] in clouds close to zero temperature. In the future, it should be equally possible to apply similar techniques to 2D gases at non-zero temperature in order the study the effects of disorder on the BKT transition.

Second, atomic gases exhibit comparatively slow dynamics, with characteristic timescales ranging from milliseconds to seconds; this allows real-time studies of dynamical effects, such as transient regimes following a quench of some Hamiltonian parameter. In the context of BKT physics, one could, for example, study the emergence of superfluidity after suddenly crossing the critical point by changing the strength of interactions in the gas. Moreover, theoretical analogies between the evolution (decay) of coherence in real space and in time have been proposed.[59] In a bi-layer system such as that introduced in Sec. 9.4.2, one can dynamically control the strength of tunnel-coupling between the two planes, so that the phases of their wave functions are initially locked together and then suddenly decoupled. By studying the interference of the two planes at various times after the quench, one could then study the evolution of their relative phases under the influence of thermal fluctuations. In the superfluid regime, the integrated interference contrast is expected to decay algebraically at long times, as $t^{-\zeta}$ with $\zeta \propto T$, in analogy with the spatial decay of the first order correlation function $g_1$ in an equilibrium 2D gas.

# References

1. F. S. Dalfovo, L. P. Pitaevkii, S. Stringari and S. Giorgini, Theory of Bose–Einstein condensation in trapped gases, *Rev. Mod. Phys.* **71**, 463 (1999).
2. M. Lewenstein, A. Sanpera, V. Ahufinger, B. Damski, A. S. De and U. Sen, Ultracold atomic gases in optical lattices: Mimicking condensed matter physics and beyond, *Adv. Phys.* **56**, 243 (2007).
3. I. Bloch, J. Dalibard and W. Zwerger, Many-body physics with ultracold gases, *Rev. Mod. Phys.* **80**, 885 (2008).
4. S. Giorgini, L. P. Pitaevskii and S. Stringari, Theory of ultracold atomic Fermi gases, *Rev. Mod. Phys.* **80**, 1215 (2008).
5. N. R. Cooper, Rapidly rotating atomic gases, *Adv. Phys.* **57**, 539 (2008).
6. V. L. Berezinskii, Destruction of long-range order in one-dimensional and two-dimensional system possessing a continous symmetry group — ii. quantum systems, *Soviet Physics JETP* **34**, 610 (1971).
7. J. M. Kosterlitz and D. J. Thouless, Ordering, metastability and phase transitions in two dimensional systems, *J. Phys. C: Solid State Phys.* **6**, 1181 (1973).
8. D. J. Bishop and J. D. Reppy, Study of the superfluid transition in two-dimensional $^4$He films, *Phys. Rev. Lett.* **40**, 1727 (1978).
9. C. Chin, R. Grimm, P. Julienne and E. Tiesinga, Feshbach resonances in ultracold gases, *Rev. Mod. Phys.* **82**, 1225 (2010).
10. N. D. Mermin and H. Wagner, Absence of ferromagnetism or antiferromagnetism in one- or two-dimensional isotropic Heisenberg models, *Phys. Rev. Lett.* **17**, 1133 (1966).
11. P. C. Hohenberg, Existence of long-range order in one and two dimensions, *Phys. Rev.* **158**, 383 (1967).
12. R. Grimm, M. Weidemüller and Y. B. Ovchinnikov, Optical dipole traps for neutral atoms, *Adv. At. Mol. Opt. Phys.* **42**, 95 (2000).
13. Z. Hadzibabic and J. Dalibard, Two-dimensional Bose fluids: An atomic physics perspective, *Rivista del Nuovo Cimento* **34**, 389 (2011).
14. A. Görlitz, J. M. Vogels, A. E. Leanhardt, C. Raman, T. L. Gustavson, J. R. Abo-Shaeer, A. P. Chikkatur, S. Gupta, S. Inouye, T. Rosenband and W. Ketterle, Realization of Bose–Einstein condensates in lower dimensions, *Phys. Rev. Lett.* **87**, 130402 (2001).
15. P. Cladé, C. Ryu, A. Ramanathan, K. Helmerson and W. D. Phillips, Observation of a 2D Bose gas: From thermal to quasicondensate to superfluid, *Phys. Rev. Lett.* **102**, 170401 (2009).
16. N. L. Smith, W. H. Heathcote, G. Hechenblaikner, E. Nugent and C. J. Foot, Quasi-2D confinement of a BEC in a combined optical and magnetic potential, *J. Phys. B* **38**, 223 (2005).
17. S. P. Rath, T. Yefsah, K. J. Günter, M. Cheneau, R. Desbuquois, M. Holzmann, W. Krauth and J. Dalibard, Equilibrium state of a trapped two-dimensional Bose gas, *Phys. Rev. A* **82**, 013609 (2010).
18. C. Orzel, A. K. Tuchmann, K. Fenselau, M. Yasuda and M. A. Kasevich, Squeezed states in a Bose–Einstein condensate, *Science* **291**, 2386 (2001).
19. S. Burger, F. S. Cataliotti, C. Fort, P. Maddaloni, F. Minardi and M. Inguscio,

Quasi-2D Bose–Einstein condensation in an optical lattice, *Europhys. Lett.* **57**, 1 (2002).
20. M. Köhl, H. Moritz, T. Stöferle, C. Schori and T. Esslinger, Superfluid to Mott insulator transition in one, two, and three dimensions, *J. Low Temp. Phys.* **138**, 635 (2005).
21. O. Morsch and M. Oberthaler, Dynamics of Bose–Einstein condensates in optical lattices, *Rev. Mod. Phys.* **78**, 179 (2006).
22. I. B. Spielman, W. D. Phillips and J. V. Porto, The Mott insulator transition in two dimensions, *Phys. Rev. Lett.* **98**, 080404 (2007).
23. Z. Hadzibabic, S. Stock, B. Battelier, V. Bretin and J. Dalibard, Interference of an array of independent Bose–Einstein condensates, *Phys. Rev. Lett.* **93**, 180403 (2004).
24. Z. Hadzibabic, P. Krüger, M. Cheneau, B. Battelier and J. Dalibard, Berezinskii–Kosterlitz–Thouless crossover in a trapped atomic gas, *Nature* **441**, 1118 (2006).
25. S. Tung, G. Lamporesi, D. Lobser, L. Xia and E. A. Cornell, Observation of the presuperfluid regime in a two-dimensional Bose gas, *Phys. Rev. Lett.* **105**, 230408 (2010).
26. C.-L. Hung, X. Zhang, N. Gemelke and C. Chin, Observation of scale invariance and universality in two-dimensional Bose gases, *Nature* **470**, 236 (2011).
27. A. Ramanathan, K. C. Wright, S. R. Muniz, M. Zelan, W. T. Hill, C. J. Lobb, K. Helmerson, W. D. Phillips and G. K. Campbell, Superflow in a toroidal Bose–Einstein condensate: An atom circuit with a tunable weak link, *Phys. Rev. Lett.* **106**, 130401 (2011).
28. D. S. Petrov, M. Holzmann and G. V. Shlyapnikov, Bose–Einstein condensation in quasi-2D trapped gases, *Phys. Rev. Lett.* **84**, 2551 (2000).
29. S. K. Adhikari, Quantum scattering in two dimensions, *American J. Phys.* **54**, 362 (1986).
30. D. S. Petrov and G. V. Shlyapnikov, Interatomic collisions in a tightly confined Bose gas, *Phys. Rev. A* **64**, 012706 (2001).
31. V. Schweikhard, S. Tung and E. A. Cornell, Vortex proliferation in the Berezinskii–Kosterlitz–Thouless regime on a two-dimensional lattice of Bose–Einstein condensates, *Phys. Rev. Lett.* **99**, 030401 (2007).
32. N. V. Prokof'ev and B. V. Svistunov, Two-dimensional weakly interacting Bose gas in the fluctuation region, *Phys. Rev. A* **66**, 043608 (2002).
33. L. Kadanoff and G. Baym, *Quantum Statistical Mechanics* (Benjamin/Cummings Publishing Company, 1963).
34. D. R. Nelson and J. M. Kosterlitz, Universal jump in the superfluid density of two-dimensional superfluids, *Phys. Rev. Lett.* **39**, 1201 (1977).
35. N. V. Prokof'ev, O. Ruebenacker and B. V. Svistunov, Critical point of a weakly interacting two-dimensional Bose gas, *Phys. Rev. Lett.* **87**, 270402 (2001).
36. D. S. Fisher and P. C. Hohenberg, Dilute Bose gas in two dimensions, *Phys. Rev. B* **37**, 4936 (1988).
37. M. Holzmann and W. Krauth, Kosterlitz–Thouless transition of the quasi two-dimensional trapped Bose gas, *Phys. Rev. Lett.* **100**, 190402 (2008).

38. M. Holzmann, M. Chevallier and W. Krauth, Semiclassical theory of the quasi two-dimensional trapped gas, *Europhys. Lett.* **82**, 30001 (2008).
39. M. Holzmann, M. Chevallier and W. Krauth, Universal correlations and coherence in quasi-two-dimensional trapped Bose gases, *Phys. Rev. A* **81**, 043622 (2010).
40. Z. Hadzibabic, P. Krüger, M. Cheneau, S. P. Rath and J. Dalibard, The trapped two-dimensional Bose gas: from Bose–Einstein condensation to Berezinskii–Kosterlitz–Thouless physics, *New J. Phys.* **10**, 045006 (2008).
41. R. N. Bisset, D. Baillie and P. B. Blakie, Analysis of the Holzmann–Chevallier–Krauth theory for the trapped quasi-two-dimensional Bose gas, *Phys. Rev. A* **79**, 013602 (2009).
42. S. T. Bramwell and P. C. W. Holdsworth, Magnetization: A characteristic of the Kosterlitz–Thouless–Berezinskii transition, *Phys. Rev. B* **49**, 8811 (1994).
43. V. S. Bagnato and D. Kleppner, Bose–Einstein condensation in low-dimensional traps, *Phys. Rev. A* **44**, 7439 (1991).
44. M. Holzmann, G. Baym, J. P. Blaizot and F. Laloë, Superfluid transition of homogeneous and trapped two-dimensional Bose gases, *P.N.A.S.* **104**, 1476 (2007).
45. R. N. Bisset, M. J. Davis, T. P. Simula and P. B. Blakie, Quasicondensation and coherence in the quasi-two-dimensional trapped Bose gas, *Phys. Rev. A* **79**, 033626 (2009).
46. R. N. Bisset and P. B. Blakie, Transition region properties of a trapped quasi-two-dimensional degenerate Bose gas, *Phys. Rev. A* **80**, 035602 (2009).
47. T. Yefsah, R. Desbuquois, L. Chomaz, K. J. Günter and J. Dalibard, Exploring the thermodynamics of a two-dimensional Bose gas, ArXiv e-prints: 1106.0188 (2011).
48. L. P. Pitaevskii and A. Rosch, Breathing mode and hidden symmetry of trapped atoms in two dimensions, *Phys. Rev. A* **55**, R853 (1997).
49. P. O. Fedichev, U. R. Fischer and A. Recati, Zero-temperature damping of Bose–Einstein condensate oscillations by vortex-antivortex pair creation, *Phys. Rev. A* **68**, 011602 (2003).
50. M. Olshanii, H. Perrin and V. Lorent, Example of a quantum anomaly in the physics of ultracold gases, *Phys. Rev. Lett.* **105**, 095302 (2010).
51. P. Krüger, Z. Hadzibabic and J. Dalibard, Critical point of an interacting two-dimensional atomic Bose gas, *Phys. Rev. Lett.* **99**, 040402 (2007).
52. A. Polkovnikov, E. Altman and E. Demler, Interference between independent fluctuating condensates, *Proc. Natl. Acad. Sci. USA* **103**, 6125 (2006).
53. S. P. Rath and W. Zwerger, Full counting statistics of the interference contrast from independent Bose–Einstein condensates, *Phys. Rev. A* **82**, 053622 (2010).
54. N. R. Cooper and Z. Hadzibabic, Measuring the superfluid fraction of an ultracold atomic gas, *Phys. Rev. Lett.* **104**, 030401 (2010).
55. I. Carusotto and Y. Castin, Non-equilibrium and local detection of the normal fraction of a trapped two-dimensional Bose gas, ArXiv e-prints: 1103.1818 (2011).
56. L. Sanchez-Palencia and M. Lewenstein, Disordered quantum gases under control, *Nature Phys.* **6**, 87 (2010).

57. G. Roati, C. D'Errico, L. Fallani, M. Fattori, C. Fort, M. Zaccanti, G. Modugno, M. Modugno and M. Inguscio, Anderson localization of a non-interacting Bose–Einstein condensate, *Nature* **453**, 895 (2008).
58. J. Billy, V. Josse, Z. Zuo, A. Bernard, B. Hambrecht, P. Lugan, D. Clément, L. Sanchez-Palencia, P. Bouyer and A. Aspect, Direct observation of Anderson localization of matter waves in a controlled disorder, *Nature* **453**, 891 (2008).
59. A. A. Burkov, M. D. Lukin and E. Demler, Decoherence dynamics in low-dimensional cold atom interferometers, *Phys. Rev. Lett.* **98**, 200404 (2007).

Chapter 10

# Vortex Physics in the Quantum Hall Bilayer

H. A. Fertig

*Department of Physics, Indiana University, Bloomington, Indiana 47405, USA*

Ganpathy Murthy

*Department of Physics and Astronomy, University of Kentucky, Lexington, Kentucky 40506, USA*

There exists a strong analogy between the quantum Hall bilayer system at total filling factor $\nu = 1$ and a thin film superfluid, in which the groundstate is described as a condensate of particle-hole pairs. The analogy draws support from experiments which display near dissipationless transport properties at low temperatures. However dissipation is always present at any accessible temperature, suggesting that in a proper description, unpaired vortex-like excitations must be present. The mechanism by which this happens remains poorly understood. A key difference between the quantum Hall bilayer and simpler thin-film superfluids is that the vortices, more properly called merons in the former context, are *charged* objects. We demonstrate that a model in which disorder induces merons in the groundstate, through coupling to this charge, can naturally explain many of the observed imperfect superfluid properties.

## 10.1. Introduction

The integer quantum Hall effect is usually associated with the physics of non-interacting electrons in a strong magnetic field. Interactions, by contrast, are generally associated with the fractional quantum Hall effect (FQHE). An intermediate situation exists, however, in quantum Hall systems with discrete degrees of freedom: situations in which the spin is not completely frozen due to the strong magnetic field (as is often the case in GaAs systems where the effective $g$ factor is quite small), semiconductor systems where electrons may reside in different valleys, or multilayer systems. The simplest of these latter systems, the quantum Hall bilayer at total filling factor $\nu = 1$, has been heavily studied for a number of years.[1] While supporting an integrally quantized Hall effect, the system displays a number of phenomena that can only be explained as being due to interactions in the system.

Even the observation of the quantized Hall effect in this system[2] is a signal of strong interaction effects: when the layers are unbiased so that each has filling factor 1/2, one does not expect a FQHE for the layers individually, which typically requires fractional filling with an odd denominator. Quantization of the Hall conductance becomes possible in this system because of interlayer coherence. For example, if the tunneling between the layers is very strong, then non-interacting states involve symmetric and antisymmetric combinations of the wavefunctions in the two wells, and filling a single Landau level in the lower energy of these will yield a $\nu = 1$ quantized Hall effect. The remarkable physics of this system, however, occurs not in this limit of strong tunneling, but rather when the tunneling is very small — much smaller than any other energy scale in the problem, including the temperature. Despite the near absence of such a term in the Hamiltonian, the Hall conductance remains quantized at $\sigma_{xy} = e^2/h$. As this situation is not possible for non-interacting electrons, this is clearly an interaction effect.

Two major pictures, closely related to one another, have been developed to describe the quantum Hall bilayer system at $\nu = 1$. One of these is in terms of quantum ferromagnetism. In this description, one identifies an electron residing in the "top" layer as having a pseudospin-up degree of freedom, while residing in the lower layer is interpreted as pseudospin-down. Because the electrostatics of the system strongly favor equal populations of the two layers, it is energetically favorable for the spin state of the electrons in this language to reside in the $x - y$ plane, even when the interlayer tunneling is negligibly small. The effective theory is then that of an easy-plane ferromagnet. Because of exchange interactions, low energy states of this system involve having the effective electron spins in the same spatial neighborhood pointing parallel to one another. The resulting system has an effective spin stiffness, and a low-energy excitation spectrum identical to that of an XY model.[3]

Like other XY systems, this coherent state of the quantum Hall bilayer should support vortex excitations. Because the spin degree of freedom can tilt out of the plane in this system (indicating a region where there is a charge imbalance between the layers), the cores of these vortices have an out-of-plane polarization which may point up or down. Such vortex-like objects are known as *merons*. Remarkably, the constraints on wavefunctions residing in a single Landau level cause such spin textures to be charged objects.[4,5] In particular, this means that merons can be incorporated into the groundstate by doping away from $\nu = 1$, or simply by sufficiently strong disorder, which may affect the doping locally. In the limit of slowly varying spin configurations, the relation between spin and charge is given by[1,4-6]

$$q(\mathbf{r}) = -\frac{1}{8\pi}\varepsilon_{\mu\nu}\mathbf{n}\cdot\partial_\mu\mathbf{n}\times\partial_\nu\mathbf{n}, \qquad (10.1)$$

where $\mathbf{n}(\mathbf{r})$ is a unit vector indicating the local direction of the spin.

An effective energy functional for this pseudospin system which captures both the spin-charge coupling and the symmetry-breaking physics has the form[7]

$$E[\mathbf{n}] = \frac{\rho_E}{2}\int d\mathbf{r}(\nabla n^\mu)^2 + \frac{1}{2}\int d\mathbf{r}d\mathbf{r}'q(\mathbf{r})v(\mathbf{r}-\mathbf{r}')q(\mathbf{r})$$

$$+ \int d\mathbf{r}V(r)q(\mathbf{r}) - \frac{\Delta_{SAS}}{4\pi\ell_0^2}\int d\mathbf{r}[n^x(d\mathbf{r})-1]$$

$$+ \Gamma\int d\mathbf{r}(n^z)^2 - \frac{e^2d^2}{16\pi\kappa}\int\frac{d\mathbf{q}}{4\pi^2}qn^z_{-\mathbf{q}}n^z_{\mathbf{q}}$$

$$+ \frac{\rho_A - \rho_E}{2}\int d\mathbf{r}(\nabla n^z)^2. \qquad (10.2)$$

The first three terms of the energy are SU(2) invariant contributions. The leading gradient term is the only one that appears in the nonlinear $\sigma$ model for Heisenberg ferromagnets, and $\rho_E$ is the spin stiffness in the $\hat{x}-\hat{y}$ plane. The second term describes the SU(2) invariant Hartree energy corresponding to the charge density associated with spin textures in quantum Hall ferromagnets. $v(r)$ is the Coulomb interaction screened by the dielectric constant $\kappa$ of the host semiconductor. The third term incorporates interactions of the charge density with an external potential, for example, due to disorder. The fourth term describes the loss in tunneling energy when electrons are promoted from symmetric to antisymmetric states; here $\Delta_{SAS}$ is the single-particle splitting between symmetric and antisymmetric states.

The last three terms are the leading interaction anisotropy terms at long wavelengths. The term proportional to $\Gamma$ implements the electrostatic energy cost of having a net charge imbalance between the layers. The $(\nabla n^z)^2$ term accounts for the anisotropy of the spin stiffness. Pseudospin order in the $\hat{x}-\hat{y}$ plane physically corresponds to interlayer phase coherence so that $\rho_A - \rho_E$ will become larger with increasing interlayer separation $d$. The sum of the first and seventh terms in Eq. (10.2) gives an XY-like anisotropic nonlinear $\sigma$ model. However, this gradient term is not the most important source of anisotropy at long wavelengths. The fifth term produces the leading anisotropy, and is basically the capacitive energy of the double-layer system. The sixth term appears due to the long-range nature of the Coulomb interaction; its presence demonstrates that a naive gradient expansion of the anisotropic terms is not valid. ($\mathbf{n}_q$ is the Fourier transform of the unit vector field $\mathbf{n}$.) Equation (10.2) can be rigorously derived from the Hartree–Fock approximation in the limit of slowly varying spin textures,[8] and explicit expressions are obtained for $\rho_E$ (which is due in this approximation entirely to interlayer interactions), $\rho_A$ (due to intralayer interactions), and $\Gamma$. Quantum fluctuations will alter the values of these parameters from those implied by the Hartree–Fock theory.

Because Eq. (10.2) is an energy functional for an easy-plane ferromagnet, it supports vortex excitations. In the clean limit, a single vortex–antivortex pair is logarithmically confined when $\Delta_{SAS}$ is zero. The out-of-plane degree of freedom allows either vorticity to carry positive or negative electric charge. In the case where the pair also has oppositely oriented polarization vectors at their centers, the total charge of the pair is $\pm e$, and is known as a bimeron. Again in the clean limit, this is a finite energy object that can cause dissipation in electric transport, and should give rise to the energy gap that one measures in the quantized Hall effect.[7]

An interesting limit is $d \to 0$, in which the two layers are essentially upon one another. Although this cannot be achieved with real double layers, the model becomes equivalent to what one expects for the electron system when the spin degree of freedom is active. (One can consider the situation in which both real spin and pseudospin are active degrees of freedom, leading to many possible states of the system.[9]) In this case, $\Delta_{SAS}$ must be reinterpreted as the Zeeman coupling of the system. In this situation, bimerons become equivalent to *skyrmions*, spin-textured objects with Pontryagin index (essentially the spatial integral of Eq. (10.1)) given by $\pm 1$, encoding the fact that this excitation has spins pointing in all possible directions on the Bloch sphere precisely once. For clean real spin systems, it is clear that the lowest energy charged excitations are skyrmions[5,10] when the Zeeman coupling is sufficiently small. Experimentally, signatures of skyrmions in the quantum Hall system would be expected in the behavior of the activation energy as a function of Zeeman coupling and also in an anomalous decay of the electron spin polarization as the system is doped away from $\nu = 1$. Although both these behaviors have been observed,[11,12] it is important to

## 10.2. Exciton Condensation and Two-Dimensional Superfluidity

Returning to the energy functional Eq. (10.2), one can describe the physics of this system in a very different analogy than that of the quantum ferromagnet. For sufficiently large $\Gamma$, out of plane fluctuations will be strongly suppressed, and one is justified in ignoring $n_z$ as a dynamical degree of freedom. To lowest order in gradients, writing $n_x + i n_y = e^{i\theta}$, the energy functional takes the simple form

$$E_{SF} = \int d\mathbf{r} \left[ \frac{\rho_s}{2} (\nabla \theta)^2 - \frac{\Delta_{SAS}}{4\pi \ell_0^2} \cos\theta \right] \quad (10.3)$$

in the absence of an external potential $V(\mathbf{r})$. For $\Delta_{SAS} = 0$ (i.e. negligible tunneling), this has exactly the form expected for a two-dimensional thin film superfluid, with $\theta$ the condensate wavefunction phase, and $\rho_s$ an effective two-dimensional "superfluid stiffness." In this case, one expects the system to have a linearly dispersing "superfluid mode" which is analogous to the spin wave of an easy-plane ferromagnet. The presence of such a mode has been verified in microscopic calculations using the underlying electron degrees of freedom.[3] This suggests the possibility that one might observe some form of superfluidity in this system. To see exactly what this means, it is convenient to consider momentarily a wavefunction for the groundstate of the system in terms of the electron degrees of freedom,

$$|\Psi_{ex}\rangle = \Pi_X [u_X + v_X c_{T,X}^\dagger c_{B,X}] |\text{Bot}\rangle \quad (10.4)$$

where $|\text{Bot}\rangle$ represents the state in which all the single particle states of the bottom layer in the lowest Landau level, created by $c_{B,X}^\dagger$, have been filled. For a state with uniform density and equal populations in each well, $u_X = v_X = 1/\sqrt{2}$. More generally, one can represent an imbalanced state, obtained physically with an electric field applied perpendicular to the bilayer, by taking $u_X = \sqrt{\nu_T}$ and $v_X = \sqrt{\nu_B}$, with $\nu_T + \nu_B = 1$. The constants $\nu_T$ and $\nu_B$ represent the filling fractions in each of the layers, and the situation where $\nu_T \neq \nu_B$ turns out to be quite interesting, as we will discuss below.

Equation (10.4) is an excellent trial wavefunction for the groundstate when the layer separation $d$ is not too large.[14] It shows that in the ideal (clean) limit, this system has a coherence much akin to that of a superconductor, and that the condensed objects in the groundstate are *excitons*, particle-hole pairs with each residing in a different layer. The structure of these condensed objects suggests that a superflow from them will carry electrons in one of the layers, and holes in the other. Superfluidity in this system then should be expected in *counterflow* transport, where electric current in each layer runs in opposite directions.

Remarkably, something akin to this occurs in experiments where electrical contact is made separately with each layer,[15] and current is injected in opposite directions for the two layers. By measuring the voltage drop in a single layer along the direction of current, one may learn about dissipation in this exciton flow. In the experiments[15] the dissipation is non-vanishing at any temperature $T > 0$, but apparently extrapolates to zero as $T \to 0$.

There are also transport experiments which use the fact that $\Delta_{SAS}$, although very small (typically several tens of microKelvin), is not zero. When the last term in Eq. (10.3) is included, the energy functional has a form very similar to that of a Josephson junction, suggesting that this system supports a Josephson effect.[16] In tunneling experiments, where one separately contacts to each layer such that the current must tunnel between them, the tunneling $I$–$V$ is nearly vertical near zero interlayer bias,[17] which appears very similar to a Josephson $I$–$V$ characteristic.

A further analogy with thin-film superfluids and Josephson junctions may be seen in nonlinear transport. In particular, Josephson junctions support a critical current above which dissipation sets in. This critical current should scale as $\sqrt{\Delta_{SAS}}$ in the model of Eq. (10.3). Behavior reminiscent of this has been found experimentally,[18] although in the quantum Hall bilayer one should note carefully that the critical current separates a low from a high dissipation regime, rather than a zero from a non-zero dissipation regime. Recent theoretical work has attempted to model this behavior, both for clean systems[19–21] and dirty ones.[22] Importantly, these give very different behaviors for how the critical current scales with respect to the sample area. The latter of these seems to agree best with experiment.

These results look strikingly reminiscent of expectations for exciton superfluidity. However, it is important to recognize that they clearly are *not* genuine superfluid behavior. If the condensate could truly flow without dissipation, one would expect zero dissipation at any finite temperature below the Kosterlitz–Thouless transition, where vortex–antivortex pairs unbind. In experiment this truly dissipationless flow appears to emerge, if at all, only in the zero temperature limit. Similarly, the Josephson effect should be truly dissipationless, whereas in experiment, there is always a measurable tunneling resistance at zero bias. The superfluidity in this system is *imperfect*. What kind of state can be nearly superfluid in this way?

A further fundamental issue is raised by the role of $\Delta_{SAS}$. Equation (10.3) has the form of an XY model with a symmetry-breaking field. This system was studied years ago in a seminal work by José et al.,[23] who demonstrated that the Kosterlitz–Thouless vortex unbinding mechanism is destabilized by such a perturbation. One way to understand this is in terms of a single vortex–antivortex pair. In the presence of the symmetry-breaking term, the constituents of this are connected by a string of overturned spins of width $\sim \sqrt{\rho_s/\Delta_{SAS}}$ in the groundstate.[1] This leads to a linear confinement of the vortices, rather than the logarithmic confinement which allows their unbinding at a finite temperature.

A renormalization group study of this system which focuses on the entropy of such strings suggests that vortices *can* unbind in these systems, although the unbinding cannot be interpreted as a thermodynamic phase transition.[24] More directly relevant to the quantum Hall bilayer system is strong evidence that disorder can overcome the string tension and nucleate unbound vortices in the groundstate.[25–27] The presence of these at arbitrarily low temperature in this system has profound physical implications, and may explain a number of the "imperfect superfluid" properties it exhibits. We explain this in more detail in the following sections.

## 10.3. Puddles from Potential Fluctuations: A Periodic Model

One important way in which skyrmions and merons of the $\nu = 1$ quantum Hall system are different than those of more standard ferromagnets is that they carry charge. In an environment with an electric potential, they will be attracted towards the extrema of the potential, and if the potential fluctuations are strong enough, they can be induced in the groundstate, even if the system is nominally precisely at $\nu = 1$. This latter situation is, in fact, quite ubiquitous in quantum Hall systems, as was shown some time ago by Efros.[28] Quantum wells and heterostructures obtain their electrons from donor atoms, leaving behind charged ions which are themselves disordered. Although separated spatially from the two-dimensional electron gas — which is the reason these systems have such high mobility — the ions present an effective disorder potential that varies slowly on the scale of the inter-electron separation. Efros demonstrated that the magnitude of this potential fluctuation is very strong, so that it is screened nonlinearly by the quantum Hall fluid. Large puddles of positively and negatively charged quasiparticles are induced in the ideal uniform density groundstate. Although these puddles locally spoil the incompressibility of the quantum Hall fluid, Efros demonstrated that a percolating strip separating the puddles supports the incompressible groundstate, and that the presence of this is sufficient to observe the quantized Hall conductance.

In building up a model of this physics, our first interest is in the effects of the puddles themselves, independent of the precise way in which they may be arranged in real samples. As a first step, we ignore that disordered nature of the puddles, and imagine a system in which they are periodically arranged. In analogy with superfluid systems,[29] and one-component quantum Hall systems near a plateau transition,[30] this likely captures some of the nonperturbative effects of disorder. Once a phase transition in which vortices are nucleated has been obtained, one may then add disorder or other perturbations and examine their relevance/irrelevance.[31] This approach has been very fruitful in the past.

As a first attempt, one of us (GM) and Subir Sachdev[6] examined the Composite Boson theory[4] near a putative Superfluid/Mott Insulator transition.[29] The primary difference between the neutral system and this one is that the Composite Boson is charged and is minimally coupled to both the external gauge field and the statistical field which attaches flux. There are two natural phases: A Bose-condensed Higgs phase in which all excitations are massive and the system can be shown to have a quantized Hall conductivity, and a Mott Insulating phase in which all conductivities vanish.[6]

In the large-$N$ approximation, one can integrate out the bosons, leaving behind an effective theory of gauge fluctuations. The phase transition turns out to be second-order in the large-$N$ limit, with the critical point having the conductivities[6]

$$\sigma_{xx}^*(0,\omega) = \frac{e^2}{h} \frac{\pi/8}{1+(\pi/8)^2} \quad \sigma_{xy}^* = \frac{\pi}{8}\sigma_{xx}^*. \qquad (10.5)$$

The XY-angle $\theta$ also acquires an imaginary part of the self-energy at the critical point which vanishes as $\omega^7$ for zero interlayer tunneling, and $\omega^5$ when interlayer tunneling is non-zero.[6]

While this model is fully quantum, it is overly simplistic in assuming that only a single phase transition exists between the uniform superfluid and a Mott Insulator. Recently, in collaboration with Jianmin Sun and Noah Bray-Ali, we have analyzed a somewhat more realistic model.[27] We assume that fermion excitations are high in energy, and that the real spin is fully polarized, leaving dynamics only in the pseudospin. The system can then be modeled purely in terms of (pseudo-)spins, which we put on a square lattice with lattice constant $a = l_0\sqrt{2\pi}$, so that there is one electron per site. Additionally, there is a periodic potential with period $L = Na$ in each direction coupling to the pseudospin textures through their charge. For the discrete lattice system, this assumes a form that involves a triplet of neighboring spins,[32]

$$\delta Q_{123} = \frac{e}{2\pi} \tan^{-1}\left(\frac{\mathbf{n}_1 \cdot \mathbf{n}_2 \times \mathbf{n}_3}{1 + \mathbf{n}_1 \cdot \mathbf{n}_2 + \mathbf{n}_2 \cdot \mathbf{n}_3 + \mathbf{n}_3 \cdot \mathbf{n}_1}\right). \tag{10.6}$$

The usual spin-stiffness, planar anisotropy, and interlayer tunneling terms are present. Finally, to model the Coulomb interaction between pseudospin textures, we introduce a Hubbard interaction for the charge on each plaquette $\square$. Thus, the Hamiltonian is

$$\mathcal{H} = -J \sum_{\langle \mathbf{rr}' \rangle} (n_x(\mathbf{r})n_x(\mathbf{r}') + n_y(\mathbf{r})n_y(\mathbf{r}')) + \frac{\Gamma}{2} \sum_{\mathbf{r}} (n_z(\mathbf{r}))^2$$

$$- h \sum_{\mathbf{r}} n_x(\mathbf{r}) - V \sum_{\square} f(X,Y) \delta Q_{\square} + H_U[\delta Q^2]. \tag{10.7}$$

Here $V$ is the strength of the periodic potential living on the dual lattice, and $f(X,Y)$ its functional form. We choose the simple form $f = \sin(2\pi X/L)\sin(2\pi Y/L)$, resulting in each unit cell having four puddles, two each being positive and negative. In what follows, we will present a few results for $J = \Gamma$, and measure all energies in units of $J$. We find the ground states (really saddle point configurations of the spins) numerically by simulated annealing. For small $V$, the uniform ferromagnetic state is the ground state, but as $V$ increases, the ground state nucleates merons/antimerons in each of the puddles. An example is shown Fig. 10.1. Note that the spins tilt out of the plane near the centers of the vortices, as expected in this type of model. This results in a local charge density associated with the vortex. It is the coupling of this charge to the potential that results in the vortices being nucleated. With increasing strength of the periodic potential, we expect more merons and antimerons to be nucleated by the potential. This is illustrated in Fig. 10.2.

The phase transitions between these ground states are generically first-order, though occasionally they can become weakly first-order or even second-order. Various physical quantities, such as the spin stiffness (without vortex corrections) and spin wave velocity, also exhibit jumps at these transitions An example of this is illustrated in Fig. 10.3.

There is an important qualitative difference between the models with and without the periodic potential as a function of interlayer coupling. In

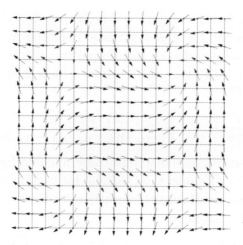

Fig. 10.1. The ground state configuration for a 16 × 16 unit cell with the strength of the periodic potential (in units where $J = 1$) being $V = 3.0$, and the Hubbard interaction is $U = 8.0$. The lengths of the spins denote their planar projection. Blue arrows tilt upward out of the plane; red ones tile downward. Note a vortex/antivortex at the center of each puddle.

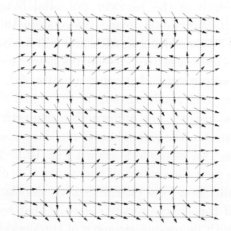

Fig. 10.2. The ground state configuration for $V = 7.0$ and $U = 8.0$.

the clean model, with no potential, the interlayer tunneling strength $h$ is a relevant perturbation.[23] At low temperature one would expect all vortices to be linearly confined,[24] so that no vortices would appear in the ground-state. Naively one would expect that systems with weak and strong $h$ at the microscopic scale will behave similarly at large length scales. However, in the model with a periodic potential, as $h$ increases, the system will undergo transitions in which merons/antimerons are lost from the puddles, ending at

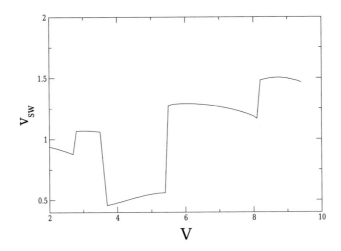

Fig. 10.3. The spinwave velocity of the $G$-mode as a function of $V$. Note that it never vanishes, but that it can vary by a factor of three, and have significant discontinuities at transitions.

very large $h$ in a uniform ferromagnetic state. We note this is consistent with simulations of disordered bilayer systems, for which coupling to a simplified model of the vortex spin texture also nucleates vortices in the groundstate.[25] Since the weak and strong $h$ systems are in different phases, there is no reason to expect similar behavior from them. Indeed, experimentally one sees the puzzling features described above only in systems with a tiny $h$.

Beyond the groundstate configuration, one may compute the collective modes of this model. One finds that when the transition is weakly first-order there is a new, quadratically dispersing mode which becomes low in energy, and can be even lower in energy than the linearly dispersing Goldstone mode (or $G$-mode). We call this new mode the $Q$-mode. Examining the wavefunctions of the $Q$-mode reveals that it represents vortex motion within the puddle. Interestingly, an analog of this kind of motion was also observed for vortices in a disordered XY model Langevin dynamics simulation,[25] and could have important consequences for dissipation in this system. An example of the low-energy part of the collective mode dispersions is shown in Fig. 10.4.

Within this simplified periodic model, the existence of these phase transitions has important physical consequences. One is the strong suppression of the Berezinskii–Kosterlitz–Thouless transition temperature $T_{KT}$ near the $V_0$ corresponding to a transition between different meron numbers in the puddles. A simple picture for this comes from considering the first such transition, from the state without merons to one with a checkerboard pattern

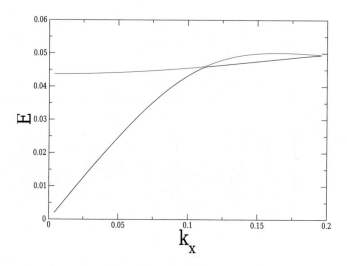

Fig. 10.4. The lowest-lying modes for $V = 4.3$ and $U = 18$. The system is very close to a ground state transition. Note the almost gapless $Q$-mode.

of merons and antimerons. Integrating out the spin-waves, the difference in ground state energies can be modeled as arising from a Coulomb gas energy of the form

$$E[\{m(\mathbf{R})\}] = \frac{1}{2} \sum_{\mathbf{R} \neq \mathbf{R}'} \mathcal{K}_s m(\mathbf{R}) m(\mathbf{R}') \log\left(\frac{|\mathbf{R} - \mathbf{R}'|}{\xi}\right)$$
$$- E_c(V) \sum_{\mathbf{R}} (m(\mathbf{R}))^2, \qquad (10.8)$$

where $E_c(V)$ is a core energy that depends on the potential strength, $m(\mathbf{R})$ is the vorticity at the dual lattice site $\mathbf{R}$, and $\xi$ is a cutoff which can be tuned so that the groundstate goes through the transition seen to occur at $V_c$ in the simulation. Near the transition the difference in ground state energies behaves as $\alpha(V - V_c)$. Attributing this difference to the difference in vortex core energies, we can extract $E_c \approx \frac{\alpha}{4}|V - V_c|$. Note that the vortex and the absence of a vortex interchange roles in this model as one goes through the transition. As the core energy of the vortices vanishes, they proliferate and disorder the system. One can solve the renormalization group equations to find $T_{\rm KT}$ as one varies $V$. We find that $T_{\rm KT}$ is typically suppressed by an order of magnitude compared to Hartree–Fock estimates, as shown in Fig. 10.5.

We can also consider the suppression of interlayer tunneling near a ground state transition with a nearly gapless $Q$-mode. It is important to note that

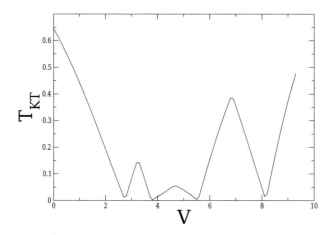

Fig. 10.5. $T_{KT}$ in units of $J/k_B$ as a function of $V$ for a $16 \times 16$ unit cell at $U = 8$. Dramatic decreases in the transition temperature $T_{KT}$ separating the low-temperature ferromagnetic phase from high-temperature paramagnetic phase occur due to changes in the topological density of the ground-state at critical potential strengths.

the XY angle $\theta$ couples to both the $G$- and $Q$-modes, and for small deviations, can be written as a linear combination of them. To see the effect of the $Q$-mode, one may decompose the tunneling term as

$$h \cos(\theta) \simeq h(e^{i\theta_G} e^{i\theta_Q} + \text{c.c.}) \tag{10.9}$$

and integrate out the $Q$-mode at non-zero temperature $T$. Assuming a quadratic dispersion $\omega_Q(\mathbf{k}) = E_{Q0} + \gamma \mathbf{k}^2$, after integrating the $Q$-mode one obtains a renormalized $h$

$$h_R \approx h e^{-Tl^2 \log \frac{\Lambda}{E_{Q0}}/\gamma}. \tag{10.10}$$

Here $\Lambda$ is a cutoff, and it can be seen that as the gap $E_{Q0}$ of the $Q$-mode vanishes, $h_R \to 0$.

One of the most important open questions is whether, and how much, of the phenomenology described above survives in the system with true disorder. Our expectation is that the qualitative difference between weak and strong tunneling will survive, because vortices will surely be induced in the groundstate by the disorder. One might also expect disorder to smooth the first-order transitions into second-order ones with a gapless mode at the transition.[33] In such a case, near a transition, one expects the suppression of interlayer tunneling to survive as well. Furthermore, an important effect of disorder is to create large rare regions in which the system is close to critical (the Griffiths phase[34]), and thus even for a *generic* disorder strength one

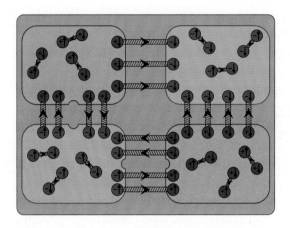

Fig. 10.6. Representation of coherence network. Links and nodes separate puddles of merons (circles). Meron charge and electric dipole moments indicated inside circles, as are strings of overturned phase connecting meron–antimeron pairs. Reproduced from Ref. 26.

expects low-energy modes to exist in the system. However, whether interlayer tunneling is thus suppressed for generic disorder strength is not clear.

## 10.4. Dissipative Transport in a Puddle-Network Model

Whether one considers an ordered or disordered model, the merons and antimerons flooding the puddles should have a high density due to the singular nature of the bare potential fluctuations.[28] Inside the puddles this will spoil interlayer coherence. The coherence however may remain strong in the regions separating the puddles, even though some meron–antimeron pairs will likely straddle them. Thus one forms a network structure for the regions where the coherence is strong, and these should dominate the "superfluid" properties of the system. A schematic picture of the system is illustrated in Fig. 10.6.

The key assumption in this model is that with such dense puddles, merons are able to diffuse independently through the system. This is supported by a renormalization group analysis, which suggests there exists a state in which disorder enters as an effective temperature, so that one would likely be above any meron–antimeron unbinding transition for such strong disorder.[26] Motion of the merons is then limited by energy barriers for them to cross the coherent links between puddles. The tendency for dissipation to vanish only in the zero temperature limit in counterflow transport now becomes very natural. When condensed excitons flow down the system, these produce a force on the merons perpendicular to that current.[35] The resulting meron current is limited by the activation energy to hop over the coherent links, and vanishes rapidly but only completely when the temperature drops to zero. This meron current induces a voltage drop in the direction of the exciton current via the Josephson relation, so that the counterflow current becomes dissipative. True superfluid response in this model can only occur at zero temperature.

Tunneling currents in this model should also be dissipative.[26] Since the current flows into (say) the top layer on the left and leaves via the bottom layer on the right, the current in the system must be decomposed into a sum of a symmetric "co-flow" and antisymmetric counterflow (CF). The former is likely carried by edge currents which are essentially dissipationless in the quantum Hall state. To obtain the correct current geometry, the CF current must point in opposite directions at the two ends of the sample. Thinking of the network as a Josephson array, the current of excitons — i.e. CF current — is proportional to $\nabla\theta$. To inject current at the edges, one may imagine an idealized situation in which there are highly coherent busbars[36] at either edge of the sample, each with a uniform superfluid phase angle that may rotate as a function of time, Josephson coupled to links at the very edges of the network. In order to inject CF currents in opposite directions at each end of the sample, the phase angle in these busbars should be rotated in the *same* direction. This means the phase angle throughout the system will tend to rotate at a uniform rate, which is limited by the $\frac{\Delta_{SAS}}{4\pi\ell_0^2}\cos\theta$ term in Eq. (10.3). This is most effective at the nodes of the network, where the coherence is least compromised by the disorder-induced merons.

A typical node, which can be modeled with a single phase angle $\theta$, is taken to be governed by a Langevin equation

$$\Gamma\frac{d^2\theta}{dt^2} = \sum_{\text{links}} F_{\text{link}} - \gamma_0\frac{d\theta}{dt} - h\sin\theta + \xi(t). \tag{10.11}$$

The quantities $F_{\text{link}}$ represent the torque on an individual "rotor" (i.e. $\theta$ variable) due to its neighbors, transmitted through the links. $\Gamma$ is the effective moment of inertia of a rotor, proportional to the capacitance of the node, $h = \frac{\Delta_{SAS}}{4\pi\ell_0^2}$, $\xi$ is a random (thermal) force, and $\gamma_0$ is the viscosity due to dissipation from the other node rotors in the system. For a small driving force, the node responds viscously, and the resulting rotation rate has the form $\gamma\dot\theta = \sum F_{\text{link}}$. The Josephson relation $V = \frac{\hbar}{e}\frac{d\theta}{dt}$ then implies that the viscosity $\gamma$ is proportional to the tunneling *conductance* $\sigma_T$ of the system.

For $k_B T \gg h$ one may show the viscosity for an individual node to be[37]

$$\gamma = \gamma_0 + \Delta\gamma = \gamma_0 + \sqrt{\frac{\pi}{2\Gamma}} \frac{h^2}{(k_B T)^{3/2}}. \qquad (10.12)$$

As each node contributes the same amount to the total viscosity, the *total* response of the system to the injected CF current obeys

$$I_{CF} \propto N_{\text{nodes}} \Delta\gamma \frac{e^2 V_{\text{int}}}{\hbar} = \sigma_T V_{\text{int}}. \qquad (10.13)$$

Note that because the nodes respond viscously, the tunneling conductance is proportional to the area of the bilayer. This is a non-trivial prediction of the model discussed here, which has been confirmed in experiment.[38] The proportionality of the tunneling conductance to $\Delta_{\text{SAS}}^2$ is another non-trivial prediction which appears to be consistent with experimental data, and which contrasts with the result one expects in the absence of disorder, for which $\sigma_T \propto \Delta_{\text{SAS}}$.

## 10.5. Drag Experiments and Interlayer Bias

When an electric field is applied perpendicular to the layers, the density in the two layers becomes imbalanced. The effect of this can be incorporated into the model, Eq. (10.2), by replacing $\Gamma \int d\mathbf{r}(n^z)^2$ with $\Gamma \int d\mathbf{r}(n^z - n_0)^2$, with $n_0 = \nu_T - \nu_B$. The imbalance has interesting consequences for merons: since the pseudospin field $\mathbf{n}$ does not drop back into the $\hat{x} - \hat{y}$ plane as one moves out from the center of meron, the spin-charge relation Eq. (10.1) indicates that the four types of merons will have four different charges. These charges are specifically given by $q_{s,T(B)} = -s\sigma\nu_{B(T)}$, where $s = \pm 1$ is the vorticity of the meron, and the $T(B)$ subscript reflects the layer in which the magnetization at the core of the meron — its polarization — resides. The index $\sigma$ indicates a sign associated with the polarization: $\sigma = 1$ for $\mathbf{n}$ in the meron center oriented in the top layer, $\sigma = -1$ for merons where it resides in the bottom layer.

The connection between polarization and charge has very interesting consequences for drag measurements in the bilayer system. In these experiments, a current is driven through only a single layer, and the voltage drop in either the drive layer or the drag layer is measured. As above, one expects this voltage drop to be related to the motion of the merons. This, in turn, is limited by the activation barriers the merons must overcome to move through the system.

Within the coherence network model, the activation barrier for merons to hop across incompressible strips will clearly depend on the relative orien-

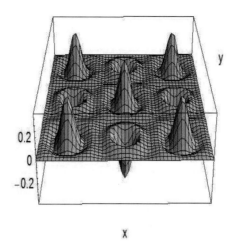

Fig. 10.7. Out-of-plane pseudospin of a Hartree–Fock groundstate with merons and antimerons in the groundstate. These are added by doping the system slightly away from $\nu = 1$.

tation of the meron polarization and the applied bias. There are two reasons for this. The first is that for a given vorticity $s$, the charge of the meron depends on the polarization $\sigma$, both in sign and magnitude. Since the barrier the meron has to cross is one in which there is no screening, while within the puddles the disorder potential is relatively well-screened, we expect a relatively high electrostatic potential barrier to be present. Clearly a lower charge meron will have an easier time crossing this. Beyond this, because the strip is incompressible, the precise details of the pseudospin will be relevant in determining the meron energy when it is inside a strip.

One can estimate the energy barrier for such a crossing with a Hartree–Fock (HF) method. It is possible to stabilize a checkerboard lattice of merons and antimerons from states residing in the lowest Landau level[9] within the HF approximation. An example of the $n_z$ content of such a state is illustrated in Fig. 10.7. One may add an external potential to the Hamiltonian producing this state in the form of a narrow barrier potential, either sitting on top of just one of the two types of merons, or between them, in each case allowing the HF state to relax locally to its lowest energy. The difference in energy between the two configurations yields an estimate of the energy barrier for a particular type of meron to cross an incompressible strip.[39] Such studies clearly show the physics expected, that the activation energy is sensitive to the interlayer bias.

For this reason, one expects the quantum Hall bilayer will contain objects with different possible activation energies. However, naively one would think that at low temperature, transport will be dominated by only the smallest of these, so that a measurement of resistance will reveal an activation energy that is symmetric around zero bias. Moreover, the measured value should be independent of whether the measurement is made in the drag or drive layer. The experimental results, however, are strikingly different than this. The activation energy as measured in the drive layer is highest when the density is biased into the drive layer, and decreases monotonically as the imbalance is changed so that more density is transferred to the drag layer. In the drag layer, the measured voltage drops turn out to be much smaller than in the drive layer, and are symmetric, and increase as the layer is imbalanced.[40]

A careful analysis of the situation requires one to determine voltage drops in individual layers. This is more information than what one obtains using the Josephson relation for the interlayer phase angle as above, which tells one just the interlayer voltage. This can be accomplished[39] using a "composite boson" description of the $\nu = 1$ quantum Hall state.[4,41] In this, one models the electrons as *bosons*, each carrying a single magnetic flux quantum in an infinitesimally thin solenoid. The Aharonov–Bohm effect then implements the correct phase (minus sign) when two of these objects are interchanged.[41] By orienting the flux quanta opposite to the direction of the applied magnetic field, on average the field is canceled, and in mean-field theory the system becomes a collection of bosons in zero field. The quantum Hall state is then equivalent to a Bose condensate of these composite bosons. But for the coherent bilayer state, there is an additional sense in which the bosons are condensed: they carry a pseudospin with an in-plane ferromagnetic alignment.

Because merons carry physical charge, they will carry a quantity of magnetic flux proportional to this charge. In analogy with a thin-film superconductor,[35] this means that a net current in the bilayer (i.e. a co-flow) creates a force on the meron perpendicular to the current. This has to be added to the force due to a counterflow component. Together, these yield a net force which may be shown to be[39]

$$\mathbf{F}_T = \frac{es}{2}\Phi_0[(1+\sigma)\mathbf{J}_B - (1-\sigma)\mathbf{J}_T] \times \hat{z}, \qquad (10.14)$$

where $\mathbf{J}_{T(B)}$ is the current density in the top (bottom) layer. As is clear from this expression, merons with only one of the possible polarizations are subject to a force in a drag experiment, since one of the two current densities vanish.

The force $F_{s,\sigma}$ on merons of vorticity $s$ and polarization $\sigma$ will cause them to flow with a velocity $u_{s,\sigma} = \mu_{s,\sigma}F_{s,\sigma}$ where $\mu_{s,\sigma}$ is an effective mobility, which we expect to be thermally activated, with a bias dependence of the activation energy as discussed above. The resulting motion of the vortices induces voltages in two ways. The first is through the Josepshon relation for the interlayer phase, yielding the relation[39]

$$\Delta V = \Delta V_T - \Delta V_B = -\frac{2\pi h}{e} y_0 \sum_{s,\sigma} n_{s\sigma} s u_{s\sigma} \qquad (10.15)$$

for the voltage drops $\Delta V_\sigma$ between two points a distance $y_0$ apart along the direction of electron current, in layer $\sigma$, where $n_{s,\sigma}$ is the meron density. The second is due to the effective magnetic flux moving with the merons, which induces a voltage drop between electrons at different points along the current flow that is independent of the layer in which they reside. This contribution is given by[39]

$$(\nu_T \Delta V_T + \nu_B \Delta V_B) = -\frac{h}{e} y_0 \sum_{s,\sigma} n_{s\sigma} q_{s\sigma} u_{s\sigma}. \qquad (10.16)$$

In a drag geometry we have, for example, $J_B = 0$ and $\mathbf{J}_T = \frac{I}{W}\hat{y}$, with $I$ the total current and $W$ the sample width. Combining Eqs. (10.15) and (10.16), we obtain $\Delta V_B = 0$ and

$$\frac{\Delta V_T}{I} = \frac{y_0}{W} h \Phi_0 (n_{1,-1} \mu_{1,-1} + n_{-1,-1} \mu_{-1,-1}). \qquad (10.17)$$

Notice the final result depends on the mobility of *only* merons with polarization $\sigma = -1$. It immediately follows that the voltage drop in the drive layer is asymmetric with respect to bias, precisely as observed in experiment.

In order to explain the voltage drop in the drag layer ($\Delta V_L \neq 0$) we must identify how forces on the $\sigma = +1$ merons might arise. A natural candidate for this is the attractive interaction between merons with opposite vorticities, which in the absence of disorder, binds them into pairs at low meron densities. Assuming that driven merons crossing incompressible strips will occasionally be a component of these bimerons, a voltage drop in the drag layer will result. The mobility of such bimerons is limited by the energy barrier to cross an incompressible strip. These strips are likely to be narrow compared to the size scale of the constituents of the bimeron,[39] so we expect the activation energy to be given approximately by the maximum of the activation energies for merons of the two polarizations $\sigma = \pm 1$. This leads to a drag resistance much smaller than that of the drive layer, with an activation energy that is *symmetric* with respect to, and increasing with, bias. These are the behaviors observed in experiment.[40] Figure 10.8

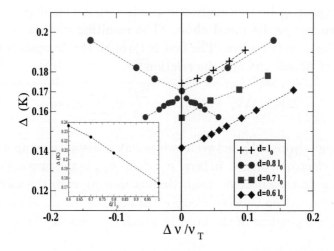

Fig. 10.8. Expected activation energy $\Delta$ as a function of relative density imbalance between the layers, $\Delta\nu/\nu_T$. For voltage drop measured in the drive layer, the results are asymmetric around $\Delta\nu = 0$, as shown explicitly for $d = 0.8\ell_0$ (see text). For measurements in the drag layer, the data is reflected around $\Delta\nu = 0$; the measured result would follow the higher of the two activation energies, yielding a result symmetric in bias. Inset: Activation energy at zero bias versus layer separation. Reproduced from Ref. 39.

illustrates the expected activation energy from a microscopic model implementing this physics[39] as a function of relative density imbalance between the layers, $\Delta\nu/\nu_T$, where $\Delta\nu = \nu_T - \nu_B$ and $\nu_T = 1$, for different layer separations. For voltage drop measured in the drive layer, the results are asymmetric around $\Delta\nu = 0$, as shown explicitly for $d = 0.8\ell_0$. For the drive layer, the large circles represent the activation energy found from a voltage drop measurement when it has a higher density than the drag layer, and the small circles the activation energy when it has the smaller density. Thus a measurement on the drive layer while continuously adjusting the density imbalance from positive through zero to negative results in a line from the large circles to the small circles as zero is crossed, yielding an asymmetric result. For measurements in the drag layer, the result would follow the higher of the two activation energies, yielding a result symmetric in bias.

This result followed from the precise cancellation between the counterflow current force on the vorticity of merons of a particular polarization, and the Lorentz force associated with meron charge and its associated effective flux. The experiments thus provide indirect evidence that the meron charges vary in precisely the way one expects from the spin-charge relation, Eq. (10.1), verifying this unique property of quantum Hall ferromagnetism.

## 10.6. Conclusion

Quantum Hall bilayers have natural descriptions in the languages of ferromagnetism and thin film superfluidity. Both these descriptions suggest the system should have vortex-like excitations which can play a key role in the dissipative properties of the system. Real bilayer systems are inevitably subject to disorder, which is likely to be strong, and to induce such vortices in the groundstate, forming a "coherence network" which surrounds puddles of merons. We have discussed how the motion of merons accounts for the deviation of the system from perfect superfluid behavior, and how the model naturally explains dissipation at finite temperatures in counterflow, tunneling, and drag geometries. To better understand quantum effects in this system, one can consider a simpler model in which the puddles are periodically arranged, and study the low energy collective modes of the system. In this case we found a series of first order zero temperature transitions separating states of different meron occupations in the puddles, and argued that the Kosterlitz–Thouless temperature should drop to zero at these transitions. Approaching these transitions, the collective mode spectrum may develop a low energy mode associated with vortex motion which can greatly suppress the effect of tunneling on the system.

There is much yet to understand about this rich and fascinating system, including how quantum effects impact the zero temperature state and its transport properties, whether there is any attainable limit in which true counterflow superfluidity could be observed, and whether the competing effects of tunneling and disorder can lead to further exotic states. We anticipate this system will remain a subject of keen interest for years to come.

## Acknowledgments

The authors have benefited from discussions and collaborations with many colleagues in the course of the research described here. We would like, in particular, to thank Noah Bray-Ali, Luis Brey, René Côté, Jim Eisenstein, Allan MacDonald, Kieran Mullen, Bahman Roostaei, Subir Sachdev, Steve Simon, Jianmin Sun, and Joseph Straley. The work described here was supported by the NSF through Grant Nos. DMR-1005035 (HAF) and DMR-0703992 (GM).

## References

1. For an introduction to this system, see article by A. H. MacDonald and S. M. Girvin in S. Das Sarma and A. Pinczuk, *Perspectives in Quantum Hall Effects* (Wiley, New York, 1996).

2. G. S. Boebinger, H. W. Jiang, L. N. Pfeiffer and K. W. West, *Phys. Rev. Lett.* **64**, 1793 (1990).
3. H. A. Fertig, *Phys. Rev. B* **40**, 1087 (1989).
4. D. Lee and C. Kane, *Phys. Rev. Lett.* **64**, 1313 (1990).
5. S. Sondhi, A. Karlhede, S. A. Kivelson and E. H. Rezayi, *Phys. Rev. B* **47**, 16419 (1993).
6. G. Murthy and S. Sachdev, *Phys. Rev.* **101**, 226801 (2008).
7. L. Brey, H. A. Fertig, R. Côté and A. H. MacDonald, *Phys. Rev. B* **54**, 16888 (1996).
8. K. Moon et al., Phys. Rev. B **51**, 5138 (1995).
9. See, for example, J. Bourassa, B. Roostaei, R. Côté, H. A. Fertig and K. Mullen, *Phys. Rev. B* **74**, 195320 (2006).
10. H. A. Fertig, L. Brey, R. Côté and A. H. MacDonald, *Phys. Rev. B* **50**, 11018 (1994).
11. S. E. Barrett, G. Dabbagh, L. N. Pfeiffer, K. W. West and R. Tycko, *Phys. Rev. Lett.* **74**, 5112 (1995).
12. A. Schmeller, J. P. Eisenstein, L. N. Pfeiffer and K. W. West, *Phys. Rev. Lett.* **75**, 4290 (1995).
13. G. Murthy, *Phys. Rev. B* **64**, 241309 (2001).
14. A. H. MacDonald, H. A. Fertig and L. Brey, *Phys. Rev. Lett.* **76**, 2153 (1996).
15. M. Kellogg, *et al.*, *Phys. Rev. Lett.* **93**, 036801 (2004); E. Tutuc *et al.*, *Phys. Rev. Lett.* **93**, 036802 (2004).
16. X. G. Wen and A. Zee, *Phys. Rev. B* **47**, 2265 (1993).
17. I. B. Spielman, *et al.*, *Phys. Rev. Lett.* **84**, 0031 (2000).
18. L. Tiemann, Y. Yoon, W. Dietsche, K. von Klitzing and W. Wegscheider, *Phys. Rev. B* **80**, 165120 (2009).
19. J. J. Su and A. H. MacDonald, *Phys. Rev. B* **81**, 184523 (2010).
20. D. V. Fil and A. S. I. Shevchenko, *J. Phys. Condens. Matter* **21**, 215701 (2009).
21. M. Abolfath, A. H. MacDonald and L. Radzihovsky, *Phys. Rev. B* **68**, 155318 (2003).
22. P. R. Eastham, N. R. Cooper and D. K. K. Lee, ArXive:1003:5192.
23. J. V. José, L. P. Kadanoff, S. Kirkpatrick and D. R. Nelson, *Phys. Rev. B* **16**, 1217 (1977).
24. H. A. Fertig, *Phys. Rev. Lett.* **89**, 035703 (2002).
25. H. A. Fertig and J. P. Straley, *Phys. Rev. Lett.* **91**, 046806 (2003).
26. H. A. Fertig and G. Murthy, *Phys. Rev. Lett.* **95**, 156802 (2005).
27. J. Sun, G. Murthy, H. A. Fertig and N. Bray-Ali, *Phys. Rev. B* **81**, 195314 (2010).
28. A. Efros, *Solid State Commun.* **65**, 1281 (1988); A. Efros, F. Pikus and V. Burnett, *Phys. Rev. B* **47**, 2233 (1993).
29. M. P. A. Fisher, P. B. Weichmann, G. Grinstein and D. S. Fisher, *Phys. Rev. B* **40**, 546 (1989).
30. W. Chen, M. P. A. Fisher and Y.-S. Wu, *Phys. Rev. B* **48**, 13749 (1993); X.-G. Wen and Y.-S. Wu, *Phys. Rev. Lett.* **70**, 1501 (1993); L. P. Pryadko and S.-C. Zhang, *Phys. Rev. B* **54**, 4953 (1996); J. Ye and S. Sachdev, *Phys. Rev. Lett.* **80**, 5409 (1989).

31. A. W. W. Ludwig, M. P. A. Fisher, R. Shankar and G. Grinstein, *Phys. Rev. B* **50**, 7526 (1994).
32. S. Sachdev and K. Park, *Ann. Phys.* **298**, 58 (2002).
33. Y. Imry and S.-K. Ma, *Phys. Rev. Lett.* **35**, 1399 (1975); M. Aizenman and J. Wehr, *Phys. Rev. Lett.* **62**, 2503 (1989); K. Hui and A. N. Berker, *Phys. Rev. Lett.* **62**, 2507 (1989).
34. R. B. Griffiths, *Phys. Rev. Lett.* **23**, 17 (1969).
35. A. J. Leggett, *Quantum Liquids* (Oxford, New York, 2006).
36. E. Granato, J. M. Kosterlitz and M. V. Simkin, *Phys. Rev. B* **57**, 3602–3608 (1998).
37. W. Dieterich, I. Peschel and W. R. Schneider, *Z. fur Physik B* **27**, 177 (1977).
38. A. D. K. Finck, A. R. Champagne, J. P. Eisenstein, L. N. Pfeiffer and K. W. West, *Phys. Rev. B* **78**, 075302 (2008).
39. B. Roostaei, K. Mullen, H. A. Fertig and S. Simon, *Phys. Rev. Lett.* **101**, 046804 (2008).
40. R. D. Wiersma *et al.*, *Phys. Rev. Lett.* **93**, 266805 (2004).
41. S. C. Zhang, T. H. Hansson and S. Kivelson, *Phys. Rev. Lett.* **62**, 82 (1989).

# Index

2D Ising model, 36, 51
2D melting, 37
2D-Coulomb gas Hamiltonian, 18
$\beta$-function RG equation, 83
$\mathbb{Z}_p$ model, 107
$\mathbb{Z}_p$ symmetry, 94, 96, 115, 122, 129
$q$-deformed boson, 96, 103, 125
$\mathbb{Z}_p$ broken symmetry phase, 118
2D XY-duality transformation, 77
2D charge–anticharge plasma, 261

Abrikosov vortex lattice, 203
Aharonov–Bohm effect, 259
Aharonov–Casher effects, 259
Anderson localization, 319
array of Josephson junctions, 201, 202
atomic gases, 297

Berezinskii–Kosterlitz–Thouless transition, 93, 96, 98, 118, 123, 257
bilayer, 326
BKT theory, 87
bolometer, 279
bond-algebraic, 93, 113, 126
Bose–Einstein condensation, 297, 309
Burgers vector, 12

cascade relaxation, 275
charge confinement, 262
charge conjugation symmetry, 100, 116
charge lattice, 211
charge pinning, 260
charge-phase duality, 259
charge-vortex duality, 241, 245, 250
circulant matrices, 109
clock handle, 108
clock model, 107, 110, 112, 114, 117, 119, 128
composite boson, 332, 342
condensed fraction, 311
controlled-NOT gate, 129

Cooper pair condensate, 257
correlation functions, 299, 306, 317
Coulomb energy, 258
Coulomb gas, 212, 216, 217
Coulomb interaction in a thin film, 266
counterflow, 339
creep, 11
critical atom number, 309, 310
critical chemical potential, 308
critical phase, 94, 96, 116, 118, 121
critical phase space density, 307, 308

Debye–Waller factor, 41
decimation RG transformation, 80
density fluctuations, 307
dielectric constant, 266
dipole excitations, 217, 218, 274
dipole moment, 215
dipole potential, 300
disclinations, 13
discrete Fourier transform, 109, 116, 124
discrete vortex, 96
discrete vortices, 94, 121, 132
discrete winding number, 94, 96, 122
dislocations, 11
disorder-induced SIT, 256
disordered phase, 98, 118
disordered state, 96
domain excitations, 217, 218
domain wall, 94, 96, 121, 130
drag, 340
dual variables, 105
duality, 93, 96, 102, 106, 111, 125, 127
duality transformation, 75, 212, 217

Efros, 331
electrostatically-driven SIT, 280
emergent U(1) symmetry, 116
equation of state, 304
essential singularity, 98, 118, 123
exciton, 329, 330, 338

extended universality, 123

Fano–Feshbach resonances, 303
field effect transistor, 279
flux quantum, 202
Fourier transform, 95, 101, 106, 123
Frank constant, 45
frustrated XY model, 201
frustration, 203, 214
fully frustrated XY model, 201, 204, 206–208, 210, 211

gauge invariance, 80, 88
gauge invariant phase difference, 202, 205
Gaussian approximation, 84, 126
Gaussian model, 79
golden mean, 230

harmonic potential, 298, 309
Hartree–Fock, 328, 336
Hartree–Fock approximation, 305
helicity modulus, 207, 208, 219
hexatic liquid crystal, 14
honeycomb lattice, 226–229

ideal Bose gas, 309
inhomogeneity, 180–182, 185

Josephson arrays, 238, 239, 241, 245, 247–250
Josephson coupling energy, 258
Josephson junction array, 135, 146, 150, 201–203, 258
Josephson junctions, 201
Josephson scaling relation, 23

kink unbinding, 222, 224, 225
Kosterlitz–Thouless stability criterion, 210
Kramers and Wannier duality, 75

Landau, L., 70
Langevin, 335, 339
laser trapping, 300
lattice Coulomb gas model, 95, 102, 106
lattice Green's function, 215
layered superconductors, 177
local density approximation, 308
long-range order, 71

macroscopic Coulomb blockade, 262

magnetic field, 185, 186, 188, 190, 192, 194
magnetization, 193
matter-wave interference, 315
Mermin–Wagner theorem, 72, 298
meron, 326, 342, 345
Migdal–Kadanoff RG, 80
momentum distribution, 315
multicritical point, 35

one-dimensional Ising model, 2, 4, 5, 19, 20
one-dimensional Ising model with $1/r^2$ interaction, 74
Onsager, L., 70, 76
order parameter, 257

paraconductivity, 163, 183
Peierls argument, 94, 120, 129
Peierls, R., 70
Peter–Weyle Fourier transform theorem, 76
phase diagram of the $p$-clock model, 118
phase fluctuations, 315
phonons, 316
Poisson summation formula, 78
polarization catastrophe, 269
polyhedral group, 94, 115
Pontryagin index, 328
Potts model, 118
pseudospin, 326, 332

quantized Hall effect, 326, 328
quantum Hall effect, 325
quantum phase transition, 256
quantum rotor model, 95, 102
quasi-condensate, 307
quasi-long-range order, 98, 118, 206
quench, 319

relation between BKT transition temperatures, 120
relaxation, 274
renormalization-group equations, 175, 178
renormalized spin stiffness, 209

scale invariance, 305, 313
scattering amplitude, 303
Schur matrix, 109
self-dual clock model, 112, 117
self-dual temperature, 94, 96, 119, 120

self-duality, 114, 123
self-generated bosonic environment, 283
sine-Gordon model, 172–174, 177, 193
skyrmion, 9, 328
solid-on-solid model, 95, 126
spin–spin correlation, 206, 210
spin-charge coupling, 327
spinwave approximation, 206
spontaneous symmetry breaking, 98, 129
staggered magnetization, 211, 221
superconducting films, 135, 136, 138, 139, 146
superconducting vortices, 237, 238, 240, 241, 245, 246, 249
superconductivity, 256
superconductor–insulator transition (SIT), 256
superfluid, 329, 331
superfluid density, 306
superfluid films, 53
superfluid-density jump in conventional films, 176, 182
superfluid-density jump in cuprates superconductors, 178, 179, 181
superinsulating state, 257
superinsulation, 257
superinsulator, 263

Thomas–Fermi regime, 305
time-of-flight, 313, 314
topological excitations, 94, 96, 118, 121
topological invariant, 96, 122

transfer matrix, 95, 107, 112, 117
transfer operator, 95, 98, 126, 128
tunneling, 274
tunneling current, 276
twisted boundary conditions, 207
two-point correlation function, 117, 123

$U(1)$ symmetry, 94, 96, 100, 115, 117, 122
uncertainty principle, 257
uniformly frustrated XY model, 201–203, 229
universal jump, 318
universal jump in helicity modulus, 210, 220
universality hypothesis, 71

vector Potts model, 107
Villain approximation, 95, 102
Villain function, 212, 228
Villain model, 94, 102, 112, 126
Villain model approximation, 80
Villain model, generalized, 83
vortex excitations, 208
vortex glass, 230
vortex–antivortex confinement, 261
vortex-core energy, 171, 174
vortices, 317

Weyl algebra, 95, 109

XY model, 94, 96, 98, 100, 102, 105, 115, 117, 118, 121, 123, 125, 201, 304